Tropical alpine environments are those regions within the tropics occurring between the upper limit of the closed canopy forest at about 3500–3900 m, and the upper limit of plant life at around 4600–4900 m. Plants growing here have evolved distinct forms to cope with a hostile environment characterized by cold and drought, with the added hazard of fire. Giant rosette plants, supported by woody stems, are by far the most characteristic aspect of tropical alpine vegetation, differentiating these communities from temperate alpine and arctic areas. Tussock grasses, prostrate shrubs and cushion plants also abound.

Using examples from all over the tropics, this fascinating account examines the unique form, physiology and function of tropical alpine plants. It will appeal to anyone interested in tropical vegetation and plant physiological adaptations to hostile environments.

TROPICAL ALPINE ENVIRONMENTS:
Plant form and function

Alan Smith was born March 31, 1945 and died August 26, 1993 at the age of 48.

TROPICAL ALPINE ENVIRONMENTS

Plant form and function

Edited by

PHILIP W. RUNDEL
Department of Biology, University of California, Los Angeles

ALAN P. SMITH
Formerly of Smithsonian Tropical Research Institute, Panama City, Panama

F. C. MEINZER
Hawaiian Sugar Planters Association, Aiea, Hawaii

CAMBRIDGE UNIVERSITY PRESS
Cambridge, New York, Melbourne, Madrid, Cape Town, Singapore, São Paulo

Cambridge University Press
The Edinburgh Building, Cambridge CB2 8RU, UK

Published in the United States of America by Cambridge University Press, New York

www.cambridge.org
Information on this title: www.cambridge.org/9780521420891

First published 1994
This digitally printed version 2008

A catalogue record for this publication is available from the British Library

ISBN 978-0-521-42089-1 hardback
ISBN 978-0-521-05411-9 paperback

Contents

Contributors

E. Beck
Botanisches Institut der Universität Bayreuth
Lehrstuhl Pflanzenphysiologie
Universitätstrasse 30
D-8580 *Bayreuth, Germany*

P. E. Berry
Missouri Botanical Garden
P.O. Box 299
St Louis, Missouri 63166, USA

R. N. Calvo
Department of Biology
University of Miami
P.O. Box 249118
Coral Gables, Florida 33124, USA

S. Carlquist
4539 Via Huerto
Santa Barbara, California 93110, USA

D. A. DeMason
Department of Botany and Plant Sciences
University of California
Riverside, California 92521, USA

G. Goldstein
Department of Botany
University of Hawaii
3190 Maile Way
Honolulu, Hawaii 96822, USA

R. Gonzalez
Harvard University Herbarium
Cambridge, Massachusetts 02138, USA

R. J. Hnatiuk
National Forest Inventory
Bureau of Resource Sciences
John Curtin House
22 Brisbane Avenue
Barton, ACT 2600, Australia

J. E. Keeley
Department of Biology
Occidental College
Los Angeles, California 90041, USA

L. L. Loope
Haleakala National Park
P.O. Box 369
Makawao, Hawaii 96768, USA

K. R. Markham
DSIR Chemistry
Petone, New Zealand

A. C. Medeiros
Haleakala National Park
P.O. Box 369
Makawao, Hawaii 96768, USA

F. C. Meinzer
Hawaiian Sugar Planters Association
P.O. Box 1057
Aiea, Hawaii 96701, USA

G. A. Miller
Andean/Southern Cone Programs
The Nature Conservancy
1815 North Lynn Street
Arlington, Virginia 22209, USA

W. A. Pfitsch
Department of Ecology
Hamilton College
1987 Upper Burford Circle
St Paul, Minnesota 55108, USA

F. Rada
Departamento de Ecología Vegetal
Universidad de los Andes
Mérida, Venezuela

H. Rehder
Institut für Botanik und Mikrobiologie
Technische Universität München
Arcistrasse 21
D-8000 München, Germany

P. W. Rundel
Laboratory of Biomedical and Environmental Sciences and
Department of Biology
University of California
Los Angeles, California 90024, USA

The late A. P. Smith

M. S. Witter
Laboratory of Biomedical and Environmental Sciences
University of California
Los Angeles, California 90024, USA

T. P. Young
The Louis Calder Center
Fordham University
Armonk, New York 10504, USA

Preface

With current attention on global problems of biodiversity and climate change, environmental interest in tropical ecosystems has increased tremendously. Very often, nevertheless, tropical biology is focused on lowland humid forests. High mountain systems, however, are also an important feature of tropical landscapes. Compared to lowland tropical forests, there has been surprisingly limited interest in the ecology of organisms in these tropical systems.

Scientific interest in the flora of tropical alpine regions goes back to the middle of the eighteenth century when Joseph de Jussieu and Charles La Condamine collected plants and mapped the high mountain areas of Ecuador as part of a five-year expedition of the French Académie des Sciences. The most vivid early scientific accounts of tropical alpine environments, nevertheless, came from travels of Alexander von Humboldt in South America and Mexico at the beginning of the nineteenth century. Accompanied by a capable young botanist named Aimé Bonpland, Humboldt travelled extensively through the high páramos of Colombia, Ecuador, and northern Peru. The patterns of vegetation zonation which he observed on this trip had a great impact on his thinking, and helped lead to the founding of the modern science of biogeography. The roots of modern ecology can also be traced to this experience which demonstrated to Humboldt the importance of interrelationships between climate, soils and biotic communities. Humboldt saw clearly that the peculiar vegetation of the páramos of the northern Andes was unlike any alpine community in temperate mountain ranges.

The middle of the twentieth century brought renewed interest in the biogeography of tropical alpine environments. The most influential of all of these geographers was Carl Troll whose 1959 monograph, *Die tropischer Gebirge. Ihre dreidimensionale klimatische und pflanzengeographischen*

Zonierung, remains a classic today. Troll described the global distribution and nature of tropical alpine climates with clarity and detail in a series of papers over a 35-year career. His studies stimulated other researchers, as seen in published symposia edited by Troll in 1968 (*Geo-Ecology of the Mountainous Regions of the Tropical Americas*, Ferd Dümmler Verlag, Bonn; 1968), and in 1978 with Wilhelm Lauer (*Geoecological Relations Between the Southern Hemisphere Zone and Tropical Mountains*, Franz Steiner, Wiesbaden; 1978). It was in this same period that Olov Hedberg published his monograph on plant ecology in the high African volcanoes, *Features of Afroalpine Plant Ecology* (Acta Phytogeographica Suecica 49: 1–144; 1964). The drier plant communities of the central Andean puna had been described earlier in 1945 by A. Weberbauer (*El Mundo Vegetal de los Andes Peruanos*, Ministerio de Agricultura, Lima).

The last decade has been characterized by a steady growth in biological interests in the remarkable features of tropical alpine ecosystems. The diversity of tropical alpine organisms and their patterns of biogeography were featured in a recent volume edited by F. Vuilleumier and M. Monasterio, *High Altitude Tropical Biogeography* (Oxford University Press, Oxford; 1986). Human impacts on the biodiversity and vegetation of tropical Andean páramos were presented in considerable detail in another new book edited by H. Balslav and J. L. Luteyn, *Páramo: An Andean Ecosystem under Human Influence* (Academic Press, London; 1992).

In this book we present a diversity of perspective on a new theme, plant form and function in tropical alpine environments. In particular, chapters focus on physiological plant ecology, plant population biology and demography, and impacts of herbivory and disturbance. It is our hope that the studies described here will stimulate increased interest in the ecology of these fascinating and complex ecosystems.

Philip W. Rundel, Los Angeles
Alan Smith, Panama City
Frederick C. Meinzer, Honolulu

1

Introduction to tropical alpine vegetation

ALAN P. SMITH

The general term 'tropical alpine' refers to regions within the tropics occurring between the upper limit of continuous, closed-canopy forest (often around 3500–3900 m) and the upper limit of plant life (often around 4600–4900 m: Hedberg 1951, 1964; Beaman 1962; Troll 1969; Wade & McVean 1969; Wardle 1971; Van der Hammen & Ruiz 1984; Vuilleumier & Monasterio 1986; see Figure 1.1) and is used in preference to regional terms such as '*páramo*' and '*jalca*' in the moist Andes from Venezuela to Northern Peru, '*puna*' in the drier central Andes, and 'Afroalpine' and 'moorland' in Africa. No clear lower boundary can be defined where natural timberline has been eliminated by man, as in many areas of the Andes and Papua New Guinea (Wade & McVean 1969; Hope 1976; Ellenberg 1979; Ruthsatz 1983), or where the forest is patchy or absent due to low rainfall, as on the north slope of Mount Kenya (Coe 1967; Figure 1.2) and the western slopes of the Peruvian Andes (Weberbauer 1911). In these cases alpine species merge gradually with species of montane pasture, savanna or desert.

Physiognomy of tropical alpine vegetation varies greatly with climatic and edaphic factors; however, certain trends are held in common by many New and Old World tropical alpine areas (see, for example, Hedberg 1964; Coe 1967; Cuatrecasas 1968; Wade & McVean 1969; J. Smith 1977, 1980; Cleef 1978), suggesting convergent evolution (Hedberg & Hedberg 1979; Halloy 1983; Smith & Young 1987).

Near the treeline, tussock grasses and erect shrubs with small leathery evergreen leaves often dominate the vegetation. With increasing elevation, tussock grasses and shrubs generally decrease in relative importance, and the vegetation is increasingly dominated by giant rosette plants (perennial, large-leaved rosettes, supported by unbranched or little-branched woody stems, with dead leaves typically retained on the stem for many years).

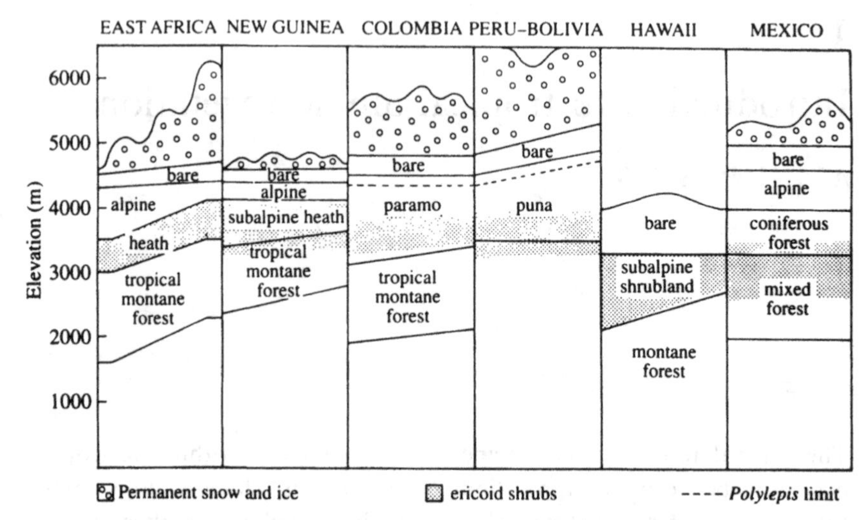

Figure 1.1. A diagrammatic summary of vegetation distribution on some New and Old World mountains. (Adapted from Troll 1968, Figure 16.)

Figure 1.2. Dry shrub-grass community on the north slope of Mount Kenya at 3300 m. Shrubs include *Helichrysum, Artemisia* and *Phillipia.* On slopes of Mount Kenya with higher rainfall, closed-canopy forest can occur at this elevation.

Figure 1.3. *Puya raimondii* (Bromeliaceae) at 4250 m, near Huancayo, Peru; individual at left is dispersing seeds.

Prostrate shrubs, non-tussock grasses and small perennial herbs are also common at intermediate elevations. Giant rosettes are by far the most characteristic aspect of tropical alpine vegetation, differentiating most of these communities from temperate alpine and arctic areas. The form has evolved repeatedly in the Andes (*Puya*, Bromeliaceae, Figure 1.3; *Espeletia*, Asteraceae, Figures 1.4–1.7; *Lupinus*, Leguminosae, Figure 1.8), Africa (*Senecio* and *Carduus*, Asteraceae; *Lobelia*, Campanulaceae, Figures 1.9–1.12), New Guinea (*Cyathea*, Figures 1.13 and 1.14), Hawaii (*Argyroxiphium*, Asteraceae, Figure 1.15) and the Canary Islands, just outside the Tropics (*Echium*, Boraginaceae, Figure 1.16). The growth form is absent from the dry volcanic peaks of Mexico, and from Mount Kinabalu (4101 m, lat. 6° N) in Borneo – a peak dominated by exposed rock. Giant rosettes

Figure 1.4. Venezuelan Andes (Mucubají). Lateral moraine at 3600 m, with dense stands of the giant rosette *Espeletia schultzii* (Asteraceae). Rosettes are interspersed with tussock grasses (mostly *Muehlenbeckia erectifolia*) and shrubs of *Hypericum*.

appear to play a particularly important role in the vegetation of the comparatively mesic alpine areas of Africa (Hedberg 1964; Coe 1967; Smith & Young 1982) and the northern Andes (Monasterio 1979; Smith 1981) where they can create 'savanna woodland' communities somewhat similar in physiognomy to palm savannas of the dry lowland tropics. These giant rosette communities should probably not be considered as merely extensions of montane forest into the alpine zone. In the northern Andes and on Mount Kenya giant rosettes are typically either absent or short in stature near treeline, and tend to increase in stature, both within and between species, with increasing elevation (A. Smith 1980), dramatically reversing the elevational trend seen in forest tree stature.

Isolated pockets of true closed canopy forest composed of *Polylepis* (Rosaceae) can occur well above the general treeline in the Andes at elevations up to 4300 m, typically occupying sheltered talus slopes (Figure 1.17).The presence of such forest pockets has led to speculation that forest may once have extended to far greater elevations than at present, but clear evidence is lacking (A. Smith 1977). Similar pockets of forest occur

Figure 1.5. A population of *Espeletia timotensis* (Asteraceae) at 4200 m, Piedras Blancas, Venezuela, with Dr José Cuatrecasas.

on Mount Wilhelm, Papua New Guinea at elevations of up to 3900 m (Wade & McVean 1969) and in this case probably are the result of human interference. Stands of the arborescent *Senecio keniodendron* (Asteraceae) on Mount Kenya can sometimes form a nearly closed canopy on mesic talus slopes at elevations of 4000–4200 m (Figure 1.18). The factors responsible for tropical treeline and for these isolated forest patches above treeline are in large part unknown (Walter & Medina 1969; Wardle 1971; A. Smith 1977).

Giant rosettes can extend to over 4600 m in Africa and to over 4500 m in the Andes, but are restricted to considerably lower elevations in other tropical alpine areas. At a given elevation, the relative importance of

Figure 1.6.

Figure 1.7.

Figure 1.8. *Lupinus alopecuroides* (Leguminosae), a monocarpic rosette species at 4150 m, Purace, Colombia.

tussock grasses versus giant rosettes generally shifts in favour of grasses as rainfall decreases; thus, the dry, fire-prone north slope of Mount Kenya and the analogous puna region of the central Andes have extensive cover by tussock grasses, with giant rosettes often restricted to comparatively rocky steep sites. Experimental studies on *Espeletia* in Venezuela suggest that tussock grasses can significantly reduce growth, reproduction and seedling establishment (Smith 1984). Fire is a major source of mortality both for *Espeletia* in the northern Andes (Smith 1981) and for *Senecio* on Mount Kenya (Smith & Young, Chapter 15).

At elevations above approximately 4200 m in the Andes and on Mount Kenya a combination of reduced rainfall and increased nocturnal soil frost heaving typically creates extensive areas of bare soil, resulting in 'alpine desert' (Monasterio 1979; Perez 1987a; see Figure 1.19). Extremely low rainfall can result in true alpine desert vegetation, as in parts of the

Figure 1.6. Purace, Colombian Andes. Foreground: *Espeletia hartwegiana* (Asteraceae) and scattered *Puta hamata* (Bromeliaceae) in a bog at 3400 m.

Figure 1.7. Purace, Colombian Andes at 3400 m; *Espeletia hartwegiana* (right) and *Puya hamata* (left).

Figure 1.9.

Figure 1.10.

Figure 1.11. Mount Kenya, 4250 m; *Carduus keniophyllum* (Asteraceae), a giant rosette species particularly common near hyrax colonies.

Peruvian and Bolivian Andes, where cactus species occur well above 4000 m (Weberbauer 1911; Herzog 1923).

Cushion plants can sometimes dominate the vegetation above 4000 m in the northern and central Andes, but appear to be less important in other tropical alpine areas (Rauh 1988). They may occur in desert-like puna habitats of both Peru and Bolivia (Weberbauer 1911; Hodge 1926) and in the moister páramo regions from northern Peru to Venezuela, where they appear to occur commonly on seepage slopes. The ecology of this growth form has been largely ignored in the tropics.

At the upper limit of plant life, low herbs, grasses and lichens dominate, and giant rosettes and shrubs generally drop out. The factors controlling the upper limit of plant life have not been analysed, although work has

Figure 1.9. Teleki Valey, Mount Kenya at 4200 m. A. *Lobelia telekii*, B. *Lobelia keniensis*, C. *Senecio brassica*, D. *Senecio keniodendron*. The dominant ground cover is *Alchemilla johnstonii*.

Figure 1.10. Teleki Valley, Mount Kenya. A topographic gradient spanning *c.* 4000–4150 m elevation, illustrating characteristic zonation of *Senecio brassica* (light-colored vegetation toward valley bottom) and *Senecio keniodendron* (upper slopes).

Figure 1.12.

Figure 1.13.

Figure 1.14. Mount Wilhelm, Pindaunde Valley, 3490 m; *Cyathea atrox.*

been done on *Draba chionophila* (Brassicaceae), one of the highest elevation species in the Venezuelan Andes (see Pfitsch, Chapter 8; also Perez 1987b). Glaciers are receding throughout the tropics, suggesting that plants may be colonizing progressively higher elevations (Coe 1967; Hastenrath 1985).

The tendency to divide tropical alpine areas into discrete elevational

Figure 1.12. Mount Elgon, 4000 m; a stand of *Senecio baratipes*, interspersed with tussock grasses and shrubs of *Helichrysum.*

Figure 1.13. Mount Wilhelm, Pindaunde Valley, Papua New Guinea, 3490 m. Treefern in foreground is *Cyathea gleichenoides.* Trees in background are primarily *Podocarpus compactus* (Podocarpaceae) and *Rapanea vaccinoides* (Myrtaceae).

Figure 1.15. Haleakala Crater, Maui, Hawaii, at 2180 m; *Argyroxiphium sandwicense* (Asteraceae).

zones (e.g. subpáramo, páramo, superpáramo; Cuatrecasas 1968; ericaceous zone, alpine zone: Hedberg 1964; see also Figure 1.1) appears to have little objective basis in actual community structure. Rather, communities appear to vary continuously along climatic and edaphic gradients, with few distinct discontinuities. However, very few quantitative data on spatial variation in community structure are available for tropical alpine areas (Hamilton & Perrott 1981; Baruch 1984).

Tropical alpine floras typically contain elements derived from the surrounding montane forests as well as from both South and North Temperate floras (Hedberg 1961; Clayton 1976; Cleef 1979; Smith 1982). Generic affinities with floras of temperate latitudes appear to increase

Figure 1.16. El Teide Volcano, Tenerife, Canary Islands, at 2100 m; *Echium wilprettii* (Boraginaceae) a subtropical monocarpic giant rosette species that converges on *Argyroxiphium*.

with increasing elevation (see, for example, Gadow 1907–1909; Wade & McVean 1969). Species richness is low compared to tropical lowland forest, but comparable to that for temperate alpine and arctic tundra communities (Wade & McVean 1969). There is no obvious latitudinal gradient in species richness among high mountain and tundra communities, at least on a local ('alpha diversity') scale (Hanselman 1975). This provides a striking contrast to the clear latitudinal gradient in local species richness found among lowland forest communities. There is great variability in taxonomic richness among tropical alpine communities. The alpine flora of the northern Andes is much richer (over 300 genera) than

Figure 1.17.

Figure 1.18.

Figure 1.19. Mount Kenya, Hobley Valley at approximately 4500 m, illustrating 'alpine desert'.

the floras of the smaller and more fragmented alpine areas of Africa (103 genera) and Papua New Guinea (107 genera) (Cleef 1979). Simpson (1974) demonstrated that variation in floristic richness from region to region within the northern Andes is positive correlated with habitat area; she also suggested that variations in habitat area and isolation during and

Figure 1.17. Venezuelan Andes, 4100 m; an isolated patch of closed-canopy *Polyepis sericea* (Rosaceae) forest on a talus slope. In foreground, *Espeletia moritziana* interspersed with shrubs of *Hypericum* and *Arcytophyllum.*

Figure 1.18. Mount Kenya, Teleki Valley, 4200 m; a closed-canopy stand of *Senecio keniodendron* on a talus slope. The tussock grassland in the foreground occurs on mineral soil.

since the Pleistocene may have influenced current patterns of species richness (see also Hooghiemstra 1984). Janzen (1967) has suggested that there is less environmental overlap between the top and bottom of elevational gradients on tropical mountains than between the top and bottom of temperate latitude gradients with similar elevational ranges. This could result in decreased gene flow along tropical gradients, potentially facilitating speciation. A. Smith (1975) and Huey (1978) present data consistent with the hypothesis.

Plant growth rates appear to be generally low in tropical alpine areas, although few data are available (Hedberg 1969; Sturm 1978; Beck *et al.* 1980; A. Smith 1980, 1981; Smith & Young, 1982; Young, Chapter 14). Growth rates appear to be reduced by dry season drought stress, especially during unusually dry years (Smith 1981) and by both intra- and interspecific competition (A. Smith 1980, 1984).

Herbivory appears to be an important influence on plant population biology and plant community structure on Mount Kenya (Young and Smith, Chapter 18). Few data are available for the Andes; herbivory on *Espeletia schultzii* in Venezuela appears to be extensive only near the lower limit on its range (A. Smith 1980); tussock grasses and herbs are commonly browsed by rabbits and small rodents in the Venezuelan Andes (A. P. Smith, personal observations); herbivory by larger mammals such as deer, mountain tapir and vicuña has been greatly reduced by hunting (Kofford 1957).

Rates of succession in tropical alpine communities appear to be low (Janzen 1973). An experimental study in the Venezuelan Andes (W. Pfitsch and A. Smith, unpublished data) suggest that at comparatively low elevations (3100 m) succession on bare soil involves invasion by early successional specialist species, which disappear later in succession and are replaced by late successional specialists. Early successional specialists decline in importance with increasing elevation: at 4200 m succession takes the form of a gradual accumulation of species which persist in mature community. At and above 3500 m succession can be greatly slowed by soil frost heaving and erosion, which eliminate seedlings before they can get established. At 3600 m several 4 m quadrats which were cleared of vegetation in 1973 were still bare in 1984 (W. Pfitsch and A. Smith, unpublished data) suggesting that higher elevation communities on tropical mountains can be quite fragile (cf. Ruthsatz 1983).

The ratio of aboveground to belowground biomass appears to be unusually high in tropical alpine areas: 10:1 in alpine grassland of Papua New Guinea (Hnatiuk 1878) and 2.5:1 to 1:1 for the Venezuelan Andes

(Smith & Klinger 1985). These values are higher than those found for arctic and temperate alpine communities (Smith & Klinger 1985), and may reflect both the absence of large underground storage organs and the more extensive development of aboveground support tissues in many tropical alpine plants. Extensive data on nutrient cycles, hydrology, energy flow and trophic structure are not available for tropical alpine sites (see Rehder, this volume and Korner 1989); tropical alpine 'ecosystems' thus remain poorly known, despite much quantitative information on component species.

References

Baruch, Z. (1984). Ordination and classification of vegetation along an altitudinal gradient in the Venezuelan paramos. *Vegetatio* **55**, 115–26.

Beaman, J. H. (1962). The timberlines of Iztaccihuatl and Popocatepetl, Mexico. *Ecology* **43**, 377–85.

Beck, E., Scheibe, R., Senser, M. & Muller, W. (1980). Estimation of leaf and stem growth of unbranched *Senecio keniodendron* trees. *Flora* **170**, 68–76.

Clayton, W. D. (1976). The chorology of African mountain grasses. *Kew Bulletin* **31**, 273–88.

Cleef, A. (1978). Characteristics of neotropical paramo vegetation and its subantarctic relation. *Erdwissenschaftlichen Forschung* **11**, 365–90.

Cleef, A. (1979). The phytogeographical position of the neotropical vascular paramo flora with special reference to the Colombian Cordillera Oriental. In *Tropical Botany*, ed. K. Larsen and L. B. Holm-Nielsen, pp. 175–84. New York: Academic Press.

Coe, M. J. (1967) The ecology of the alpine zone of Mount Kenya. *Monographiae Biologicae 17*. The Hague: Junk.

Cuatrecasas, J. (1968). Páramo vegetation and its life forms. *Colloquium Geographicum* **9**, 163–86.

Ellenberg, H. (1979). Man's influence on tropical mountain ecosystems of South America. *Journal of Ecology* **67**, 401–16.

Gadow, H. (1907–1909). Altitude and distribution of plants in Southern Mexico. *Journal of the Linnean Society, Botany* **38**, 429–40.

Halloy, S. (1983). The use of convergence and divergence in the interpretation of adaptation in high mountain biota. *Evolutionary Theory* **6**, 232–55.

Hamilton, A. & Perrott, R. (1981). A study of altitudinal zonation in the montane forest belt of Mt. Elgon, Kenya/Uganda. *Vegetatio* **45**, 107–25.

Hanselman, D. P. (1975). Species diversity in tundra environments along a latitudinal gradient from the Andes to the Arctic. MS thesis, Duke University, Durham NC.

Hastenrath, S. (1985). *The Glaciers of Equatorial East Africa*. Boston: D. Reidel.

Hedberg, O. (1951). Vegetation belts of the East African mountains. *Svensk Botanik Tidskrift* **45**, 140–202.

Hedberg, O. (1961). The phytogeographical position of the Afroalpine flora. *Recent Advances in Botany 1961*, pp. 914–19.

Hedberg, O. (1964). Features of Afroalpine plant ecology. *Acta Phytogeographica Suecica* **49**, 1–144.
Hedberg, O. (1969). Growth rate of the East African giant senecios. *Nature* **222**, 163–4.
Hedberg, I. & Hedberg, O. (1979). Tropical-alpine life forms of vascular plants. *Oikos* **33**, 297–307.
Herzog, T. (1923). *Die Pflanzenwelt der bolivischen Anden und ihres ostlichen Vorlandes.* Leipzig: Verlag Wilhelm Engelman.
Hnatiuk, R. (1978). The growth of tussock grasses on an equatorial high mountain and on two sub-Antarctic islands. In *Geoecological Relations between the Southern Temperate Zone and the Tropical Mountains*, ed. C. Troll and W. Lauer. pp. 159–90. Erdwissenschaftliche Forschung 11. Wiesbaden: Franz Steiner.
Hodge, W. (1926). Cushion plants of the Peruvian puna. *Journal of the New York Botanic Garden* **47**, 133–41.
Hodge, W. (1960). Yareta – fuel umbellifer of the Andean puna. *Economic Botany* **14**, 113–18.
Hooghiemstra, H. (1984). Vegetational and climatic history of the high plain of Bogota, Colombia: a continuous record of the last 3.5 million years. Dissertations Botanicae, Band 79. Cramer.
Hope, G. (1976). The vegetational history of Mt. Wilhelm, Papua New Guinea. *Journal of Ecology* **64**, 627–64.
Huey, R. (1978). Latitudinal pattern of between-altitude faunal similarity: mountain passes may be 'higher' in the tropics. *American Naturalist* **112**, 225–9.
Janzen, D. (1967). Why mountain passes are higher in the tropics. *American Naturalist* **101**, 233–49.
Janzen, D. (1973). Rate of regeneration after a tropical high elevational fire. *Biotropica* **5**, 117–22.
Korner, C. (1989). The nutritional status of plants from high altitudes. *Oecologia* **81**, 379–91.
Kofford, C. (1957). The vicuna and the puna. *Ecological Monographs* **27**, 153–219.
Monasterio, M. (1979). El páramo desertico en el altiandino de Venezuela. In *El Medio Ambiente Páramo*, ed. M. L. Salgado-Labouriau, pp. 117–76. Caracas, Venezuela: IVIC.
Perez, F. L. (1987a). Needle-ice activity and the distribution of stem-rosette species in a Venezuelan paramo. *Arctic and Alpine Research* **19**, 135–53.
Perez, F. L. (1987b). Soil moisture and the upper altitudinal limit of giant paramo rosettes. *Journal of Biogeography* **14**, 173–86.
Rauh, W. (1988). *Tropische Hochgebirge Pflanzen.* Berlin: Springer-Verlag.
Ruthsatz, B. (1983). Der Einflusz des Menschen auf die Vegetation semiarider bis arider tropischen Hochgebirge am beispiel der Hochanden. *Berichte der Deutschen Botanische Gesellschaft* **96**, 535–76.
Simpson, B. (1974). Glacial migrations of plants: island biogeographical evidence. *Science* **185**, 698–700.
Smith, A. (1975). Altitudinal seed ecotypes in the Venezuelan Andes. *American Midland Naturalist* **94**, 247–50.
Smith, A. (1977). Establishment of seedlings of *Polylepis sericea* in the paramo zone of the Venezuelan Andes. *Bartonia* **45**, 11–14.
Smith, A. (1980). The paradox of plant height in an Andean giant rosette species. *Journal of Ecology* **68**, 63–73.

Smith, A. (1981). Growth and population dynamics of *Espeletia* (Compositae) in the Venezuelan Andes. *Smithsonian Contributions to Botany* **48**, 1–45.
Smith, A. (1984). Postdispersal parent–offspring conflict in plants: antecedent and hypothesis from the Andes. *American Naturalist* **123**, 354–70.
Smith, A. & Young, T. (1982). The cost of reproduction in *Senecio keniodendron*, a giant rosette species of Mt. Kenya. *Oecologia* **55**, 243–7.
Smith, A. & Young, T. (1987). Tropical alpine plant ecology. *Annual Review of Ecology and Systematics* **18**, 137–58.
Smith, J. (1975). Notes on the distributions of herbaceous angiosperm species in the mountains of New Guinea. *Journal of Biogeography* **2**, 87–101.
Smith, J. (1977). Vegetation and microclimate of east- and west-facing slopes in the grassland of Mt. Wilhelm, Papua New Guinea. *Journal of Ecology* **65**, 39–53.
Smith, J. (1980). The vegetation of the summit zone of Mount Kinabalu. *New Phytologist* **84**, 547–73.
Smith, J. (1982). Origins of the tropical alpine flora. *Monographiae Biologicae* **42**, 287–308.
Smith, J. & Klinger, L. (1985). Aboveground:belowground phytomass ratios in Venezuelan paramo vegetation and their significance. *Arctic and Alpine Research* **17**, 189–98.
Sturm, H. (1978). *Zur Ökologie der Andinen Paramoregion.* The Hague: Dr W. Junk.
Troll, C. (ed.) (1968). Geo-ecology of the mountainous regions of tropical America. *Colloquium Geographicum* **9**.
Van der Hammen, T. & Ruiz, P. M. (1984). *Studies on Tropical Andean Ecosystems*, Vol. 2. Berlin: J. Cramer.
Vuilleumier, F. & Monasterio, M. (1986). *High Altitude Tropical Biogeography.* New York: Oxford University Press.
Wade, L. & McVean, D. (1969). *Mt Wilhelm Studies I. The alpine and subalpine vegetation.* Research School of Pacific Studies Publication BG/1. Canberra: Australian National University.
Walter, H. & Medina, E. (1969). Die Bodentemperatur als ausschlagenbenden Faktur für die Gliederung der subalpinen und Stufe in den Anden Venezuelas. *Berichte der Deutschen Botanische Gesellschaft* **82**, 275–81.
Wardle, P. (1971). An explanation for Alpine timberline. *New Zealand Journal of Botany* **9**, 371–402.
Weberbauer, A. (1911). *Die Pflanzenwelt der peruanischen Anden.* Leipzig: Verlag Wilhelm Engelmann.

2
Tropical alpine climates

PHILIP W. RUNDEL

Historical introduction

Ecological interest in climates of the tropical alpine regions of the world dates back perhaps to the early travels of La Condamine in Ecuador in the mid-19th century, but more dramatically to the remarkable explorations of Alexander von Humboldt in the northern Andes, Central America and Mexico from 1799 to 1804. His contributions to science from these travels, which encompassed geography, biology, geology, climatology, anthropology and other subjects, filled 30 volumes (von Humboldt 1807–39), and had a tremendous influence on the intellectual and economic development of Latin America in the 19th century. No less an authority than Simon Bolivar once remarked, 'Baron Humboldt did more for the Americas than all of the conquistadors' (Von Hagen 1948). The scientific studies of von Humboldt served as the foundation of the modern science of biogeography, and his climatological observations in the Andes and on the Mexican volcanoes played a major role in the development of his ideas.

Another notable advance in scientific knowledge of alpine climatology in tropical mountain regions came not from a typical scientist at all but from the noted European alpinist, Edward Whymper. Whymper, who had been the first man to scale the Matterhorn in the Swiss Alps in 1865, came to Ecuador in 1879 to attempt climbs of Chimborazo and other high volcanoes of that region. He was spectacularly successful not only in these ascents, but in the wealth of ecological and climatological data which he collected and published (Whymper 1892).

Modern studies of tropical alpine environments had their beginning with the work of the great biogeographer, Carl Troll. In a large body of work, beginning with his first publication in 1941 and ending with his death in 1975, Troll presented the global distribution and significance of tropical alpine environments in great detail and clarity. His 1959 monograph on the phytogeography and climatology of tropical mountain ranges remains a classic today.

Climatic characteristics

Temperature cycles in tropical alpine environments

Tropical alpine climates differ sharply from those of temperate alpine regions in many traits, particularly with respect to seasonal and diurnal patterns of temperature change. Much of this difference results from the latitudinal effect on annual change in solar radiation at the outer surface of the atmosphere. Low tropical latitudes are regions of net heat gain and have relatively constant levels of solar irradiance over annual cycles. Over a 12-month cycle at the Equator, for example, the daily maximum irradiance is only 13% higher than the minimum level. This ratio remains relatively small up to the latitudes of the Tropic of Cancer and Tropic of Capricorn where it reaches about 60%, but thereafter increases much more sharply toward higher latitudes (List 1971). The global pattern of irradiance results in a condition where tropical latitudes are characterized by small seasonal changes in temperature, but strong diurnal patterns of change are present at high elevations. Indeed, diurnal ranges of temperature change are commonly 3–10 times greater than seasonal changes in tropical alpine environments. This contrasts markedly from alpine environments in high latitudes where seasonal temperature changes are large and diurnal changes relatively small.

This striking pattern of seasonal versus diurnal temperature range is one of the most significant factors characterizing tropical alpine environments. Tropical and temperate differences in alpine temperature pattern can be seen in Figure 2.1. The Zugspitze in the Alps has a mean diurnal temperature range of about 5 °C, compared with about 13 °C in the seasonal range of mean monthly temperature. Tropical alpine sites show the reverse pattern of temperature change. Mucubají in the Venezuelan Andes and Pindaunde in Papua New Guinea, two tropical alpine sites near the Equator, have mean diurnal temperature ranges of 8–9 °C, but only about 2 °C range in mean seasonal range in temperature. Quito, in the Andes very near to the Equator in Ecuador, has only a 0.5 °C difference between mean July and January temperature, but a 9 °C range in mean diurnal temperature. Diurnal ranges in temperature for tropical alpine

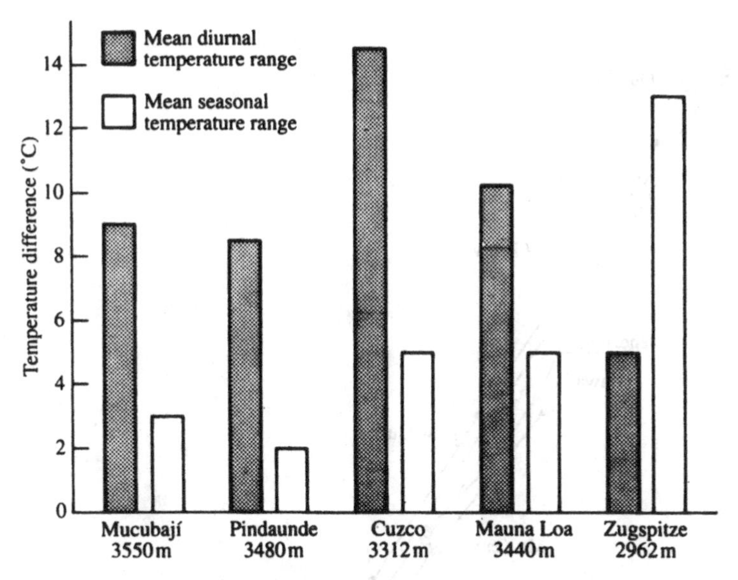

Figure 2.1. Mean diurnal and seasonal temperature ranges for four tropical alpine and one temperate alpine sites. Data are presented for Mucubají, Venezuela (Azócar & Monasterio 1980), Pindaunde, Papua New Guinea (Hnatiuk *et al.* 1976), Cuzco (Johnson 1976), Mauna Loa, Hawaii (Doty & Mueller-Dombois 1966), and the Zugspitze in the Alps (Troll 1959).

environments generally increase with greater distance from the Equator but the tropical pattern of relatively small seasonal change in temperature remains the same. Cuzco at lat. 14° S in Peru and Mauna Loa in the Hawaiian Islands at lat. 21° N have mean diurnal changes of about 15 and 10 °C, respectively (Figure 2.1). Diurnal ranges of temperature in the puna of Bolivia at about lat. 18° S commonly reach 25 °C, while seasonal changes in mean monthly temperature remain small. With increasingly higher latitudes in the Southern and Northern Hemispheres tropical alpine environments grade into temperate alpine environments as seasonal temperature ranges increase to match and then exceed diurnal ranges (Paffen 1967).

Tropical alpine environments have been characterized as having summer every day and winter every night. This extreme diurnal shift in temperature and the frosts that it brings provides one of the most critical elements of environmental stress faced by plants in these habitats. The frequency of frosts is a key selective force in adaptation to high tropical environments. While cold tolerance is certainly an important component of plants' survival in this zone, additional stress comes from the nightly freeze–thaw

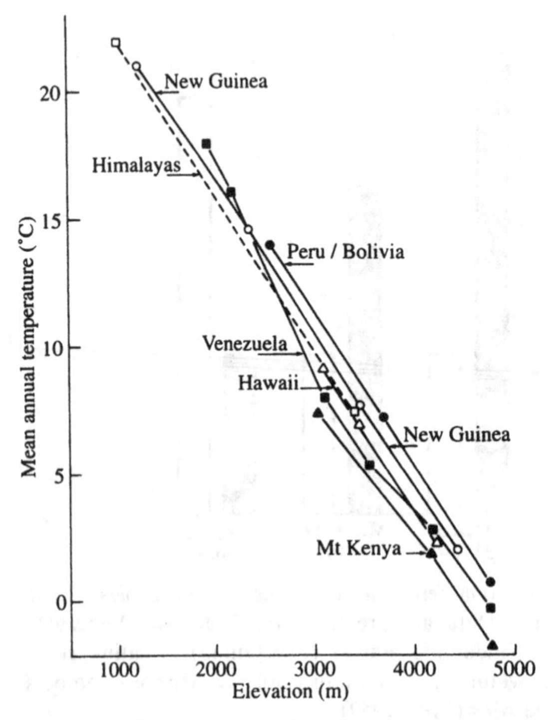

Figure 2.2. Elevational gradient for mean annual temperature in tropical mountain ranges. A broken line is shown with comparative values for the temperate Himalaya Range. Lines connect continuous gradients at the same site. Data from Troll (1959); Doty & Mueller-Dombois (1966); Coe (1967a,b); Hnatiuk *et al.* (1976); Johnson (1976); and Azócar & Monasterio (1980).

cycles which cause dynamic soil movements. Solifluction and frost heaving provide an inhospitable soil environment for woody plant roots.

Temperature gradients with elevation

Lapse rates of mean annual temperature with elevation are remarkably similar in different mountain regions of the world, both tropical and temperate. Temperature gradients from the Himalayas in the temperate zone are virtually identical to those present in New Guinea, the northern and central Andes, East Africa, and the Hawaiian Islands (Figure 2.2). The lapse rate in mean annual temperature for all of the sites is about 0.6 °C per 100 m elevation. Among the tropical alpine regions, however, the mean low temperature of the coldest month differs among regions. Such mean low temperatures are from 3 to 4 °C warmer in New Guinea

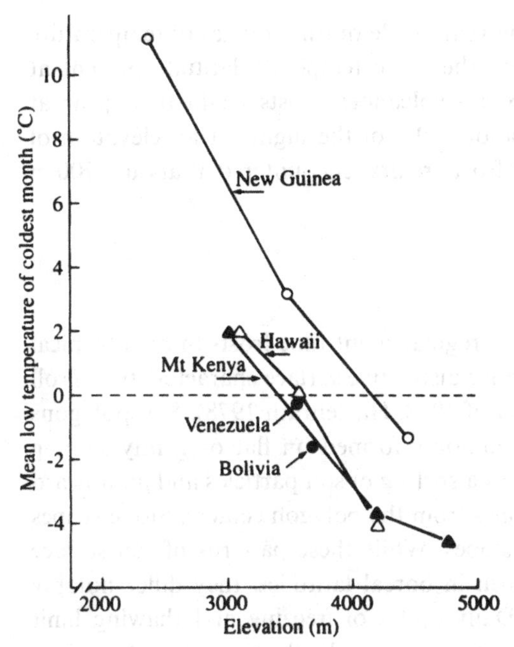

Figure 2.3. Elevational gradient for mean low temperature of the coldest month of the year in tropical mountain ranges. Lines connect continuous gradients at the same site. Data from sources cited in Figure 2.2.

than in the Andes, East Africa, or the Hawaiian Islands (Figure 2.3). While the continental positions of the Andes and East African volcanoes might suggest colder extreme temperatures due to a continental effect, it is not clear why gradients for New Guinea should differ markedly from those present on Mauna Loa in the Hawaiian Islands. Limited climatic data available from the New Guinea Highlands suggest that heavy cloud cover and high humidity may be important factors in these relatively warm gradients in temperature (Smith 1977a, 1980b).

Patterns of elevational zonation in frost regimes have been described in detail for the high Andes of southern Peru and the high volcanoes of Mexico (Troll 1968; Lauer & Klaus 1975). Night-time freezing temperatures begin to occur above about 3000 m elevation at Volcán El Misti in Peru, and by 4000 m such diurnal frosts are a regular event on 80–90% of the nights of the year. Days with permanent frost begin to appear above 4700 m elevation and at 5800 m such conditions are present throughout the year. Even at 4700 m elevation, the typical pattern of tropical temperature cycles remains. Seasonal variation in mean maximum temperature varies only about 3 °C over the year, while diurnal ranges of temperature are 6–10 °C (Troll 1968). With the more temperate latitude present at Orizaba and other high Mexican volcanoes, frosts begin to appear at about 1800 m and occur on about 90% of the nights at an elevation of 4500 m. Days with permanent frost are first encountered at about 4700 m (Lauer & Klaus 1975).

Solifluction

Soil movements associated with regular night-time frosts in high tropical alpine habitats frequently produce distinctive surface characteristics (Troll 1944; Löffler 1975; Furrer & Graf 1978; Hastenrath 1978). Soil polygons with stoney perimeters are commonly formed on flat or gently sloping terrain where frost action causes a sorting of soil particles and an upward and outward movement of stones from the polygon centers. Stone stripes typically form across steeper slopes. While these patterns of soil surface superficially resemble those seen in boreal latitudes, they differ notably in the depth of solifluction. Daily cycles of freezing and thawing limit the soil movements to a few centimeters in depth, much less deep than solifluction patterns in high latitudes. The resultant soil polygons are small, compared with those in high latitudes (Troll 1944, 1968).

Temperature effects on vegetation zonation

Gradients of temperature change with elevation, particularly those that influence patterns of freeze–thaw cycles, have pronounced effects on patterns of vegetation zonation in tropical mountains. The upper limit of forest vegetation in East Africa, and in the northern Andes, corresponds to the boundary of the frost-free zone at about 3000 m (Troll 1973). Sarmiento (1986) has pointed out that even small differences in frost frequency can apparently determine the dividing line between forest and alpine páramo habitats on moist Andean slopes. Two neighboring stations in Ecuador, Quito at 2818 m and Izobamba at 3058 m, have almost identical climatic regimes, differing only in the total absence of frosts at Quito and their occasional occurrence at Izobamba. Nevertheless, the area of Quito originally supported montane forest vegetation, while nearby Izobamba has an alpine páramo. A very different pattern of vegetation influence is present on the high volcanoes of Mexico, however. Here the lower frost line serves to delineate mixed forest from coniferous forest, not montane forest from páramo (Lauer & Klaus 1975).

The environmental constraints of treelines are not simply temperature alone, of course, but rather a complex mix of interacting forces (Stevens & Fox 1991). One notable difference in environmental conditions between tropical and temperate treelines is the role played by snowcover. Temperate zone trees in their *krummholz* form of growth at high elevations are characteristically covered through much of the winter by deep snow which protects plant tissues from the extremes of low temperature and the desiccating effects of wind. In contrast, seasonal snow seldom accumulates below the zone of perennial snow or ice cover in tropical mountain regions because of the lack of seasonality in temperature and the relatively warm daytime conditions present. Large glacial tongues of ice are rare below the line of permanent frost in tropical mountains, where ablation can be a daily event throughout the year (Troll 1968).

Differences in seasonality of temperature conditions are also responsible for distinctive patterns of interface between forest and alpine vegetation zones which separate temperate and tropical mountain ranges. In temperate and boreal latitudes, strong seasonal changes in temperature produce cold air drainage in valley systems and lead to higher timberlines on adjacent slopes. Snow accumulation and associated avalanche movements down valleys reinforce this pattern. With an absence of seasonality in tropical mountain ranges, however, this pattern of interface between zones is frequently altered. Diurnal climate shifts buffer valley habitats from extremes of temperature, producing conditions with frost frequencies and higher ambient humidities. These conditions promote a higher elevation distribution of trees in valleys than on adjacent slopes.

Precipitation

Patterns of precipitation in tropical alpine regions are complex and much more difficult to generalize than patterns of temperature discussed above. Seaonality of precipitation is highly variable depending on the region. This variation is largely influenced by position on windward or leeward slopes, by latitude, and by the relative movement of the intertropical convergence zone (ITCZ). In the northern Andes, for example, a series of distinctive precipitation regimes is present. These include areas with strongly unimodal patterns of seasonality peaking in June–July and other regions with strong bimodal distribution of rainfall with peaks in April–May and again in October–November. Further south in the Andean Cordillera, the biseasonality of rainfall becomes increasingly less pronounced and the June–July–August period of drought intensifies. These

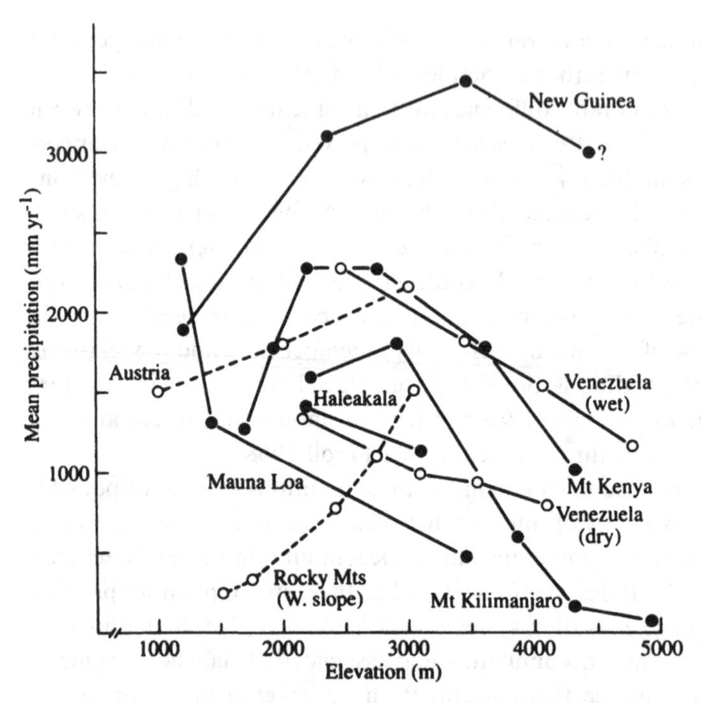

Figure 2.4. Elevational gradients for mean annual precipitation in tropical (solid lines) and temperate mountain ranges (dashed lines). Data from the same sources as Figures 2.1 and 2.2.

South American patterns of precipitation are discussed in more detail below.

Elevational gradients of precipitation are likewise highly variable in tropical alpine environments, owing to the influence of tradewind inversions and the relative exposure and continental influence on alpine regimes in each region. Unlike annual precipitation in temperate mountain ranges which generally increases predictably with increasing elevation, as shown for gradients in the Austrian Alps and west slope of the Rocky Mountains in Figure 2.4, precipitation commonly reaches peak levels at intermediate elevation on tropical mountains and declines steadily toward the peak. A relatively low tradewind inversion in the Hawaiian Islands produces peak rainfall amounts at about 1600 m elevation (Price 1983). Maximum precipitation in the Venezuelan Andes falls at 2000–2500 m (Monasterio & Reyes 1980; Sarmiento 1986), and at 2200–3000 m on the East African high volcanoes. The peak level on Mount Wilhelm in New Guinea appears to be higher at about 3400 m (Figure 2.4). There are exceptions to this pattern, however, in which precipitation does not appear to decline at higher elevations on the slope. Such conditions have been reported for the llanos-facing exposures of the Andes near Mérida in Venezuela (Sarmiento 1986) and on Mount Kinabalu in North Borneo (Smith 1980a).

Total precipitation is another highly variable climatic trait of tropical alpine environments. Local rainshadow effects in the Cordillera Oriental of Colombia produce pronounced gradients in annual rainfall over short distances, ranging from lows of about 700 mm to highs of 3250 mm (Weischet 1969). The summit of Mount Kilimanjaro in Tanzania is thought to receive less than 100 mm rainfall per year (Hedberg 1964), making it the driest tropical alpine site. Oruro at 3705 m elevation in central Bolivia near the southern margin of the tropical alpine zone receives less than 300 mm precipitation per year, while the summit of Mauna Kea at 4205 m elevation has a mean annual precipitation of about 350 mm. At the other extreme, alpine areas on Mount Wilhelm in New Guinea appear to receive nearly 3500 mm per year (Figure 2.4), while Mount Kinabalu in North Borneo may receive as much as 5000 mm (Smith 1980a).

Regional patterns of climate

The tropical Andes

The climatology of tropical alpine environments is much better studied in the Andean Cordillera than in any other region. No other area of the world possesses such a major mountain range cutting across tropical latitudes, nor so many climatological stations at high elevation. The Andes extend from about lat. 9° N in northern Columbia southward to Tierra del Fuego at a latitude of about 56° S. The southern edge of the tropical Andean zone is generally taken to be about lat. 17–18° S in southern Bolivia.

Two distinct environmental regions can be distinguished in the alpine zone of the tropical Andes. The páramo region of Colombia, Venezuela and Ecuador is characterized by relative moist conditions (with a few notable exceptions) and highly stable seasonal patterns of change in both mean maximum and mean minimum monthly temperatures. Precipitation in the Northern Hemisphere winter and spring comprises no more than 60% of total precipitation. In contrast, puna habitats in high elevations of Peru and Bolivia are generally much more arid than páramo regions. While mean monthly maximum temperatures show little change seasonally

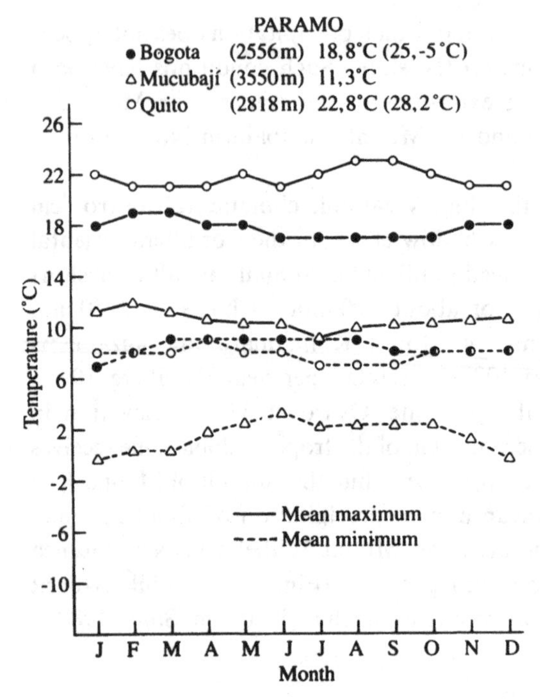

Figure 2.5. Seasonal patterns of mean maximum and mean minimum monthly temperature for three stations in the páramo zone of the northern Andes. For each station, data are shown for elevation, mean annual maximum and minimum and extreme maximum and minimum. Data from Azócar & Monasterio (1980) and Johnson (1976).

except near the limits of the subtropical zone in southern Bolivia, a seasonal decline in mean minimum temperatures during the Southern Hemisphere winter becomes increasingly pronounced with increasing latitude.

Typical examples of temperature seasonal regimes for Andean páramo sites is shown in Figure 2.5. Quito has a wet páramo climate with a mean monthly maximum temperature of 22 °C and a means monthly minimum of 8 °C. Despite a lower elevation, Bogotá has a mean maximum temperature 4 °C lower, but a nearly identical mean minimum. Mucubají at 3550 m elevation in the Andes of Venezuela has mean monthly maximum and minimum temperatures of 11 and 3 °C, respectively (Azócar & Monasterio 1980). Frosts do not occur at Quito, but temperatures have reached as low as −5 °C in Bogotá (Johnson 1976).

For puna stations, mean monthly maximum temperatures show the same steady range of values seen for páramo stations (Figure 2.6), with

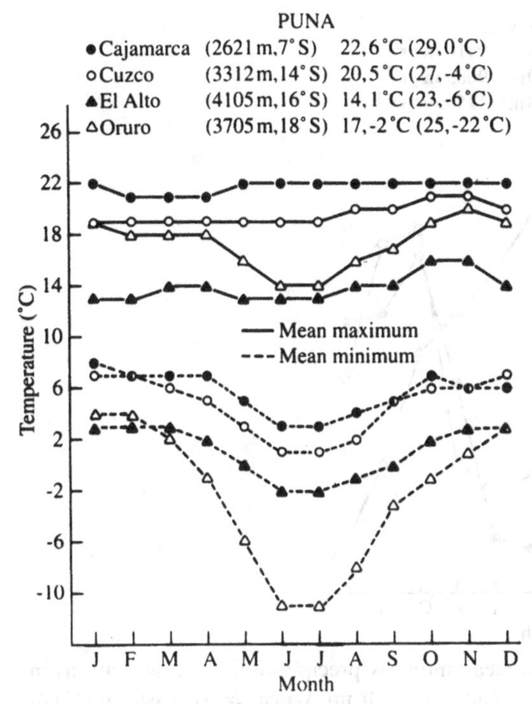

Figure 2.6. Seasonal patterns of mean maximum and mean minimum monthly temperatures for four stations of the puna zone of the central Andes. For each station, data are shown for elevation, mean annual maximum and minimum and extreme maximum and minimum. Data from Johnson (1976).

elevation as the primary variable predicting maximum temperature. A winter decline in mean monthly minimum temperatures centred on June and July is sharply evident at all of the puna stations and becomes more pronounced moving southward toward higher latitudes. At Oruro in central Bolivia at lat. 18° S, a sharp decline in winter maximum temperatures becomes apparent. Temperatures commonly reach −15 °C in the altiplano of southern Peru (Winterhalder & Thomas 1978) and −22 °C at Oruro (Figure 2.8, below). Sarmiento (1986) reports temperatures as low as −30 °C in the altiplano of Bolivia. With these winter lows, daily ranges in temperature for puna stations (16–25 °C) are commonly much greater than for páramo stations (7–14 °C). This is true even during the warmer season of the year from October to March (Figure 2.6). Daily ranges of temperature in the Bolivian altiplano can reach as much as 30–40 °C (Johnson 1976; Sarmiento 1986), making these among the highest reported anywhere in the world (Paffen 1966).

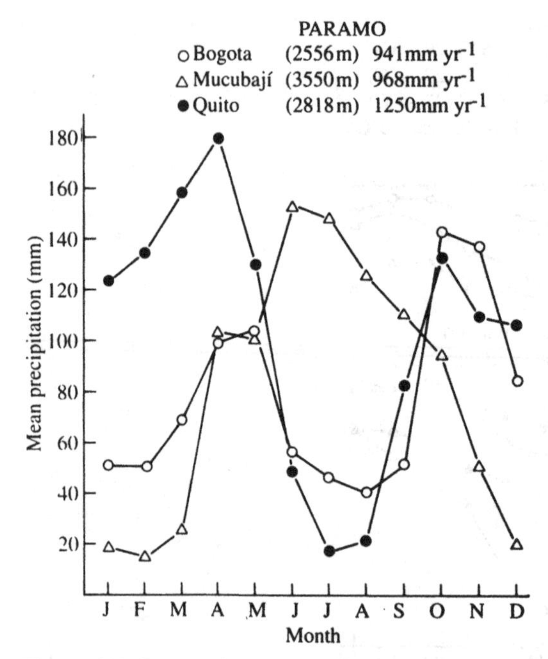

Figure 2.7. Seasonal patterns of mean monthly precipitation for three stations in the páramo zone of the northern Andes. Data from Azócar & Monasterio (1980) and Johnson (1976).

The páramo region of the northern Andes is exceedingly complex in the patterns of precipitation amount and seasonality that are present. These patterns have been reviewed in detail by Sarmiento (1986). The region from central Colombia to Venezuela generally experiences a marked bimodal pattern of rainfall with two dry seasons, as seen in Figure 2.7 for Bogotá. There is an exception to this pattern, however, in the slopes facing the Venezuelan llanos where there is a unimodal distribution of precipitation with a dry season from November to March. This type of precipitation regime is illustrated by Mucubají (Figure 2.7). Further south, the more pronounced dry season shifts markedly to mid-year, and the bimodal pattern of rainfall is almost lost. Quito provides a good example of this type of regime (Figure 2.7).

Moving southwards from the páramo habitats of Ecuador into the puna of the antiplano region of Peru and Bolivia, there is a distinctive gradient in several characteristics of rainfall. First, mean annual precipitation drops steadily from the 1300 mm present around Quito to 700–800 mm in central and northern Peru to 560 mm at La Paz and finally below 300 mm

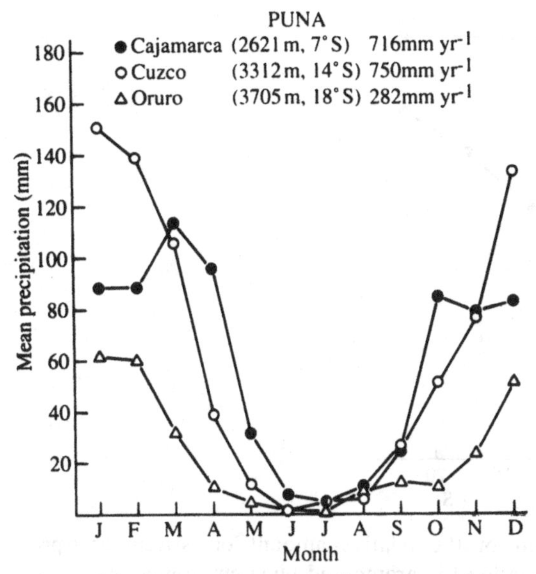

Figure 2.8. Seasonal patterns of mean monthly precipitation for three stations in the puna zone of the central Andes. Data from Johnson (1976).

at Oruro (Johnson 1976). This gradient of increasing aridity also brings with it an increasing seasonality of precipitation and duration of the dry season in mid-year. Cajamarca in northern Peru still retains elements of a biseasonal pattern of rainfall like that present in Quito, but this pattern is totally lost in Cuzco in central Peru (Figure 2.8). The mid-year dry season increases in length from 3 months with less than 20 mm rainfall in Cajamarca to 4 months in Cuzco and finally 7 months at Oruro. This gradient of increasing seasonality in rainfall serves to separate páramo from puna climates. Additionally, páramo climates commonly have 60% or less of their annual precipitation falling in the 6 months from November to April, but this proportion increases to 80–90% or more in the puna (Figure 2.9). Finally, the increasing aridity of puna habitats also brings increasing interannual variability in rainfall. Drought years occur with increasing frequency, increasing the coefficient of variation in mean annual precipitation moving from north to south (Johnson 1976).

Central America and the Mexican Highlands

Unlike South America, there is no extensive cordillera through Central America and Mexico that reaches above the forest line. Nevertheless there

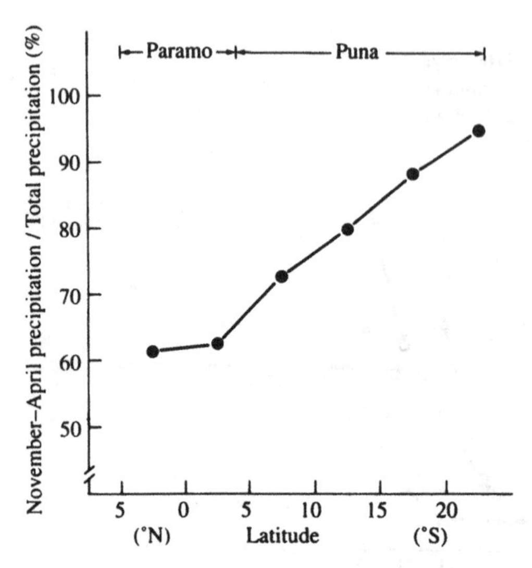

Figure 2.9. Latitudinal gradient of the relative amount of November–April precipitation in the Andes in relation to páramo and puna environments.

are three mountain areas that reach well above 3000 m elevation and present tropical alpine habitats (Hastenrath 1968; Lauer 1968). The Cordillera Talamanca in Costa Rica and western Panama and the adjacent Cordillera Central in Costa Rica have several peaks which rise above 3000 m, with Volcán Chirripó reaching 3820 m in the former range and Volcán Irazú at 3422 m in the latter. Treeline in the area commonly occurs at about 3100 m, but the boundary has been obscured by past human actions. The Costa Rican peaks contain the northern phytogeographic limit of páramo taxa from Colombia (Weber 1959). Rainfall in the páramo zone of the Cordillera de Talamanca is high and seasonal in distribution, with four relatively dry months from January to April.

The Guatemalan Highlands in northern Central America is a second region with tropical alpine habitats. Two volcanoes, Tacana and Tajumulco, reach above 4000 m elevation. Good climatic records are lacking for this area, but timberline is known to be high at about 3800–3900 m (Lauer 1968).

The Mexican Highlands present a transitional area from tropical to temperate alpine condition. The high plateau in eastern Mexico at 2000–2500 m gives rise to a number of very high peaks, most notably Ixtaccíhuatl (5285 m), Popocatépetl (5455 m) and Pico de Orizaba (5675 m). Precipitation regimes are highly seasonal in these mountains, with only about 10% of the annual total falling in winter (Garcia 1970).

Rainfall peaks at about 2000–2300 m and declines at higher elevations. Studies of diurnal patterns of temperature change on the slopes of the high volcanoes have shown that there is generally a decrease in the range of diurnal temperature change as both high and low temperatures drop with increasing elevation (Lauer & Klaus 1975). Frost hours per day were estimated to change from less than one at 3000 m to a full 24 hours at 5000 m. This lower frost limit is well below the forest line and serves roughly to separate mixed from coniferous forests. Freezing temperatures appeared to be seasonal at intermediate elevations above 3000 m, being most frequent during the dry winter months, and much rarer in April and May.

Despite the subtropical latitude of the Mexican Highlands, the vegetation structure of these mountains is strikingly boreal. Forest vegetation is dominated by species of *Pinus, Abies, Acer, Quercus, Fraxinus* and other genera of clear Holarctic origin. Above the forest line, which extends to about 4000 m elevation, the expected vegetation would be an analog of the arid puna of the Andes. Instead, there are mountain grass steppes with strong Holarctic and Nearctic affinities. This apparent paradox of a boreal vegetation occurring in an area of seemingly tropical alpine climate has led to a controversy which has existed since the early 18th century. Alexander von Humboldt considered this situation on his visit to Mexico in 1803–4 and attributed it to the continental position of Mexico and the presumed influence of cold air masses from the north. Climatic data have not supported von Humboldt's hypothesis, however, The range of mean monthly temperatures on the Mexican volcanoes is only about 5–6 °C at 3000 m and about 2.5 °C above 4000 m (Lauer & Klaus 1975). Additionally, the Mexican volcanoes lack a zone of winter snow cover below the permanent snowline, although seasonal snow accumulation may occur in the humid summer season (Troll 1949). Thus the climatic regime is clearly that of a tropical alpine region, not a boreal one. The floristic dominance of boreal elements can be attributed to geological history and the relative biogeographic isolation of the Mexican Highlands from tropical mountains to the south (Troll 1968).

East African Highlands

Two areas in East Africa include important areas of tropical alpine habitat. Best known of these are the isolated volcanic peaks lying near the Equator: Mount Kenya (5195 m) in Kenya, Mount Kilimanjaro (5899 m) in Tanzania, Mount Elgon (4324 m) at the Kenya–Uganda border, and the volcanic chains of the Ruwenzori (5119 m) and Virunga (4500 m) along the Uganda–Zaïre–Rwanda border. Further north, the Ethiopian Highlands present a broad plateau above 2000 m elevation with several chains extending above 3000 m. A number of isolated peaks exceed 4000 m, with the high point at Ras Dashan (4620 m) in northern Ethiopia.

Little is known about seasonal patterns of temperature regimes on the East African volcanoes. Data collected by Coe (1967a, b) along an elevational gradient over 30 days in December and January suggest that relatively low mean and absolute minimum temperatures were present. Mean minimum temperature over this period dropped from 1.7 °C at 3048 m to −3.9 °C at 4770 m. Absolute minimum temperatures were −1.5 °C and −8.3 °C, respectively. Klute (1920) has reported temperature data for a 46-day period at 4160 m on Mount Kilimanjaro. The mean daily temperature of 1.9 °C was almost identical to that reported for Mount Kenya by Coe (1967b).

As with temperature regimes, only limited precipitation data are available for higher elevations in the East African volcanoes. All of these peaks appear to have a bimodal pattern of precipitation, with peaks in April and November–December. Dry seasons are centered on January–February and July–August (Hedberg 1964). Total precipitation is highly variable, however, varying with both geographic position and slope exposure on a single peak. Hedberg (1964) classified the east African volcanoes along a moisture gradient on the basis of observed floristic and vegetation patterns. On this basis the Ruwenzori is the wettest range, followed closely by the Virunga volcanoes. Conditions become increasingly drier from Mount Kenya to Mount Elgon to Mount Kilimanjaro.

Exposure is of critical importance in determining annual precipitation levels on the East African volcanoes, with south and southeastern slopes receiving the highest levels. On Mount Kenya, rainfall is highest on the windward southeast slopes where annual precipitation peaks at about 2500 mm at 1400–2200 m elevation. The summit area receives about 850 mm (Hedberg 1951; Thompson 1966). On the wetter windward slopes, the principle dry season occurs from June to August, while on the drier leeward slopes where precipitation is only half as much a more pronounced drought occurs in January and February (Thompson 1966). Mount Kilimanjaro appears to be far drier than Mount Kenya. Hedberg (1964) suggests that rainfall on the windward slope peaks at about 2200 m where it comes close to 2000 mm per year, but drops sharply to levels below 200 mm above 4200 m. Dewfall in the summit area, however, may increase water availability for plant growth (Salt 1954). Sarmiento (1986) draws the parallel of Mount Kenya and its extensive areas of Afroalpine succulent vegetation with the páramos of the northern Andes, while the drier alpine slopes of Mount Kilimanjaro are more comparable to a puna habitat in the Bolivian altiplano.

Tropical alpine habitats are poorly developed in the Ethiopian Highlands. Rainfall gradients are pronounced, running from the north near the Red Sea where the plateau area receives only about 400 mm per year to the southwest margin of the highlands where annual precipitation exceeds 2500 mm. Dry seasons change from 10 months in the north (September–June) to only 2 months at the southwestern edge of the plateau (January and February). A cloud forest belt occurs from about 2000–2500 m in this latter area, but annual precipitation declines to about 1600 mm in subalpine woodlands at 3000 m elevation (Lauer 1976). Climatic data are not available for a higher zone of Afroalpine vegetation. Although there are no meteorological stations at high elevations, some impressions of alpine temperature conditions can be extrapolated from data for Addis Ababa at 2370 m. Rare frosts do occur at this elevation in winter, and the coldest mean monthly temperatures (4 °C) are present in November at the height of the dry season (Brown & Cocheme 1973).

Three isolated mountain ranges reach just above 3000 m elevation in arid north-central Africa. These include the Darfur Range in Sudan on the eastern margin of the Sahara Desert and the Ahaggar and Tibesti Ranges in the central Saharan region within Algeria and Chad, respectively. The peaks of these ranges are arid, with no more than 100–200 mm of annual rainfall (Yacono 1968). Also highly isolated is Mount Cameroon which towers to 4070 m near the coast in West Africa at about lat. 4° N. This peak appears to be one of the wettest tropical mountains in the world. Lauer (1976) has estimated that annual rainfall exceeds 8000 mm at 1000 m and declines to no less than 2000 mm at the summit.

Malaysia and New Guinea

The most extensive area of tropical alpine habitats in the Indo-Malaysian region occurs in the highlands of New Guinea which extend for more than 2000 km from west to east across the island (Smith 1980b). The highest peak is Mount Jaya (Carstensz) at 5030 m, but Mount Trikora (Wilhelmina), Mount Wisnumurti, and Mount Madala in Irian Jaya all exceed 4500 m. To the east 1000 km, in Papua New Guinea, Mount Wilhelm reaches 4510 m, and a large number of other peaks extend beyond 4000 m. Climatic regimes on these high tropical peaks have had

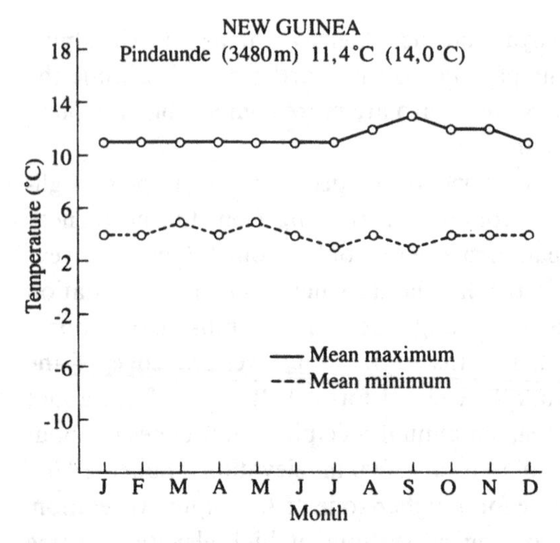

Figure 2.10. Seasonal patterns of mean maximum and mean minimum monthly temperatures for Pinaunde on Mount Wilhelm in Papua New Guinea. Data are shown for elevation, mean annual maximum and minimum and extreme maximum and minimum. Data from Hnatiuk *et al.* (1976).

only limited study, but from these data they appear to be remarkably humid and mild in temperature (Barry 1978, 1980). Allison & Bennett (1976) found a virtual lack of seasonal climatic change on Mount Jaya, and diurnal ranges in temperature were surprisingly small as well. They reported daily changes of only 3.4 °C at 3600 m and 2.7 °C at 4250 m. Studies on Mount Wilhelm likewise have indicated a remarkably equitable climate at high elevations (McVean 1968; Hnatiuk *et al.* 1976). Seasonal studies at the Pinaunde research station at 3480 m found a seasonal range of 2 °C mean monthly temperatures, with mean maximum temperatures of 11 °C and mean minimum temperatures of 4 °C (Figure 2.10). While this seasonal and daily range were not strikingly different from those present at Mucubají at a similar elevation in the Venezuelan Andes (Figure 2.1), the extremes of temperature were much more moderate. Extreme highs and lows recorded at Pinaunde were 14 and 0 °C (Hnatiuk *et al.* 1976) compared with 22.2 and −8.6 °C at Mucubají (Azócar & Monasterio 1980; Sarmiento 1986). Regular heavy cloud cover and high atmospheric humidity may contribute significantly to the moderate temperature conditions. It is not surprising that freezing temperatures are rare below 3500 m. Exceptional frosts extending as low as 1600 m in valleys have occurred, however, associated with rare events of extended drought and thus clear skies (Brown & Powell 1974).

Annual precipitation is notably high throughout the New Guinea alpine region, but diminishes from west to east. Rainfall seasonality is pronounced in the Eastern Highlands of Papua New Guinea. There is a brief dry season in June and July, but each still receives about 100 mm of monthly precipitation. Annual rainfall reaches a peak of 3400 mm at about 3500 m, falling only slightly to about 2900 mm at the summit (Hnatiuk *et al.* 1976). Snowfall is frequent above 4000 m, and may occur in any months of the year. In contrast, the Western Highlands of New Guinea in Irian Jaya appear to be much less seasonal in precipitation and total annual rainfall is probably higher than in Papua New Guinea (Allison & Bennett 1976).

The upper limit of forests in the New Guinea Highlands is notably high, extending to 3800–3900 m (Smith 1977a, b, 1980b). This is 600–700 m higher than typical limits in other tropical alpine regions. Mild temperatures, low frost frequency, and humid conditions all undoubtedly act to favor this high forest line.

Numerous islands in Indonesia (Sumatra, Java, Bali and Lombok) have peaks which exceed 3000 m in elevation, but these are too low to have true alpine habitats in this region. Mount Kinabalu in Northern Borneo, however, reaches 4101 m, extending well into the alpine zone. Rainfall is thought to be exceedingly high on the mountain, due perhaps to its proximity to the coast. Smith (1977b, 1980a) has suggested that annual precipitation continues to rise with elevation, with more than 3000 mm at 3350 m and perhaps as much as 5000 mm at the summit. Rainfall occurs throughout the year, with the wettest conditions occurring from November to January. Despite the high rainfall, páramo-like communities are not present on Mount Kinabalu. This surprising absence is due to the high treeline present and the bare rocky summit area with little soil development.

Hawaiian Islands

Three high volcanic peaks are present in the Hawaiian Islands. Haleakala on Maui reaches to 3094 m, while the younger Mauna Loa and Mauna Kea on the Big Island extend to 4169 and 4205 m, respectively. Diurnal temperature ranges at 3440 m on the slopes of Mauna Loa average about 10 °C, similar to the range of 8–12 °C at the summit of Mauna Kea (Figure 2.11). Seasonal ranges of mean monthly temperature

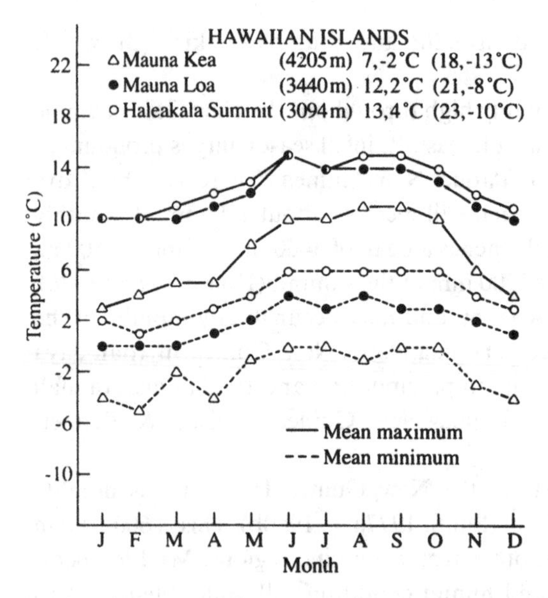

Figure 2.11. Seasonal patterns of mean maximum and mean minimum monthly temperatures for the high elevation Hawaiian volcanoes. Data are shown for elevation, mean annual maximum and minimum and extreme maximum and minimum. The data presented are for the summits of Mauna Kea and Haleakala, and the north slope of Mauna Loa (Doty & Mueller-Dombois 1966).

are about 4 °C for all the Hawaiian high volcanoes. Strong trade-wind inversions produce peak rainfall levels at relatively low elevations and these summits are arid. Mean annual precipitation on the north slope of Mauna Loa at 3440 m, in the lee of Mauna Kea, is less than 500 mm, rainfall at the summit of Mauna Kea is estimated to be about 350 mm. There is little seasonality to precipitation in these high volcanoes, with only a slight tendency toward drier conditions in June (Figure 2.12). Haleakala, however, exhibits a marked dry season in summer (Price 1983).

Conclusions

Tropical high mountain environments share the characteristic of large diurnal cycles in temperature coupled with little or no seasonality in mean temperature. It is this trait that strongly distinguishes these tropical alpine ecosystems from those of temperate and boreal mountain ranges. Large diurnal temperature shifts in temperature and the frosts that they bring have strong effects on plant vegetation patterning and zonation, both

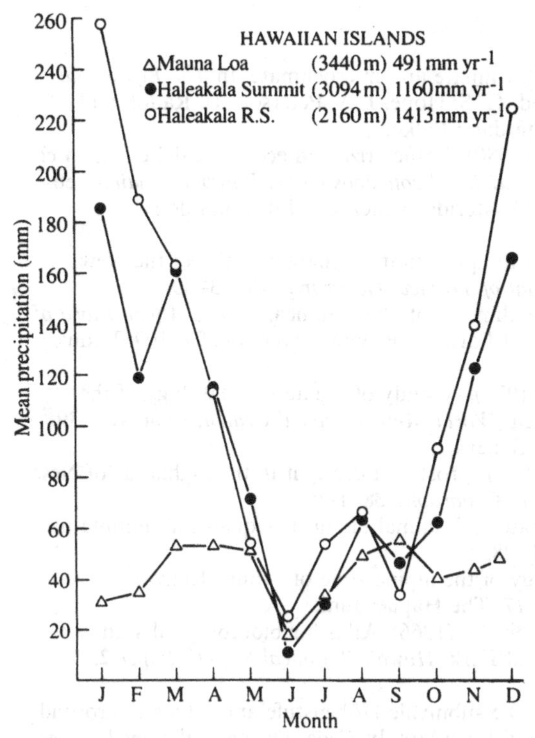

Figure 2.12. Seasonal patterns of mean monthly precipitation for two high elevation Hawaiian volcanoes. Data from Doty & Mueller-Dombois (1966) and US National Park Service (unpublished).

directly and indirectly through frost heaving in soils. Within tropical alpine environments, the frequency and duration of frosts may play a major role in delineating patterns of community distribution along elevational gradients. Precipitation is a much less predictable trait of tropical alpine environments. The amount and seasonality of precipitation varies greatly with elevation, slope orientation and geographic position in relation to oceanic influences. These variables of rainfall and the duration of annual drought conditions certainly have profound biological influence as well. While it is the macroclimatic characteristics of temperature, frost and precipitation that determine broad patterns of vegetation zonation, however, it is microclimatic conditions and gradients that influence the growth and survival of individual plant species. These microclimatic conditions are the theme of the following chapter.

References

Allison, I. & Bennett, J. (1976). Climate and microclimate. In *The Equatorial Glaciers of New Guinea*, ed. G. S. Hope, J. A. Peterson, U. Radok and I. Allison, pp. 61–81. Rotterdam: Balkema.

Azócar, A. & Monasterio, M. (1980). Caracterización ecológica del clima en el Páramo de Mucubají. In *Estudios Ecológicos en los Páramos Andinos*, ed. M. Monasterio, pp. 207–23. Mérida, Venezuela: Ediciones de la Universidad de los Andes.

Barry, R. G. (1978). Aspects of the precipitation characteristics of the New Guinea Mountains. *Journal of Tropical Geography* **47**, 13–30.

Barry, R. G. (1980). Mountain climates of New Guinea. In *The Alpine Flora of New Guinea*, Vol. I. General Part, ed. P. Van Royen, pp. 74–109. Vaduz: Cramer.

Brown, L. H. & Cocheme, J. (1973). A study of the agroclimatology of the highlands of Eastern Africa. *World Meteorological Organization No. 339*. FAO–UNESCO–WMO, Geneva.

Brown, M. & Powell, J. M. (1974). Frost and drought in the highlands of New Guinea. *Journal of Tropical Geography* **38**, 1–6.

Coe, M. J. (1967a). Microclimate and animal life in the equatorial mountains. *Zoologica Africana* **4**, 101–28.

Coe, M. J. (1967b). The ecology of the alpine zone of Mount Kenya. *Monographiae Biologicae 17*. The Hague: Junk.

Doty, M. S. & Mueller-Dombois, D. (1966). Atlas for bioecological studies in Hawaii Volcanoes National Park. *Hawaii Botanical Society Paper* **2**, 1–509.

Furrer, G. & Graf, K. (1978). Die subnivale Höhenstufe am Kilimandjaro und in den Anden Boliviens und Ecuadors. In *Geoecological Relations between the Southern Temperate Zone and the Tropical Mountains*, ed. C. Troll and W. Lauer, pp. 441–57. Wiesbaden: Franz Steiner.

Garcia, E. (1970). Los climas del Estado de Vera Cruz. *Anales Institut Biologia, Universidad National Autonoma México, Séries Botánica* **1**, 3–42.

Hastenrath, S. (1968). Certain aspects of the three-dimensional distribution of climate and vegetation belts in the mountains of Central America and southern Mexico. In *Geo-ecology of the Mountainous Regions of the Tropical Americas*, ed. C. Troll, pp. 122–30. Bonn: Dümmlers.

Hastenrath, S. (1978). On the three-dimensional distribution of subnival soil patterns in the high mountains of East Africa. In *Geoecological Relations between the Southern Temperate Zone and the Tropical Mountains*, ed. C. Troll and W. Lauer, pp. 458–81. Wiesbaden: Franz Steiner.

Hedberg, O. (1951). Vegetation belts of the East African mountains. *Svensk Botanisk Tidskrift* **45**, 140–202.

Hedberg, O. (1964). Features of Afroalpine plant ecology. *Acta Phytogeographica Suecica* **49**, 1–144.

Hnatiuk, R. I., Smith, J. M. B. & McVean, D. N. (1976). *The climate of Mt Wilhelm. Mt Wilhelm Studies 2*. Research School of Pacific Studies Publication BG/4. Canberra: Australian National University.

Humboldt, A. von (1807–39). *Voyage aux régions equinoctiales du Nouveau Continent, fait en 1799–1804*, 30 vols. Paris.

Johnson, A. M. (1976). The climate of Peru, Bolivia and Ecuador. In *Climates of Central and South America*, ed. W. Schwerdtfeger, pp. 147–218. Amsterdam: Elsevier.

Klute, F. (1920). *Ergebnisse der Forschungen am Kilimandscharo 1912.* Berlin: Dietrich Riemer.

Lauer, W. (1968). Problemas de la división fitogeográfica en America Central. In *Geo-ecology of the Mountainous Regions of the Tropical Americas*, ed. C. Troll, pp. 139–56. Bonn: Dummlers.

Lauer, W. (1973). The altitudinal belts of the vegetation in the central Mexican highlands and their climatic conditions. *Arctic and Alpine Research* **5**, A99–A113.

Lauer, W. (1976). Zur hygrischen Höhenstufung tropischer Gebirge. In *Neotropische Oekosysteme*, ed. F. Schmithüsen, pp. 169–82. The Hague: Junk.

Lauer, W. and Klaus, D. (1975). Geoecological investigations on the timberline of Pico de Orizaba, Mexico. *Arctic and Alpine Research* **7**, 15–330.

List, R. J. (1971). *Smithsonian Meteorological Tables*, 6th edn. Washington, DC: Smithsonian Institution Press.

Löffler, E. (1975). Beobachtungen zur periglazialen Höhenstufe in den Hochgebirgen von Papua New Guinea. *Erdkunde* **29**, 285–92.

McVean, D. N. (1968). A year of weather records at 3480 m on Mt. Wilhelm, New Guinea. *Weather* **23**, 377–81.

Monasterio, M. & Reyes, S. (1980). Diversidad ambiental y variación de la vegetación en los páramos de la Andes Venezolanos. In *Estudios Ecológicos en los Páramos Andinos*, ed. M. Monasterio, pp. 47–91. Mérida, Venezuela: Ediciones de la Universidad de los Andes.

Paffen, K. H. (1966). Die tägliche Temperaturschwankung als geographisches Klimacharakteristikum. *Erdkunde* **20**, 252–63.

Paffen, K. H. (1967). Das Verhältnis der tages – zur jahrzeitlichen Temperaturschwankung. *Erdkunde* **21**, 94–111.

Price, S. (1983). Climate. In *Atlas of Hawaii*, 2nd edn, ed. R. W. Armstron, pp. 59–66. Honolulu: University of Hawaii Press.

Salt, G. (1954). A contribution to the ecology of upper Kilimanjaro. *Journal of Ecology* **42**, 375–423.

Sarmiento, G. (1986). Ecological features of climate in high tropical mountains. In *High Altitude Tropical Biogeography*, ed. F. Vuilleumier and M. Monasterio, pp. 11–45. Oxford: Oxford University Press.

Smith, J. M. B. (1977a). Vegetation and microclimate of east- and west-facing slopes in the grasslands of Mt. Wilhelm, Papua New Guinea. *Journal of Ecology* **65**, 39–53.

Smith, J. M. B. (1977b). An ecological comparison of two tropical high mountains. *Journal of Tropical Geography* **44**, 71–80.

Smith, J. M. B. (1980a). The vegetation of the summit zone of Mt. Kinabalu. *New Phytologist* **84**, 547–73.

Smith, J. M. B. (1980b). Ecology of the high mountains of New Guinea. In *The Alpine Flora of New Guinea*, Vol. I, General Part, ed. P. Van Royen, pp. 11–1131. Vaduz: Cramer.

Stevens, G. C. & Fox, J. F. (1991). The causes of treeline. *Annual Review of Ecology and Systematics* **22**, 177–91.

Thompson, B. W. (1966). The mean annual rainfall of Mt. Kenya. *Weather* **21**, 48–49.

Troll, C. (1944). Strukturböden, Solifluktion und Frostklimate der Erge. *Geologisches Rundschau* **35**, 545–694.

Troll, C. (1949). Schmelzung und Verdunstung von Eis und Schnee in ihrem Verhältnis zur geographischen Verbreitung der Ablationsformen. *Erdkunde* **4**, 18–29.

Troll, C. (1959). Die tropischen Gebirge: Ihre dreidimensionale klimatische und pflanzengeographische Zonierung. *Bonner Geographische Abhandlungen* **25**, 1–93.

Troll, C. (1968). The cordilleras of the tropical Americas: aspects of climatic, phytogeographic and agrarian ecology. In *Geo-ecology of the Mountainous Regions of the Tropical Americas*, ed. C. Troll, pp. 15–56. Bonn: Dummlers.

Troll, C. (1973). The upper timberlines indifferent climatic zones. *Arctic and Alpine Research* **5**, A3–A18.

Von Hagen, V. W. (1948). *The Green World of the Naturalists*. New York: Greenberg.

Weber, H. (1959). *Los Páramos de Costa Rica y su Concatenación Fitogeográfica con los Andes Suramericanos*. San José: Instituto Geográfico de Costa Rica.

Weischet, W. (1969). Klimatologische Regeln zur Verticalverteilung der Niederschlage in den Tropengebirgen. *Die Erde* **100**, 287–306.

Whymper, E. (1892). *Travels Amongst the Great Andes of the Equator*. New York: Charles Scribner's Sons.

Winterhalder, B. P. & Thomas, R. B. (1978). Geoecology of southern highland Peru: A human adaptation perspective. *Institute of Arctic and Alpine Research Occasional Paper No.* 27.

Yacono, D. (1968). Essai sur le climat de montagne au Sahara, l'Ahaggar. *Travaux de l'Institut de Recherches Sahariennes, No.* 1.

3

Páramo microclimate and leaf thermal balance of Andean giant rosette plants

F. C. MEINZER, G. GOLDSTEIN and F. RADA

Introduction

General climatic features of tropical alpine regions have been discussed in Chapter 1 of this volume. To reiterate, the nearly complete lack of temperature seasonality in tropical alpine zones is a key feature in distinguishing them from temperate alpine zones. The Andean páramo zone and similar zones in other tropical high mountains are characterized by high inputs of solar radiation in the presence of low inputs of thermal energy. This characteristic might be expected to present special circumstances from the standpoint of regulation of leaf thermal balance, in contrast to temperate alpine and desert habitats where both solar radiation and thermal energy inputs may be seasonally high. It has been suggested that some of the prominent morphological features found in giant rosette plants (see Chapter 1) represent adaptations for regulation of thermal balance under the special microclimatic conditions encountered in tropical mountains (Hedberg 1964; Larcher 1975).

Numerous studies have dealt with the importance of characteristics such as leaf absorptance to solar radiation, leaf angle and rate of transpirational cooling as determinants of leaf temperature under a given set of environmental conditions (Mooney *et al.* 1977; Geller & Smith 1982). However, fewer studies have examined the interaction between spatial and temporal changes in environmental variables and plant features thought to be important for regulation of leaf thermal balance (Smith & Nobel 1977; Ehleringer & Mooney 1978). Ideally, this type of study would be carried out with a single species or several closely related species of similar morphology known to occur over a wide range of environmental conditions (Ehleringer *et al.* 1981; Field *et al.* 1982). The focus of this chapter will be a discussion of certain morphological and behavioral features of Andean giant rosette plants of the genus *Espeletia* and their influence on leaf thermal balance, under microenvironmental conditions that change both with elevation, and vertically within sites at a given elevation.

Elevational changes in environmental factors

General patterns of decreasing temperature with increasing elevation in the Venezuelan Andes and other tropical and temperate mountains have been described by Rundel in Chapter 2. In addition to temperature, several other factors that influence energy exchange between the leaf and its environment exhibit altitude-dependent changes. These factors affect leaf temperature through their influence on radiation, convection and evaporation, the principal mechanisms of energy exchange. Among the most important of these parameters are solar and longwave radiation, the volumetric heat capacity of the air (ρC_p), radiative resistance (r_r), and the diffusion coefficients for sensible heat (D_H) and water vapor (D_V) in air. The latter two exert a direct influence on resistances to convective and evaporative heat transfer, respectively. An appreciation of how these parameters may ultimately determine leaf temperature can be gained by examining standard energy balance equations (Monteith 1973; Campbell 1977; Jones 1992). Upon examining the equations it should become apparent that the overall effect of elevation on leaf temperature is not entirely predictable unless certain standard weather and microsite conditions are specified. Nevertheless, greater leaf-to-air temperature differences at higher elevations are predicted when simulations are performed using energy balance equations. Observations in temperate (Smith & Geller 1979; Field *et al.* 1982) and tropical mountains (Korner *et al.* 1983; F. Meinzer and G. Goldstein, unpublished observations) seem to confirm this.

The discussion in this section will be restricted to those physical properties of the atmosphere which show 'intrinsic' or consistent changes with elevation regardless of local variations in weather or microsite. Calculated changes in some of these properties along an altitude gradient in the Venezuelan Andes are shown in Fig. 3.1. While it is obvious that these factors vary dramatically between sea level and the upper limit of the páramo zone at 4600 m, it is also apparent that they change significantly within the páramo zone (*c.* 3000–4600 m). This points to the necessity of always including altitude corrections when calculations are to be performed with energy balance equations.

The increases in D_H and D_V with increasing elevation should favor

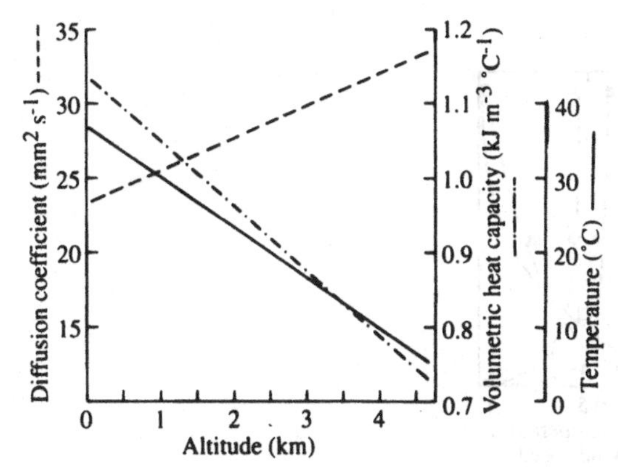

Figure 3.1. Calculated changes in mean maximum air temperature, the diffusion coefficient for sensible heat in air (D_H), and the volumetric heat capacity of the air (ρC_p) with increasing altitude in the Venezuelan Andes. The air temperature lapse rate determined from climatic data for the region, and a function describing decreasing pressure with increasing elevation were used to calculate D_H and ρC_P.

enhanced rates of convective and evaporative cooling, respectively. On the other hand, the reduction in ρC_p with increasing elevation would diminish the efficiency of convective cooling because of the reduced capacity of the air as a sink for sensible heat. 'Resistance' to radiative heat transfer,

$$r_r = \frac{\rho C_p}{4\varepsilon\sigma T^3}$$

where ε is the longwave emissivity of the leaf, σ is the Stefan–Boltzmann constant and T is temperature in °K, should decrease with increasing elevation but the rate will depend on the air temperature lapse rate. In giant rosette plants elevational changes in characteristics, such as rosette geometry, leaf boundary layer thickness, absorptance to solar radiation and stomatal opening, interact with intrinsic changes in physical parameters of the external environment to determine patterns of regulation of leaf thermal balance.

Vertical profiles within sites

Typical vertical profiles of air temperature and wind speed for two páramo sites in the Venezuelan Andes are shown in Fig. 3.2. As expected, air temperatures are lower at the higher (4200 m) site. Differences in the

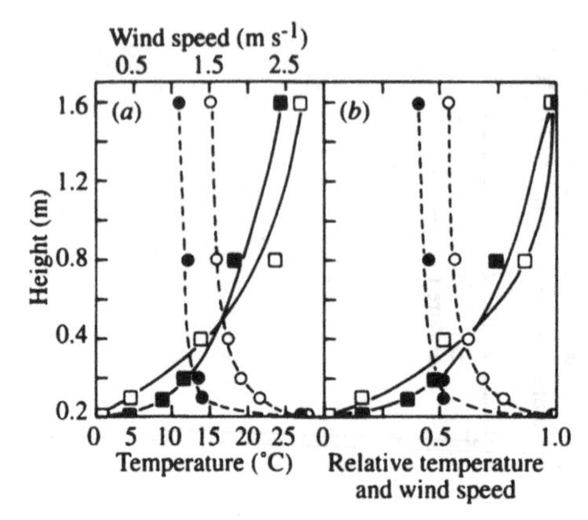

Figure 3.2. Representative vertical profiles (*a*), (*b*) of air temperature (○, ●) and wind speed (□, ■) for Piedras Blancas Páramo at 4200 m (closed symbols) and Mucubají Páramo at 3550 m (open symbols). Data were collected between 1200 and 1300 h local time on clear days during January, 1983.

density and stratification of the vegetation at the two sites (Figure 3.3) may account for the different shapes of the vertical profiles. For example, although wind speeds at 1.6 m are similar for the two sites, the more open high altitude site shows an initial steeper rise in wind speed with increasing height (Figure 3.2*b*). This is reflected in a steeper decrease in air temperature with increasing height for the 4200 m site (Figure 3.2*b*). This vertical differentiation within sites could have important implications for regulation of thermal balance in leaves of juveniles and adults of caulescent *Espeletia* species, and in adults of co-occurring acaulescent and caulescent species.

Plant characteristics and leaf thermal balance

At the leaf level *Espeletia* species exhibit significant variation in a number of features that interact to influence leaf temperature. Among the most important of these features are leaf absorptance to solar radiation, boundary layer resistance to convective and latent heat transfer, and stomatal conductance to water vapor. Boundary layer resistance to heat transfer has two components. One of these is a variable aerodynamic component which is a function of leaf dimensions and wind speed. The magnitude of the other component is fixed and determined by the thickness of leaf pubescence. High boundary layer resistance, by impeding heat loss, would tend to raise leaf temperature. On the other hand, the

Figure 3.3. (*a*) Piedras Blancas Páramo and (*b*) Mucubají Páramo near the sites where temperature and wind speed profiles shown in Fig. 3.2 were measured.

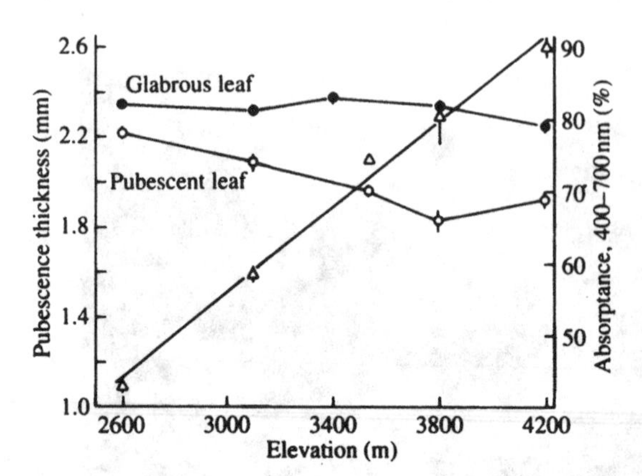

Figure 3.4. Thickness of pubescence covering the upper leaf surface (△) and absorptance to solar radiation (400–700 nm) for intact leaves (○) and leaves with their hairs artificially removed (●) in five *Espeletia schultzii* populations along an elevation gradient. Bars represent ±SE; pubescence $n = 80$; absorptance $n = 10$ (From Meinzer *et al.* 1985).

leaf pubescence also reduces leaf absorptance to solar radiation (Meinzer *et al.* 1985), which would tend to lower leaf temperature. In some *Espeletia* species, pubescence effects on boundary layer resistance outweigh those on absorptance to solar radiation, resulting in higher leaf temperatures than would occur in a glabrous leaf at high radiation loads (Meinzer & Goldstein 1985). Attenuation of photosynthetically active radiation by the pubescent layer has been predicted to exact a relatively high cost in terms of reduced carbon dioxide assimilation of *Espeletia* species under some environmental conditions (Goldstein *et al.* 1989).

Pubescence-induced temperature increases have previously been demonstrated for leaves (Wuenscher 1970) and other structures such as inflorescences (Krog 1955). In five populations of *E. schultzii* along an elevation gradient from 2600 to 4200 m, thickness of leaf pubescence increased dramatically while the associated decline in leaf absorptance to solar radiation was relatively small (Figure 3.4). These increases in fixed boundary layer thickness result in potentially higher leaf temperatures relative to air temperature at higher elevations (Meinzer *et al.* 1985). A similar pattern of increasing pubescence thickness and higher plant temperatures relative to air temperature with increasing elevation has been observed for pubescent *Puya* inflorescences in the Ecuadorian Andes (Miller 1986 and Chapter10).

Some leaf characteristics of several *Espeletia* species are summarized in Table 3.1. Their net effect on leaf temperature is not easily predicted from

Table 3.1 *Summary of leaf characteristics for several* Espeletia *species*

Species	Elevation (m)	Absorptance 400–700 nm		Pubescence thickness (mm)	Leaf width (cm)	Resistance ratio[b]
		Intact	Glabrous[a]			
E. timotensis	4200	0.70	0.83	2.3	3.7	4.7
E. lutescens	4200	0.61	0.89	1.8	3.4	4.0
E. schultzii	4200	0.69	0.79	2.6	2.7	5.9
E. spicata	4200	0.60	0.79	0.9	1.9	3.0
E. moritziana	4200	0.63	–	1.0	1.4	3.6
E. floccosa	3550	0.61	–	0.4	1.1	2.2
E. lindenii	2800	0.71	–	0.2	2.4	1.4
E. atropurpurea	3100	–	0.87	0	6.0	1

[a] Absorptances for glabrous leaves were obtained by plucking the pubescence from the leaf.
[b] Resistance ratio is that of sensible heat transfer for intact and glabrous leaves, calculated at a wind speed of 1.5 m s^{-1}.

a simple inspection of the table. It is apparent that, although pubescence invariably decreases absorptance to solar radiation with respect to a glabrous leaf, there is no simple relationship between the thickness of the pubescence and leaf absorptance. The most useful indicators in Table 3.1 for assessing the balance between the effects of pubescence on boundary layer resistance and solar radiation absorption are probably the ratio of sensible heat transfer resistance for pubescent and glabrous leaves, and the difference in absorptance between intact leaves and leaves with their pubescence removed. The possibility that leaf pubescence may have different consequences for thermal balance in different *Espeletia* species will be discussed in the next section.

There is good evidence that the basic rosette geometry of *Espeletia* species strongly influences leaf thermal balance, especially near the apical region. Previous measurements have demonstrated considerable heating in the apical region of *E. schultzii* (Larcher 1975) and this has been attributed to a parabolic heating effect in the rosette, in which the apical bud would form the focus of a paraboloid and the older leaves would form the walls (Smith 1974). When features of rosette geometry in several populations of *E. schultzii* were used to calculate the foci of hypothetical paraboloids formed by the rosettes, it was found that a portion of the apical bud always fell within the focal region (Meinzer *et al.* 1985). It is questionable, however, whether a significant proportion of the warming observed in apical leaves of Andean giant rosette plants arise from a true parabolic heating effect. Unlike a parabolic mirror which exhibits specular reflectance, the surface of pubescent *Espeletia* leaves exhibits diffuse reflectance. Furthermore, even if the rosette leaves were to form a specular surface, true parabolic heating could take place only when the sun's rays were parallel to the axis of the parabola formed by the rosette. Nevertheless, heat storage in the apical bud of species such as *E. schultzii* is promoted by the basic rosette geometry. Instead of receiving an unobstructed view of the sky and, therefore, the low effective sky temperature, the apical bud is surrounded by a longwave source radiating at a temperature often several degrees above air temperature. The amount of shortwave radiation incident on the apex would also be enhanced by reflection from the pubescent layer of the surrounding leaves.

As suggested earlier, intra- and interspecific differences in rosette height above the ground might be expected to have a significant effect on regulation of leaf temperature. Vertical air temperature profiles in páramo sites are such that temperature gradients of 10 °C between a few centimeters and 1 m above the surface are not uncommon (Figure 3.2). There is some evidence to suggest that within a given páramo site, giant rosette species may exhibit a surprising degree of homeostasis in temperature control regardless of plant height. At 3550 m *E. schultzii* and *E. floccosa* typically exhibit similar average leaf temperatures during periods of moderate to high incident solar radiation (Figure 3.5) despite the different regions of the air temperature profile in which their rosettes are situated. Thus, the nearly sessile rosettes of *E. floccosa* exhibit much smaller leaf-to-air temperature differences than do the taller (up to 1 m) rosettes of *E. schultzii.* Leaf characteristics of these two species probably account for much of the difference in regulation of leaf temperature with respect to air temperature (Table 3.1). While both species have pubescent leaves, the pubescence of *E. floccosa* is more reflective, and represent a less significant contribution to boundary layer resistance to heat transfer than that of *E. schultzii.* Simulations carried out with energy balance equations predict that under most conditions this should result in smaller leaf-to-air temperature differences for *E. floccosa.* It is interesting to note that, unlike the matted, 0.4 mm thick pubescence of its rosette leaves, pubescence of the inflorescence of *E. floccosa* is 2–3 mm thick. These inflorescences occupy a portion of the vertical profile similar to that of *E. schultzii* rosettes. Pubescence on *E. floccosa* inflorescences may have consequences similar to those suggested for Andean *Puya* species (Miller 1986 and Chapter 10).

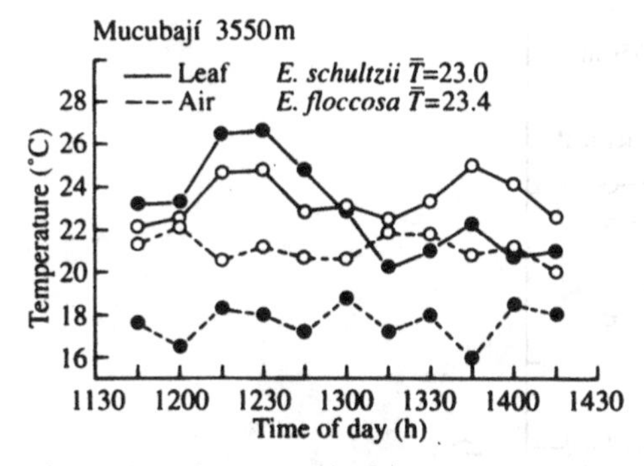

Figure 3.5. Courses of leaf temperature, and air temperature at mid-rosette height, for *Espeletia schultzii* and *E. floccosa* during a clear, warm dry season day in the Mucabají Páramo. Air temperatures were measured at 70 cm for *E. schultzii* and 15 cm for *E. floccosa.* Leaf temperatures for each species are averages of eight leaves at different positions in the rosette.

Significance of plant characteristics and patterns of regulation of plant temperature

It has been suggested that a principal consequence of variation in some of the plant features described above is to decouple plant temperatures and physiological processes from unfavorably low environmental temperatures (Meinzer *et al.* 1985). In high páramo sites environmental temperatures may frequently be suboptimal for translocation and growth. Leaf and apical bud temperatures in many *Espeletia* species are commonly 5–15 °C above air temperature (Meinzer & Goldstein 1986). Since a typical Q_{10} for translocation and leaf expansion may be about 2, temperature increases of even a few degrees could be highly significant, especially in an environment lacking temperature seasonality. Some instances in which this hypothesis has been used to interpret and predict pattern of regulation of plant temperature and variation in plant features are discussed below.

The ability of the apical bud core of *Espeletia* species to remain above 0 °C during the night is well documented (Smith 1974; Rada 1983). Nocturnal loss of stored heat is retarded by leaf nyctinasty and the presence of pubescence between successive layers of leaves which reduces heat conduction and convection from one leaf layer to the next. The principal adaptive consequence of this diurnal heat storage may be more favorable temperatures for apical growth and leaf expansion rather than protection from freezing, however. Apical bud tissue and young leaves of many *Espeletia* species have the capacity to avoid freezing by supercooling.

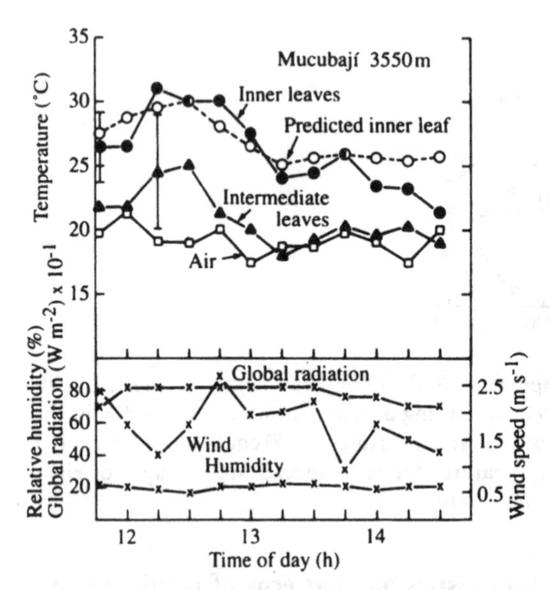

Figure 3.6. The influence of leaf position on temperature in an *E. schultzii* rosette during a portion of a clear warm way in the Mucubají Páramo. Bars represent the largest $\pm$SE ($n = 4$). Air temperature, humidity and wind speed were measured at mid-rosette height nearby (from Meinzer *et al.* 1985).

Freezing exotherms occur at temperatures well below minimum air and leaf temperatures commonly measured in the field (Goldstein *et al.* 1985; Rada *et al.* 1985).

On clear days, with their corresponding high levels of incident solar radiation, the temperature of the inner expanding and recently expanded rosette leaves is higher than that of older leaves with an intermediate position (Figure 3.6). Leaf–air temperature differences for individual inner rosette leaves are often as high as 15 °C. This heat storage in the apical region may partly explain the observation that leaf production and growth rates in *Espeletia* species from high páramos are relatively rapid considering the low mean air temperature (Estrada 1984). A similar adaptive consequence, promotion of growth, has been proposed to result from northerly orientation and elevated apical temperature in the Atacama Desert cactus *Copiapoa* (Ehleringer *et al.* 1980).

Potential increases in leaf temperature caused by high radiation loads in the apical region of *Espeletia* rosettes are further enhanced by variations in two features which exert their influence at the leaf level. Increasing leaf angle in the rosette from nearly horizontal senescent leaves to vertical expanding leaves corresponds to a gradient of increasing

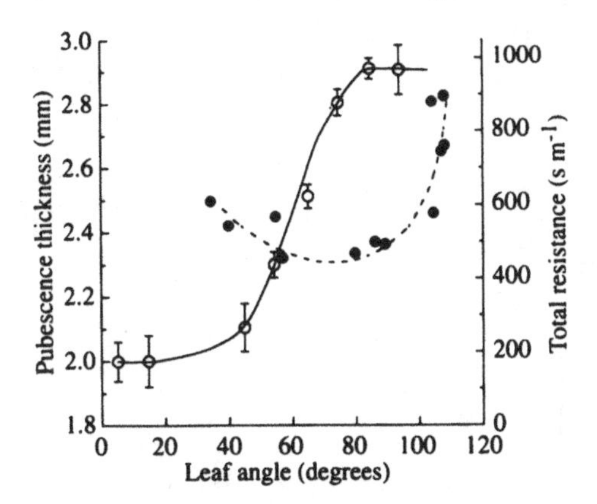

Figure 3.7. Pubescence thickness (○) and total diffusive resistance (●) in relation to leaf angle (inversely related to leaf age) in an *E. timotensis* rosette. Pubescence thicknesses at a given leaf angle are means (±SE) for 7–10 leaves. Angle is 0° for a horizontal leaf. Resistance measurements are for individual leaves (from Meinzer & Goldstein 1985).

thickness of pubescence and stomatal resistance (Figure 3.7). Energy balance equations predict that this gradient of increasing boundary layer and total diffusive resistance with decreasing leaf age would promote greater coupling between incident radiation and temperature in young leaves.

It is apparent that leaf pubescence is one of the key factors involved in regulation of leaf thermal balance of many *Espeletia* species. Leaf pubescence has been shown to be a major determinant of leaf temperature in many desert species where it typically results in reduced absorptance to solar radiation, reduced leaf temperature and thus lower transpirational losses (Smith & Nobel 1977; Ehleringer & Bjorkmann 1978; Ehleringer & Mooney 1978; Ehleringer *et al.* 1981). Thus, in desert plants, the influence of leaf hairs on solar radiation absorption predominates, while in many *Espeletia* species their influence on boundary layer resistance to heat transfer predominates. This suggests that evolution of leaf pubescence in plants of hot, arid habitats and high tropical mountains has occurred in response to dissimilar selective pressures. While both habitats are characterized by high incoming solar radiation, patterns of moisture availability and air temperature are quite different. During most of the year in Andean páramos, low temperatures, rather than lack of precipitation, limit water availability. Air temperatures are consistently low with no distinct favorable season for growth. Typical maximum air temperatures

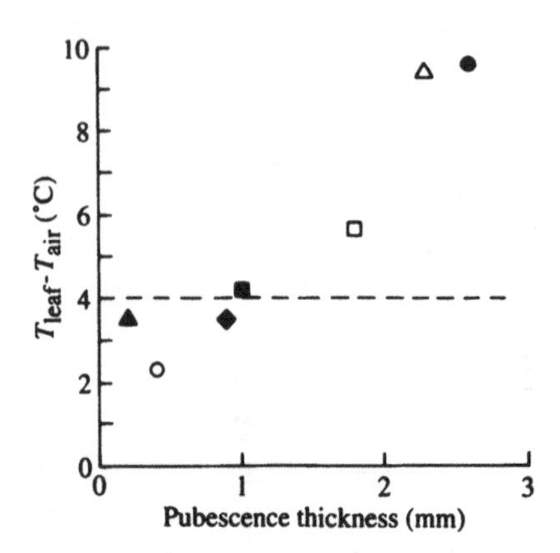

Figure 3.8. Predicted leaf-to-air temperature difference under simulated midday conditions at 3500 m for seven *Espeletia* species with different thicknesses of leaf pubescence. Leaf characteristics from Table 3.1 were used in the simulations. The broken line represents the predicted temperature difference for a completely glabrous *Espeletia* leaf. ●, *E. schultzii* 4200 m; △, *E. timotensis* 4200 m; □, *E. lutescens* 4200 m; ■, *E. moritziana* 4200 m; ◆, *E. spicata* 4200 m; ○, *E. floccosa* 3500 m; ▲. *E. lindenii* (2800 m). Stomatal resistance was held at 200 s m^{-1}, wind speed at 1.5 m s^{-1}, incident solar radiation at 800 W m^{-2}, air temperature at 15 °C, and external humidity at 3.5 g m^{-3} for all simulations.

in warm desert habitats may exceed 40 °C, at least 30 °C above maximum air temperatures in high páramos. While water availability may be limited in both habitats, it is apparent that leaf temperature variations at the higher desert temperatures will have a much greater effect on leaf–to–air vapor pressure gradient and potential rates of transpiration due to the much steeper slope of the saturation vapor pressure curve near 40 °C. The consequences of leaf pubescence in some *Espeletia* species and in desert plants are similar in the sense that in both cases, the pubescence represents a mechanism which decouples plant temperature from air temperature.

It is likely that variation in páramo microclimate, and therefore selective pressures, are sufficient to have caused variation in leaf pubescence and its consequences within the genus *Espeletia.* Figure 3.8 shows the results of a simulation in which pubescence thickness and other leaf characteristics of seven *Espeletia* species from several páramo sites were used to predict leaf–air temperature differences under identical clear, midday environmental conditions.

Since stomatal conductance was held constant for the simulation, the predicted temperature differences are due primarily to the opposing influence of pubescence on boundary layer thickness and absorptance to solar radiation. Leaf dimensions varied among species, but at the wind speed used for the simulation variations in aerodynamic boundary layer did not have a significant effect on leaf temperature. If the leaf–air temperature difference for an 'average' glabrous *Espeletia* leaf is used as a reference, it is predicted that for three of the species the presence of leaf pubescence results in lower leaf temperature relative to air temperature under conditions of high incident solar radiation. The species with the smallest predicted leaf–air temperature difference is *E. floccosa* whose leaves are located in the warmer air layers near the soil surface (Figure 3.5). The three species with the highest predicted leaf–air temperature differences are all caulescent, attaining maximum heights of 1.5–3 m and occurring at elevations up to 4200 m.

Conclusions

In the Andean páramo, variations in temperature that occur diurnally, altitudinally, and along vertical profiles within sites are much larger than seasonal temperature fluctuations. Within the páramo zone giant rosette members of the genus *Espeletia* experience a wide range of microclimates, especially with respect to temperature and wind. These conditions have led to a series of plant features that permit a substantial degree of thermoregulation mainly through heat storage and variable coupling between the plant and its temperature and radiation environment. Control of coupling between air temperature, incident radiation and leaf temperature is achieved principally through variations in leaf spectral characteristics and boundary layer resistance to convective heat transfer. This form of thermoregulation contributes to homeostasis of processes such as carbon dioxide assimilation, translocation and growth along environmental gradients.

References

Campbell, G. S. (1977). *An Introduction to Environmental Biophysics.* Berlin: Springer-Verlag.

Ehleringer, J. & Bjorkmann, O. (1978). Pubescence and leaf spectral characteristics in a desert shrub, *Encelia farinosa. Oecologia* **36**, 151–62.

Ehleringer, J. & Mooney, H. A. (1978). Leaf hairs: effects of physiological activity and adaptive value to a desert shrub. *Oecologia* **37**, 183–200.

Ehleringer, J., Mooney, H. A., Gulmon, S. L. & Rundel, P. W. (1980). Orientation and its consequences for *Copiapoa* (Cactaceae) in the Atacama Desert. *Oecologia* **46**, 63–7.

Ehleringer, J., Mooney, H. A., Gulmon, S. L. & Rundel, P. W. (1981). Parallel evolution of leaf pubescence in *Encelia* in coastal deserts of North and South America. *Oecologia* **49**, 38–41.

Estrada, C. (1984). Dinámica del crecimiento y reproducción de *Espeletia* en el páramo desertico. MS thesis, Universidad de los Andes, Mérida, Venezuela.

Field, C., Chiariello, N. & Williams. W. E. (1982). Determinants of leaf temperature in California *Mimulus* species at different altitudes. *Oecologia* **55**, 414–20.

Geller, G. N. & Smith, W. K. (1982). Influence of leaf size, orientation, and arrangement on temperature and transpiration in three high-elevation, large-leafed herbs. *Oecologia* **53**, 227–34.

Goldstein, G., Rada, F. & Azócar, A. (1985). Cold hardiness and supercooling along an altitudinal gradient in Andean giant rosette species. *Oecologia* **68**, 147–52.

Goldstein, G., Rada, F., Canales, M. O. & Zabala, O. (1989). Leaf gas exchange of two giant caulescent rosette species. *Oecologia Plantarum* **10**, 359–70.

Hedberg, O. (1964). Features of Afroalpine plant ecology. *Acta Phytogeographica Suecica* 49, 1–144.

Jones, H. G. (1992). *Plants and Microclimate*, 2nd edn. Cambridge: Cambridge University Press.

Korner, C., Allison, A. & Hilscher, H. (1983). Altitudinal variation of leaf diffusive conductance and leaf anatomy in heliophytes of montane New Guinea and their interrelation with microclimate. *Flora* **174**, 91–135.

Krog, J. (1955). Notes on temperature measurements indicative of special organization in arctic and sub-arctic plants for utilization of radiated heat from the sun. *Physiolgia Plantarum* **8**, 836–9.

Larcher, W. (1975). Pflanzenökologische Beobachtungen in der Paramostufe der Venezolanischen Anden. *Anzeiger der mathematisch-naturwissenshaftlichen Klasse der Österreichischen Akademie der Wissenschaften* **11**, 194–213.

Meinzer, F. C. & Goldstein, G. (1985). Some consequences of leaf pubescence in the Andean giant rosette plant *Espeletia timotensis*. *Ecology* **66**, 512–20.

Meinzer, F. & Goldstein, G. (1986). Adaptations for water and thermal balance in Andean giant rosette plants. In *On the Economy of Plant Form and Function*, ed. T. Givnish, pp. 381–411. Cambridge: Cambridge University Press.

Meinzer, F. C., Goldstein, G. H. & Rundel, P. W. (1985). Morphological changes along an altitude gradient and their consequences for an Andean giant rosette plant. *Oecologia* **65**, 278–83.

Miller, G. A. (1986). Pubescence, floral temperature and fecundity in species of *Puya* (Bromeliaceae) in the Ecuadorian Andes. *Oecologia* **70**, 155–60.

Montieth, J. L. (1973). *Principles of Environmental Physics*. New York: American Elsevier.

Mooney, H. A., Ehleringer, J. & Bjorkmann, O. (1977). The energy balance of leaves of the evergreen desert shrub *Atriplex hymenelytra*. *Oecologia* **29**, 301–10.

Rada, F. (1983). Mecanismos de resistencia a temperaturas congelantes en *Espeletia spicata y Polylepis sericea*. MS thesis, Universidad de los Andes, Mérida, Venezuela.

Rada, F., Goldstein, G., Azócar, A. & Meinzer, F. (1985). Freezing avoidance in Andean giant rosette plants. *Plant, Cell and Environment* **8**, 501–7.

Smith, A. P. (1974). Bud temperature in relation to nyctinastic leaf movement in an Andean giant rosette plant. *Biotropica* **6**, 263–6.

Smith, W. K. & Geller, G. N. (1979). Plant transpiration at high elevations: theory field measurements, and comparisons with desert plants. *Oecologia* **41**, 109–22.

Smith, W. K. & Nobel, P. S. (1977). Influences of seasonal changes in leaf morphology on water-use efficiency for three desert broadleaf shrubs. *Ecology* **58**, 1033–43.

Wuenscher, J. E. (1970). The effect of leaf hairs of *Verbascum thapsus* on leaf energy exchange. *New Phytologist* **58**, 65–73.

4

Comparative water relations of tropical alpine plants

F. C. MEINZER, G. GOLDSTEIN
and P. W. RUNDEL

Introduction

In tropical alpine regions drought may be the most important seasonal factor in an environment that otherwise lacks significant seasonality (see Smith, Chapter 1). In many of these regions diurnal cycles of physiological drought associated with low soil temperatures may be superimposed on the seasonal changes in soil moisture. From studies of cold temperate zone plants, particularly conifers (Kaufmann 1975, 1977; Running & Reid 1980), it is known that even in soils near field capacity, water uptake by roots may be severely impaired by low soil temperatures (0–5 °C). This high root resistance to water uptake may extend through spring and into early summer in temperate zone coniferous forests with persistent snow cover. In arctic regions, physiological drought may extend throughout the entire summer if the roots are situated above a permafrost layer (Goldstein 1981).

In the tropical alpine zone, where snow cover is not persistent and permafrost does not exist, the potential for physiological drought is nevertheless present. The risk is especially great during the early morning hours when soil temperatures in the root zone are near freezing and potential transpiration is high due to high solar radiation loads. The simultaneous occurrence of low water availability in the absolute sense on a seasonal basis, and in the physiological sense on a diurnal basis, complicates the analysis of drought resistance mechanisms in tropical alpine plants. It also provides an excellent opportunity to study adaptations to drought along both altitudinal and geographical gradients of relative importance of diurnal versus seasonal drought.

The concepts of tolerance and avoidance as components of drought resistance (Levitt 1980) are useful in understanding responses to seasonal and diurnal drought stress. Included among drought tolerance mechanisms are responses such as: (i) osmotic adjustment, which acts to maintain tissue turgor and volume in the face of declining soil water potential; and (ii) a decline in the lethal tissue water content resulting from increased tissue dehydration tolerance. Drought avoidance mechanisms include: (i) internal water storage (capacitance), which can dampen both diurnal and seasonal fluctuations in water status; (ii) temporal displacement of activity; and (iii) sensitive stomatal control of transpiration.

Studies undertaken in the Venezuelan Andes have shown that avoidance is the principal means by which giant rosette members of the genus *Espeletia* resist the effects of diurnal drought. Morning transpiration is partially compensated by removal of water from internal storage, especially from the central stem pith (Goldstein & Meinzer 1983). The effects of high root resistance to water uptake are thus largely avoided by minimizing the resistance between the water source and atmospheric sink. In tropical alpine shrubs and trees (e.g. *Polylepis*) lacking significant internal water storage, drought tolerance mechanisms might be expected to predominate. The discussion in this chapter will focus on the water relations of giant rosette plants, the tropical alpine growth form for which the most extensive water relations data are available. This growth form is well suited for comparative studies of drought resistance mechanisms, since it occurs along gradients over which the relative importance of diurnal versus seasonal drought varies.

Water relations of giant rosette plants

The striking similarity among giant rosette growth forms both within and between the South American and African tropical alpine regions belies considerable variation in diurnal patterns of regulation of water balance (Figures 4.1, 4.2). In this section the role of a number of species, site, and size-specific factors as determinants of the daily course of water balance will be discussed. When leaf water potential is used as an indicator of water balance, two extreme diurnal patterns can often be observed. In the Venezuelan páramo, adult individuals of some *Espeletia* species such as *E. lutescens* may exhibit little diurnal fluctuation in leaf water potential while the fluctuation may be quite pronounced in adults of other co-occurring species such as *E. moritziana* (Figure 4.1). These divergent patterns of leaf water balance can occur in spite of very similar daily courses of stomatal conductance and transpiration. In *Espeletia* species showing significant diurnal water potential fluctuations the daily minimum

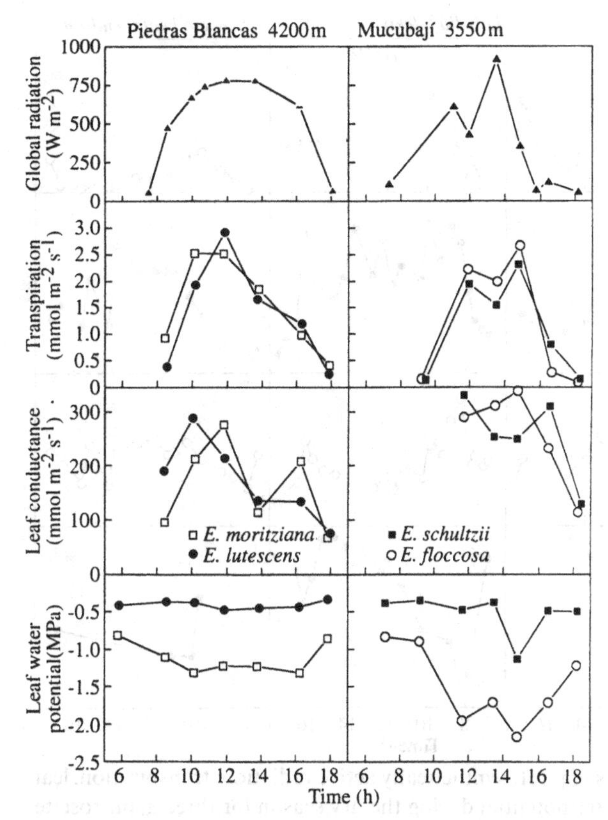

Figure 4.1. Daily courses of global radiation, transpiration, leaf conductance and leaf water potential for four *Espeletia* species in two páramo sites during the dry season (after Goldstein *et al.* 1984).

water potential generally coincides with the period of maximum transpiration, but the rate of water potential decline with increasing transpiration during the morning hours shows considerable species-to-species variation.

The ranges of leaf water potential, leaf conductance, and transpiration rates encountered during the dry season in African giant rosette species (Schulze *et al.* 1985) appear to be similar to those reported for *Espeletia* species. As in the genus *Espeletia*, giant rosette members of the African genera *Lobelia* and *Senecio* show large variation in the magnitude and rate of daily fluctuations in leaf water potential. In the African genera changes in leaf water potential appear to be less coupled to changes in transpiration rate than in the Venezuelan species. In *Lobelia telekii*, for example, leaf water potential can exhibit considerable recovery after

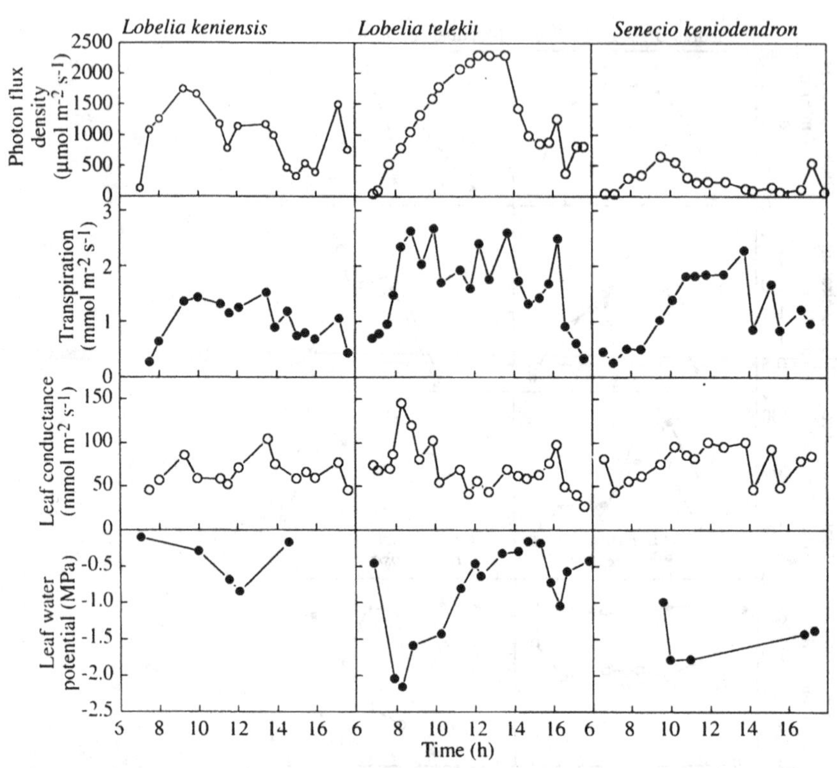

Figure 4.2. Daily courses of photosynthetically active radiation, transpiration, leaf conductance and leaf water potential during the dry season for three giant rosette species growing at 4200 m on Mount Kenya (after Schulze *et al.* 1985).

its early morning decline in spite of a more or less stable average transpiration rate during the late morning and early afternoon (Figure 4.2). Temperature-dependent changes in the hydraulic resistance in the plant, particularly in the leaves, may play an important role in determining the diurnal course of leaf water balance in *L. telekii* (Schulze *et al.* 1985).

Diurnal patterns of water balance in *Espeletia* species are greatly influenced by plant capacitance (Goldstein *et al.* 1984). A quantitative index of relative internal water storage capacity can be obtained by determining the pith volume in the stem per amount of actively transpiring leaf area served (Table 4.1). The physiological significance of this morphological index is that *Espeletia* species with the highest pith volume per unit leaf area show the lowest apparent resistance to water flow (drop in water potential/transpiration) under field conditions (Fig. 4.3). The buffering effect of pith water storage permits early stomatal opening and

Table 4.1 *Absolute and relative pith water storage capacity for adult individuals of seven* Espeletia *species*

Species	Pith volume (cm^3)	Leaf area ($\times 10^3$ cm^2)	Pith volume/ Leaf area ($\times 10^{-2}$ cm^3 cm^{-2})	Available water[a] (g)
E. lutescens	968	8.60	10.5	176
E. moritziana	268	4.85	5.7	57
E. spicata	566	9.96	5.6	160
E. schultzii	466	11.58	4.7	99
E. marcana	270	6.73	3.8	86
E. atropurpurea	29	1.98	1.8	9
E. floccosa	72	5.45	1.3	27

[a] Available water in the pith was calculated according to Goldstein *et al.* (1984).

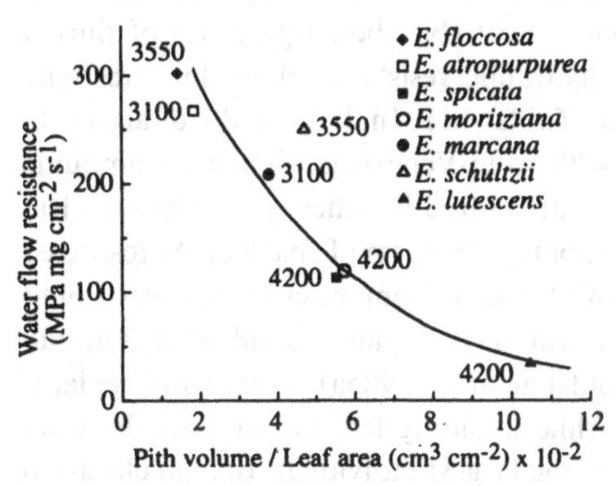

Figure 4.3. The relationship between apparent liquid flow resistance and relative pith water storage capacity for seven *Espeletia* species. Elevations of the study populations of each species are given in meters (after Goldstein *et al.* 1984).

gas exchange with little change in plant water status. Species with the largest relative water storage capacity tend to be those that occur in the highest and coldest sites where the potential for physiological drought is the greatest.

The absolute quantity of available water stored in the pith and other tissues can be estimated if the relationship between water potential and water content for the tissue is known. A conservative estimate of this available water can be obtained if the lower limit for water removal from the tissue is set at a water potential corresponding to the turgor loss point of the leaf tissue. Using this criterion, available water in the pith of adult

Table 4.2 *Water storage in pith, xylem, and leaf tissue of three large individuals of* E. lutescens *growing at 4200 m*

Plant	Height (m)	Leaf area (m^2)	Available water Pith (g)	Xylem (g)	Leaves (g)	Total (g)
1	1.90	1.32	519	382	63	964
2	1.95	1.57	722	511	87	1320
3	2.00	1.15	917	718	60	1695
Mean	1.95	1.35	719	537	70	1326

individuals ranged from 9 to 176 g among seven *Espeletia* species studied (Goldstein *et al.* 1984). An index of relative water storage capacity such as pith volume per unit leaf area appears to be a better predictor of diurnal patterns of water balance and hydraulic resistance than does absolute water storage (Figures 4,1, 4.3; Table 4.1). In large individuals of *E. lutescens*, a species common at 4200 m in Venezuela, significant amounts of physiologically available water are stored in other tissues such as leaf midribs, secondary xylem, and probably periderm (Table 4.2). At restricted transpiration rates water removed solely from internal sources could prevent leaf turgor loss for several days in an individual 2–3 m tall (Goldstein & Meinzer 1983; Goldstein *et al.* 1985a). In terms of replacement of transpirational losses while avoiding leaf turgor loss, the total internal water storage capacity of the largest individuals of high elevation *Espeletia* species exceeds water storage capacities reported for both elastic tissues such as fruits and buds (Jarvis 1975; Hinckley *et al.* 1981) and inelastic tissues such as sapwood (Richards 1973; Waring & Running 1978; Running 1980).

Characteristics of leaf tissue water relations in *Espeletia* reflect the differences in relative water storage capacity and patterns of water balance described above. As might be expected, leaf osmotic potential at zero turgor declines with decreasing relative water storage capacity and therefore greater diurnal fluctuations in leaf water potential (Figure 4.4).

It is interesting to note that these relationships between leaf turgor loss point and water storage capacity observed for *Espeletia* do not seem to fit the pattern observed in the Hawaiian giant rosette species *Argyroxiphium sandwicense* (silversword). The leaves of this species have a turgor loss point of −1.53 MPa (Robichaux *et al.* 1990), but a pith volume to leaf

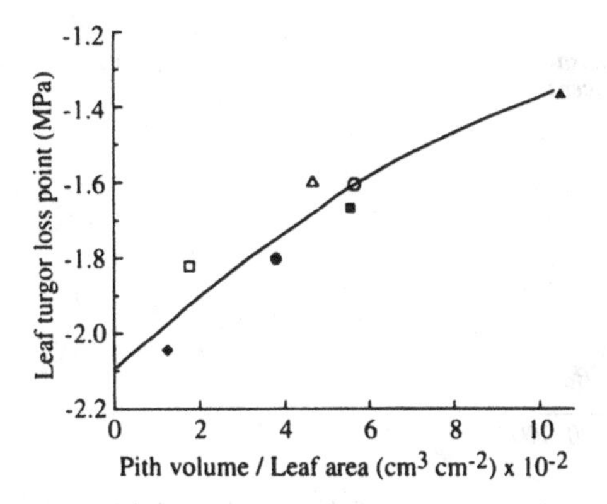

Figure 4.4. The relationship between leaf tissue water potential at the turgor loss point and relative pith water storage capacity for seven *Espeletia* species. Symbols as in Figure 4.3 (after Goldstein *et al.* 1984).

area ratio of only $0.25 \times 10^{-2}\ cm^{-2}$. An Andean *Espeletia* species with an equivalent pith volume to leaf area ratio would be expected to have a leaf turgor loss point of approximately -2.1 MPa (Figure 4.4).

Leaf tissue dehydration tolerance may also be related to patterns of capacitance and leaf water potential components. A comparison of dehydration tolerance for *E. floccosa,* a species with a low relative water storage capacity and low osmotic potential at zero turgor, and *E. lutescens,* a species with high storage capacity and osmotic potential, revealed that irreversible damage (90% resaturation attained) began to occur at water saturation deficit of 40% in *E. floccosa* and only about 20% in *E. lutescens* (Figure 4.5). For *E. lutescens* the water saturation deficit associated with irreversible damage coincided approximately with the water saturation deficit and water potential at zero turgor, but for *E. floccosa* irreversible damage began to occur below the turgor loss point. These observations suggest that diurnal drought avoidance mechanisms may predominate in higher elevation species such as *E. lutescens,* while seasonal drought tolerance mechanisms are more prominent in lower elevation species such as *E. floccosa.* In lower elevation sites in Venezuela, seasonal drought is more severe than diurnal drought. The dominant *Espeletia* species here either form sessile rosettes or have much shorter stem and therefore much lower water storage capacity than species occupying higher sites (Figure 4.3). Water storage of the magnitude found in *Espeletia,* while adequate on a diurnal basis, may offer little protection against prolonged seasonal drought. Thus, lower elevation species tend to have higher leaf tissue osmotic concentrations and therefore greater ability to

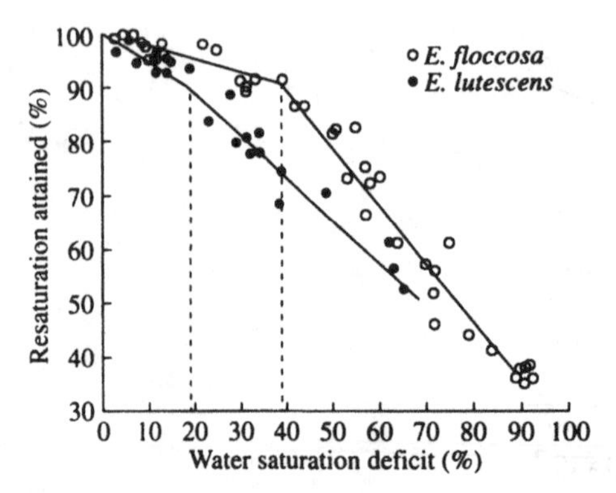

Figure 4.5. Relationship between water saturation deficit experienced by leaf tissue and the subsequent percentage resaturation attained. The critical damage point (90% resaturation) and corresponding saturation deficits are shown (data of F. C. Meinzer and G. Goldstein).

maintain turgor volume, and also exhibit greater tolerance of tissue dehydration.

Studies of intraspecific patterns of plant–water relations in the genus *Espeletia* along both environmental (Meinzer *et al.* 1985) and size (Goldstein *et al.* 1985a) gradients have helped to clarify the influence of species-specific versus habitat-specific features. For example, in five populations of *E. schultzii* distributed along an altitude gradient from 2600 to 4200 m, a 20-fold increase in pith volume per unit leaf area was observed (Figure 4.6). This was due both to an increase in plant height, and therefore pith volume, and to a decrease in total rosette leaf area. If the simultaneous reduction in total leaf area with increasing elevation had not occurred, there would have been only a 10-fold increase in available water due to greater pith volume alone. This pattern supports the idea that pith water storage in giant rosette plants is primarily a habitat-specific response to water availability limited by low temperature at higher elevations. This may partially explain the apparent paradox that height of giant rosette species, unlike temperate tree species, increases with elevation (Hedberg 1964; Mabberley 1973; Smith 1980).

In the high elevation species *E. timotensis*, relative water storage capacity varies greatly with plant height during the life of each individual (Goldstein *et al.* 1985a). Relative storage capacity shows a slow initial increase as both leaf area and stem length are increasing. Then a rapid increase in relative storage capacity with increasing height occurs as rosette area begins to stabilize. As expected, these changes have a

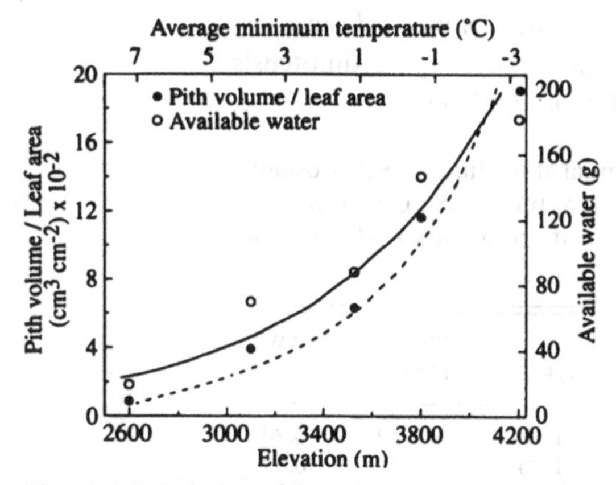

Figure 4.6. Relative pith water storage capacity and mass of available water stored in the pith for five *Espeletia schultzii* populations along an elevation gradient (after Meinzer *et al.* 1985).

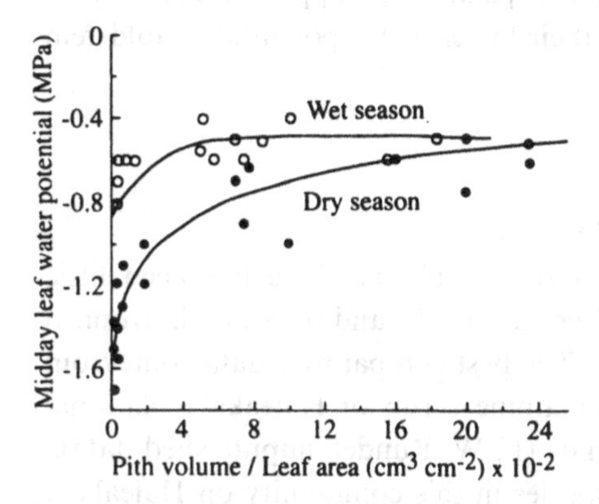

Figure 4.7. Influence of pith water storage capacity on minimum leaf water potential of *Espeletia timotensis* individuals growing at 4200 m elevation (after Goldstein *et al.* 1985a).

significant impact on plant water balance, especially during the dry season (Figure 4.7). Minimum dry season leaf water potentials may fall below the turgor loss point for individuals in the smallest size classes (cf. Table 4.3; Figure 4.7). Despite the large differences in water balance among different height classes, there is relatively little difference in leaf water relations characteristics except for a slightly higher dehydration tolerance in the smallest size classes (Table 4.3). It has been suggested that the high risk of mortality in young individuals of *E. timotensis* may be due in part

Table 4.3 *Leaf water relations characteristics for three size classes of* E. timotensis *from a population at 4200 m*

Osmotic potential at full turgor (Ψ_{π}^{100}), osmotic potential at zero turgor (Ψ_{π}^{z}), and relative water content at zero turgor (RWC_z) are shown

Height (m)	Ψ_{π}^{100} (MPa)	Ψ_{π}^{z} (MPa)	RWC_z
1.50	−1.12	−1.39	0.90
0.85	−1.06	−1.28	0.89
0.25–0.30	−1.14	−1.49	0.84

to their low relative water storage capacity and apparent inability to adjust osmotically in response to their lower water potentials (Goldstein *et al.* 1985a).

Water relations of woody plants

In comparison to studies of giant rosette plants, there has been little research on the water relations of woody shrubs and trees and in tropical alpine and subalpine communities. The best comparative data come from investigations of shrubs in the subalpine scrub of Haleakala National Park, Maui, in the Hawaiian Islands (P. W. Rundel, unpublished data).

Although all of the dominant species in this community on Haleakala are evergreen shrubs averaging 1–2 m in height, leaf morphology and structure are quite varied (Figure 4.8). Leaves vary from simple leptophylls of area less than 0.2 cm^2, as in *Styphelia tameiameiae* (Epacridaceae), to compound microphylls about 15 cm^2 as in *Sophora chrysophylla* (Leguminosae). Mean ratios of turgid to dry weight in leaves of six species studied ranged from a low of 1.96 in *Styphelia* to 2.91 in *Coprosma montana.* (Rubiaceae) (Table 4.4). These latter variations in leaf water storage are directly related to variation in the components of leaf water relations.

The osmotic potentials at full turgor and zero turgor were significantly related in the six shrub species studied (Figure 4.9), and each of these in turn was related to the turgid/dry weight ratio of the leaves (Figure 4.9). As either osmotic potential at full or zero turgor decreased (became more

Table 4.4 *Components of tissue water relations for six species of subalpine shrubs collected on 1 September 1983 in Haleakala National Park, Maui*

All samples were collected at Halemau Trailhead at 2440 m with the exception of *Dodonaea eriocarpa* which was collected at 2200 m. All values are means of replicated analyses

Species	Ψ_π^{100} (MPa)	Ψ_π^{z} (MPa)	WD_z (%)	Symplastic fraction (%)	Turgid/dry ratio
Geranium cuneatum	−1.32	−1.68	19.5	58	2.67
Coprosma montana	−1.39	−2.22	25.2	56	2.91
Sophora chrysophylla	−1.38	−1.80	14.0	42	2.68
Vaccinium reticulatum	−1.96	−2.50	18.8	67	2.22
Dodonaea eriocarpa	−1.99	−2.67	22.5	78	2.12
Styphelia tameiameiae	−2.34	−3.04	20.0	54	1.96

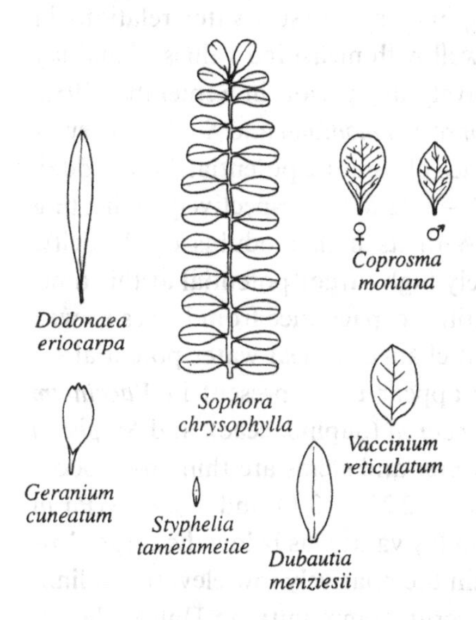

Figure 4.8. Characteristic leaf form in seven dominant shrubs in the subalpine scrub zone of Haleakala National Park, Hawaiian Islands.

negative) the turgid/dry weight ratio decreased linearly. The one species that fell away from an almost perfect fit in these relationships was *Coprosma montana.* This shrub changed in relative water content more rapidly per unit of turgor change than the other species.

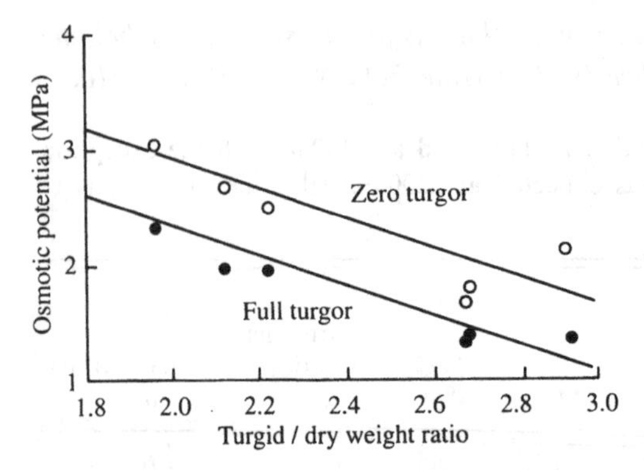

Figure 4.9. Relationship of turgid/dry weight ratio to osmotic potentials at full and zero turgor for six species of shrubs from the subalpine scrub of Haleakala National Park, Hawaii.

The wide range of values for components of tissue water relations in the Haleakala shrubs corresponded well with measured values of midday water potential at the end of a relatively dry period in September 1983. *Geranium cuneatum* (Geraniaceae), *Coprosma montana* and *Sophora chrysophylla*, with less drought-tolerant values of osmotic potential, had midday water potentials of -1.28, -1.33 and -1.72 MPa, respectively, indicating positive levels of turgor potential. With its high modulus of elasticity, *C. montana* still maintained a relatively high turgor potential at this time. All three of these species appear to utilize capacitance from leaves as well as from fleshy stems to buffer diurnal changes in leaf water potential.

Much lower levels of capacitance appear to be present in *Vaccinium reticulatum* (Ericaceae), *Dodonaea eriocarpa* (Sapindaceae), and *Styphelia tameiameiae*, where leaves hold less water and stems are thin and woody. Midday levels of water potential were -2.22, -3.14 and -2.82 MPa in these species. For *D. eriocarpa*, the midday value was below the turgor loss point. This condition may well explain the relatively low elevational limit of *D. eriocarpa* within the subalpine scrub community on Haleakala.

The range of relative water deficits at the turgor loss point for the six Haleakala shrubs ranged from 14 to 25%. This is a range comparable to that reported for evergreen hardwood forest and woodland species in Australia and New Zealand (Myers & Neales 1984; Clayton-Green 1983; Jane & Green 1983). It is much lower than that reported for more drought-tolerant chaparral shrubs in California (Davis & Mooney 1986). The symplastic water fraction in the Haleakala shrubs ranged from 42 to

78%. The significance of such a broad range is not apparent. For evergreen hardwoods in Australia and New Zealand the range reported is 51–87%.

Conclusions

Data on giant rosette plants, as in the subalpine shrubs from the Hawaiian islands, suggest that capacitance may be an important component of adaptation to diurnal drought stress. It is certainly not the only component, however. *Coprosma montana* and *Styphelia tameiameiae*, both of which reach above 3000 m to the top of Haleakala, represent opposite poles of adaptation in tissue water relations. The former species relies heavily on avoidance of diurnal drought stress through tissue water storage, while the latter tolerates diurnal drought through relatively low tissue osmotic potentials. As more tropical alpine shrubs are studied in the future, the relative importance of these two modes of drought adaptation will become better understood. *Polylepis sericea*, an alpine tree from the Venezuelan Andes, has elements of both strategies in its water relations (see Chapter 7).

Direct comparisons of tissue water relations between giant rosette growth forms and woody shrubs and trees are obviously difficult because of the very large differences of growth form involved. Nevertheless, the fundamental principles of comparing mechanisms of drought tolerance and drought avoidance are equally applicable to both groups.

Recent data comparing the water balance of sympatric populations of *Argyroxiphium sandwicense* and *Dubautia menziesii* on Haleakala in the Hawaiian Islands (Robichaux *et al.* 1990) provides an insightful look at similarities and differences in rosette and shrub growth forms. These two genera are closely related (and interfertile), but *A. sandwicense* is a monocarpic rosette plant and *D. menziesii* a shrub. Despite differences in growth form, both species exhibit very similar diurnal patterns of leaf water potential. Nevertheless, *A. sandwicense* has a transpiration rate twice that of *D. menziesii*. This surprising condition results from the higher capacitance of *A. sandwicense* as well as from its higher hydraulic conductivity. More comparative studies of this type may be very important in assessing the evolutionary significance of the giant rosette growth form found so widely in tropical alpine ecosystems.

Future research

Tropical alpine regions represent ideal natural laboratories for the study of plant water relations and adaptations to drought in different growth forms along gradients of relative importance of diurnal versus seasonal drought. Information about growth forms other than the conspicuous giant rosettes is largely lacking. Comparative studies involving 'conventional' woody and herbaceous growth forms that have temperate alpine zone counterparts would be particularly valuable in revealing features specifically suited to the unique conditions prevailing in tropical alpine zones. Even among the apparently similar African and Andean giant rosettes, recent findings concerning similarities between the two groups may only be superficial and more detailed comparative water relations data are needed.

Many questions in the area of low temperature effects on water relations of tropical alpine plants remain unresolved. For example, in growth forms with exposed stems, how is the problem of freezing-induced gas embolisms in the xylem overcome? Does root resistance to water uptake in tropical alpine plants show the same kind of temperature dependence as in temperate zone plants? And finally, studies carried out with giant rosettes in Africa (Beck *et al.* 1984) and Venezuela (Goldstein *et al.* 1985b) indicate that the relationship between frost resistance, tissue freezing, and water relations behaviour needs to be examined. African giant rosette species should be more tolerant of water deficits than their Andean counterparts because they are regularly subjected to equilibrium freezing during which the symplast may become severely dehydrated as extracellular ice formation takes place (Beck *et al.* 1984). In Andean giant rosette species, on the other hand, freezing of leaf tissue has not been observed in the field because the leaves are able to supercool without freezing to temperatures well below minimum leaf temperatures encountered in the field (Goldstein *et al.* 1985b).

References

Beck, E., Schulze, E.-D., Senser, M. & Scheibe, R. (1984). Equilibrium freezing of leaf water and extracellular ice formation Afroalpine giant rosette plants. *Planta* **162**, 276–82.

Clayton-Green, K. A. (1983). The tissue water relations of *Callitris columellaris*, *Eucalyptus melliodora* and *Eucalyptus microcarpa* investigated using the pressure-volume technique. *Oecologia* **57**, 368–73.

Davis, S. D. & Mooney, H. A. (1986). Tissue water relations of four co-occurring chaparral shrubs. *Oecologia* **70**, 527–35.

Goldstein, G. (1981). Ecophysiological and demographic studies of white spruce (*Picea glauca* (Moench) Voss) at treeline in the central Brooks Range of Alaska. PhD dissertation, University of Washington, Seattle.

Goldstein, G. & Meinzer, F. (1983). Influence of insulating dead leaves and low temperatures on water balance in an Andean giant rosette plant. *Plant, Cell and Environment* **6**, 649–56.

Goldstein, G., Meinzer, F. C. & Monasterio, M. (1984). The role of capacitance in the water balance of Andean giant rosette species. *Plant, Cell and Environment* **7**, 179–86.

Goldstein, G., Meinzer, F. C. & Monasterio, M. (1985a). Physiological and mechanical factors in relation to size-dependent mortality in an Andean giant rosette species. *Oecologia Plantarum* **6**, 263–75.

Goldstein, G., Rada, F. & Azócar, A. (1985b). Cold hardiness and supercooling along an altitudinal gradient in Andean giant rosette species. *Oecologia* **68**, 147–52.

Hedberg, O. (1964). Features of Afroalpine plant ecology. *Acta Phytogeographica Suecica* **49**, 1–144.

Hinckley, T. M., Teskey, R. O., Duhme, F. & Richter, H. (1981). Temperate hardwood forests. In *Water Deficits and Plant Growth*, Vol. VI, ed. T. T. Kozlowski, pp. 153–208. New York: Academic Press.

Jane, G. T. & Green, T. G. A. (1983). Utilization of pressure-volume techniques and non-linear squares analysis to investigate site induced stresses in evergreen trees. *Oecologia* **57**, 380–90.

Jarvis, P. G. (1975). Water transfer in plants. In *Heat and Mass Transfer in the Environment of Vegetation*, ed. D. A. de Vries and N. K. Van Alfen, pp. 369–94. Washington, DC: Script Book Co.

Kaufmann, M. R. (1975). Leaf water stress in Engelmann spruce. *Plant Physiology* **56**, 841–4.

Kaufmann, M. R. (1977). Soil temperature and drying cycle effects on water relations of *Pinus radiata*. *Canadian Journal of Botany* **55**, 2413–18.

Levitt, J. (1980). *Responses of Plants to Environmental Stresses*, 2nd edn, Vol. 2, *Water, radiation, salt and other stresses*. New York: Academic Press.

Mabberley, D. J. (1973). Evolution in the giant groundsels. *Kew Bulletin* **28**, 61–96.

Meinzer, F. C., Goldstein, G. & Rundel, P. W. (1985). Morphological changes along an altitude gradient and their consequences for an Andean giant rosette plant. *Oecologia* **65**, 278–82.

Myers, B. A. & Neales, T. F. (1984). Seasonal changes in the water relations of *Eucalyptus behriana* F. Muell. and *E. microcarpa* (Maiden) Maiden in the field. *Australian Journal of Botany* **32**, 495–510.

Richards, G. P. (1973). Some aspects of the water relations of Sitka spruce. PhD dissertation, University of Aberdeen, UK.

Robichaux, R. H., Carr, G. D., Liebman, L. & Pearcy, R. W. (1990). Adaptive radiation of the Hawaiian silversword alliance (Compositae: Madiinae): ecological, morphological and physiological diversity. *Annals of the Missouri Botanical Garden* **77**, 64–72.

Running, S. W. (1980). Relating plant capacitance to the water relations of *Pinus contorta*. *Forest Ecology and Management* **2**, 237–52.

Running, S. W. & Reid, P. C. (1980). Soil temperature influences of *Pinus contorta* seedlings. *Plant Physiology* **65**, 635–40.

Schulze, E.-D., Beck, E., Scheibe, R. & Ziegler, P. (1985). Carbon dioxide assimilation and stomatal response of afroalpine giant rosette plants. *Oecologia* **65, 207**–13.

Smith, A. P. (1980). The paradox of plant height in Andean giant rosette species. *Journal of Ecology* **68**, 63–73.
Waring, R. H. & Running, S. W. (1978). Sapwood water storage: its contribution to transpiration and effect upon water conductance through the stems of old-growth Douglas-fir. *Plant, Cell and Environment* **1**, 131–40.

5

Cold tolerance in tropical alpine plants

E. BECK

Introduction

'Summer every day and winter every night' (Hedberg 1964) is a brief but succinct characterization of the tropical alpine climate, pointing to the fact that the amplitude of the daily temperature oscillation by far exceeds that of the monthly mean values. Although cloudiness exerts a mitigating effect on the daily temperature extremes during the rainy seasons, nocturnal frost may occur throughout almost all of the year at altitudes above 4000 m. Therefore tropical alpine plants must maintain mechanisms of permanent frost hardiness which differ considerably from those providing the overwintering plants of temperate climates with seasonal frost resistance. Whereas, for example, in Norway spruce the frost-hardy state is characterized by a high proportion of unsaturated fatty acids in the membrane lipids (Senser 1982), by a shift from photosynthetic starch formation to the production of sucrose and its galactosides (Kandler *et al.* 1979), by a reduced capability of photosynthetic electron transport (Senser & Beck 1979) and by a suspension of growth activity, tropical alpine plants must combine physiological features providing frost resistance with continuously high rates of photosynthesis and growth.

On the other hand, the so-called *Frostwechselklima* (Troll 1943) of the tropical alpine regions confronts plants with a short – although daily – frost period during which the air temperature usually does not drop below −12 °C (Beck *et al.* 1982; Azócar *et al.* 1988), so that a moderate degree of frost tolerance together with an efficient mechanism to delay cooling (i.e. an efficient insulation) are sufficient to allow the survival of higher plants.

Comparing the equatorial regions of the South American Andes (the so-called páramos) with those of East Africa (the so-called Afroalpine zone: Hauman 1955) on Mounts Kilimanjaro, Kenya, Ruwenzori and

Elgon, frost as a critical ecological factor appears to be of more importance in the latter areas than in the former (Goldstein *et al.* 1985; Azócar *et al.* 1988). This may, in part, be due to the high elevation of large valley systems such as those on Mount Kenya, where cold air flowing down from the ridges accumulates to form extensive cold air lakes. Although the majority of the morphological and physiological features of páramo and Afroalpine plants appear to be similar, the lower subfreezing temperatures in many Afroalpine regions have brought about a special adaptation of some plants to tolerate regular freezing of the leaf water.

The presence of tree-like species (*Senecio*, *Espeletia*) in the tropical alpine regions, i.e. far above the timberline, impressively demonstrates the vitality of these specialized plants. However, the occurrence of even a single extraordinarily severe frost or a prolonged dry season are possible environmental factors which may seriously impair their viability. Especially during the dry season, the clear nocturnal sky occasionally gives rise to considerable radiation emission which results in leaf temperatures of up to 3 °C lower than the air temperature (Beck *et al.* 1982; Azócar *et al.* 1988); thus leaf temperatures as low as −14 °C have been recorded on Mount Kenya at an altitude of 4200 m (Beck 1987), and the insulation of the stem or the bud might not always be completely sufficient to cope with such a degree of frost.

Another problem of vascular plants subjected to a *Frostwechsalklima* arises from the rapid thawing of frozen organs upon warming by the first incident rays of sunlight in the morning. If, for instance, a leaf thaws (which usually happens within a few minutes) and transpires while the water in the stem is still frozen or the liquid water flow resistance in the pith is still too high (Goldstein & Meinzer 1983), or the volume of the pith is too small (Monasterio 1986), embolism and collapse of the water transport system to the leaves must be the consequence (Hedberg 1964; Goldstein *et al.* 1984). With respect to tropical alpine life-forms (Hedberg & Hedberg 1979) in this context, insulation or reduction of the stem on the one hand (giant rosette and acaulescent rosette plants, cushion plants) or reduction of the transpiring leaf area on the other (night bud, tussocks, sclerophyllous scrubs) take on ecological significance.

However, it should additionally be mentioned that tropical alpine plant life-forms and species are also found at lower elevations, where nocturnal frost may be the exception rather than the rule or may not occur at all. This suggests that the special features of the plants under discussion must not be interpreted solely in terms of adaptations to regularly occurring frost, but must also be regarded as evolutionary answers to the challenges posed by other environmental factors such as drought, inhibitory high light intensities, recurrent burning (Beck 1983; Beck *et al.* 1986a) or a shortage of nutrients (Monasterio 1986; Beck, Chapter 11).

Cold tolerance: mechanisms and functions

According to Levitt (1980), cold resistance encompasses mechanisms resulting in freezing avoidance and others providing for freezing tolerance. Taking into account the extreme habitat in which these plants are found, both these types of cold resistance may be expected to be relevant to tropical alpine species.

Freezing avoidance: general considerations

Freezing avoidance may be achieved in two ways: (i) by frost mitigation (Sakai & Larcher 1987), i.e. by protection of plant organs from freezing temperatures, by insulation or the provision of a thermal buffer, and (ii) by prevention of the formation of ice at subfreezing temperatures, i.e. by avoidance of ice nucleation. Both mechanisms enable the spanning of a relatively short period of moderate frost but are inappropriate as protection against severe or permanent subfreezing temperatures. With respect to the tropical *Frostwechselklima*, both types of freezing avoidance function well, since the lowest temperatures are not usually attained until just before dawn, which is immediately followed by rapid warming after sunrise (Figure 5.1). Since irradiance during the usually cloudless morning hours (1.3–1.5 kJ m^{-2} s^{-1} (Beck *et al.* 1982) and 1.05 kJ m^{-2} s^{-1} (Larcher 1975)), approaches or even exceeds the average solar constant (1.4 kJ m^{-2} s^{-1}), the rate of energy uptake by the plant after sunrise by far exceeds that of nocturnal heat loss due to radiation emission. Consequently, those mechanisms of freezing avoidance based on insulation or thermal buffering which are just sufficient to protect plant organs against moderate cooling will not prevent them from rapidly getting warm in the morning.

Freezing avoidance has been widely interpreted in the sense of protecting sensitive plant parts from freezing damage. However, more recent data may offer other physiological interpretations: it has been shown that one of the major problems faced by the caulescent giant rosette species *Espeletia timotensis* as a result of low temperatures is the maintenance of water flow through the stem. This obviously takes place predominantly via living pith cells and hence is detrimentally affected by frost (Goldstein

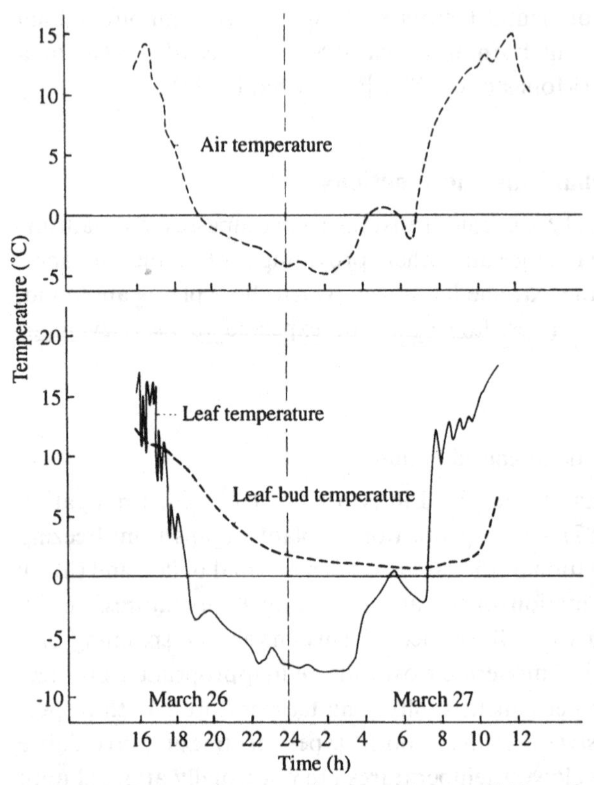

Figure 5.1. *Senecio keniensis* (Mount Kenya, 4150 m elevation): comparison of the temperature course of an adult (outer) rosette leaf with that measured inside the cone-shaped leaf bud. Reference air temperatures are shown in the top panel (from Beck *et al.* 1982).

& Meinzer 1983). Thus avoidance of subfreezing temperatures in the stem is a prerequisite for functioning water relations. Similar conclusions have been drawn with respect to other caulescent *Espeletia* (Smith 1979; Hedberg & Hedberg 1979) and *Senecio* species (Hedberg 1964).

Another plant organ which has been thought to be susceptible to freezing damage is the leaf or inflorescence bud. Since growth and leaf production take place throughout the year, resting buds are usually not to be found on tropical alpine plants. In the giant rosette plants, the leaf buds – which also encompass the apical meristem of the stem – consist of a great number (e.g. about 80 in *Senecio keniodendron*: Beck *et al.* 1982) of firmly packed, more or less revolute young leaves, which form a yellowish cone in the centre of the (terminal) leaf rosette (Figure 5.2).

Figure 5.2. Longitudinal section of a rosette of *Lobelia telekii* (Mount Kenya) showing the cone-shaped central bud where the developing leaves are firmly appressed. The spreading adult leaves are in the 'day-position' (photo: E. Beck 1979).

When these young leaves have attained about two thirds of their final size, they separate from the cone, unroll and become green and contribute to the large leaf rosette. This rosette, which consists of 50–150 adult leaves and assumes a diameter of up to 1 m or more, via nyctinastic inward bending of leaves during the cold hours, forms a so-called night-bud around the central cone thus protecting the developing leaves from freezing (Figure 5.3). In an elegant experiment Smith (1974) prevented the formation of the night-bud in *Espeletia schultzii* by removing or immobilizing the adult leaves. As a result the bud-core temperatures dropped well below freezing point and the youngest fully expanded leaves finally

Figure 5.3. Rosettes of *Senecio keniensis* (Mount Kenya) exhibiting the typical night-bud with leaves in the 'night-position'. Note the white felty indumentum on the lower leaf surface (photo: E. Beck, March 1985).

wilted and died. They did not, however, exhibit the typical necroses of freezing damage and thus may have rather been damaged by rapid warming and the resulting accelerated transpiration in the early morning, when the water supply from the stem was still limited. The idea that the protection of the leaf bud from freezing may primarily serve a purpose other than that of merely ensuring the survival of the meristems is corroborated by the finding that the developing leaves are almost as frost hardy as are the adult ones, and that even the seedlings exhibit a quite comparable degree of frost resistance (Table 5.1). Beck *et al.* (1982) have put forward the hypothesis that the effect of the formation of the night-bud may rather be to prevent a nocturnal standstill of leaf growth which would otherwise cease due to freezing.

Mechanisms of freezing avoidance

Insulation

Insulation by dead leaves

The species of the genera *Senecio* (East Africa) and *Espeletia* (South America) which are characteristic of the alpine zones all have thick and sparsely branched stems, the older parts of which are surrounded by a thick cork cortex, while the young stems or branches are protected by a dense mantle of marcescent dry leaves. Investigating *Espeletia timotensis* Monasterio (1986) found 82.5% of the total biomass accumulated in the dead leaves while the rosette of living leaves amounted only to 2.2%. In the case of *Senecio keniodendron* this mantle is about 10–15 cm thick and consists of about 1600 shrivelled brittle leaves per meter of stem length (Beck *et al.* 1980). The insulating effect of the mantle has been investigated by Hedberg (1964). Whereas in an early morning the temperature of the outer surface of the dry leaf cylinder of *S. keniodendron* was −5 °C, that of the outermost part of the pith was +3 °C. A similar insulating affect of the marcescent leaves of *E. schultzii* has also been demonstrated (Hedberg & Hedberg 1979). Interestingly, removal of the insulating material resulted in a gradual wilting of the rosette leaves, most of which then died 3–4 months later (Smith 1979). But no sudden collapse of leaf integrity was observed, as would have been the consequence of a loss of capacitance from water storage of the pith due to direct freezing injury. After formation of the cork cortex mentioned above removal of the insulating leaf mantle apparently does not give rise to frost-triggered damage.

Insulation by the formation of a night-bud: giant rosette plants

A typical feature of the giant rosette plants is the ability of the adult leaves to carry out special movements. The leaves bend upwards and inwards upon cooling and since the rosettes are composed of some 70–200 adult leaves, this movement results in the whole system taking on a globular shape which has been designated a 'night-bud' (Hedberg & Hedberg 1979; see Figure 5.3). The reverse leaf movement takes place upon warming, thus opening the rosette to its day position.

The dense layering of the night-bud effectively insulates the inner leaves and especially the core of meristematic leaves from nocturnal frosts. The insulating effect of this multilayer is further enhanced by the spongy mesophyll of the adult leaves, which resembles the structure of styrofoam (Figure 5.4). In addition, the dense packing of the developing leaves in the central cone forms a solid mass of considerable size which acts as a thermal buffer. The cooperation of heat storing and insulation ultimately results in a remarkable delay in cooling of the central core, the temperature of which rarely drops below 0 °C even on very cold nights (Figure 5.1).

In some giant rosette species the leaf surface is covered by a more or less dense tomentum of dead hairs. An extreme example for such a hairy

Table 5.1 *Frost hardiness of Afroalpine plants as determined in the field*

Method 1: leaf pieces were cooled stepwise to various final subfreezing temperatures at which they remained for 3–4 hours. After rewarming in ice they were kept in moist Petri dishes at ambient temperature and inspected during the course of 4–5 days for necroses. A limit temperature of −5 °C indicates that less than 50% of the samples had developed necroses, whereas at the next lower final temperature more than 50% of the leaf pieces showed frost damage.
Method 2: cooling and rewarming as above. After 1 h at 0 °C, a constant number of leaf discs were freshly harvested from the leaf pieces and placed in 50 ml distilled water. The conductivity of the resulting solution was followed over 6 min and compared with that stemming from untreated controls and from completely frozen samples. A sharp increase of conductivity indicated frost damage and the corresponding temperature was regarded as the limit of frost hardiness

Species	Plant part	Date	Altitude (m.a.s.l.)	Limit of frost hardiness (°C)	Method of determination	Ref. no.
Senecio keniodendron	green leaf	Oct. 1980	4200	−8	1	13
S. keniodendron	green leaf	Feb. 1983	4200	−5	1	12
S. keniodendron	seedling	Feb. 1983	4200	−5	1	*
S. keniodendron	green leaf	Mar. 1985	4200	−14	2	*
S. keniodendron	green leaf	Mar. 1985	4500	lower than −15	2	*

Senecio keniensis	green leaf	Oct. 1980	4100	−10	1	13
S. keniensis	green leaf	Mar. 1983	4100	−5	1	12
S. keniensis	seedling	Mar. 1983	4100	−8	1	*
Lobelia keniensis	green leaf	Oct. 1980	4100	lower than −20	1	13
L. keniensis	green leaf	Mar. 1983	4100	−10	1	12
L. keniensis	seedling	Mar. 1983	4100	−10	1	*
Lobelia telekii	green leaf	Oct. 1980	4200	lower than −20	1	13
L. telekii	green leaf	Mar. 1983	4200	−15	1	12
L. telekii	seedling	Mar. 1983	4200	−10	1	*
L. telekii	green leaf	Mar. 1985	4200	−14	2	*
L. telekii	green leaf	Mar. 1985	4500	−14.5	2	*
L. telekii	meristematic leaf	Mar. 1985	4200	−14	2	*
L. telekii	meristematic leaf	Mar. 1985	4500	lower than −15	2	*
L. telekii	seedling	Mar. 1985	4200	lower than −12	2	*
L. telekii	seedling	Mar. 1985	4500	−14	2	*
Haplocarpha rueppellii	green leaf	Mar. 1985	4200	−13	2	*
Ranunculus oreophytus	green leaf	Mar. 1985	4200	−14	2	*
Senecio purtschelleri	green leaf	Mar. 1985	4250	−14	2	*
Carduus chamaecephalus	green leaf	Mar. 1985	4250	−14	2	*
C. chaemaecephalus	green leaf	Mar. 1985	4500	−15	2	*

* Original data.

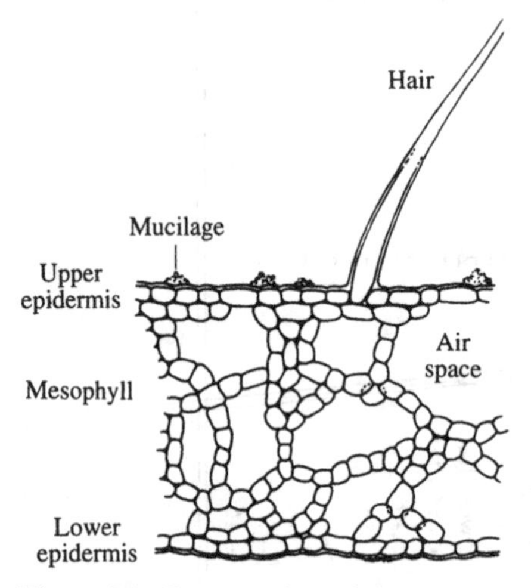

Figure 5.4. Cross section of the basal region of a leaf of *Lobelia keniensis*. The extremely large intercellular spaces of the spongy mesophyll render a structure with an insulating effect similar to that of styrofoam.

indumentum is provided by *Senecio keniensis* (formerly designated as *Senecio brassica*, Jeffrey 1986; see also Beck *et al.* 1992) the leaves of which are coated by an approximately 2 mm thick white felt on their lower surfaces. This felt layer covers the whole outer surface of the night-bud when leaves have assumed their night position. It has been suggested that the woolly indumentum might help to reduce the nocturnal radiation emission from the rosette (Hedberg 1964). This insulating effect could not, however, be demonstrated, as no temperature difference could be detected between the outer surface of the felt and the epidermis during the course of the entire night (Figure 5.5). Only upon the absorption of sunlight by the leaf did the temperature of the leaf surface exceed that of the outer surface of the indumentum. Meinzer & Goldstein (1985) studied the physiological importance of such a felt in *Espeletia timotensis* in the field. They recorded pubescence-induced increases in leaf temperature of up to 3% during daytime while in the night removal of the felt had no or at most a small reversed effect. The available data indicate that such a felt cannot protect the rosette from longwave infrared radiation emission. Similar results have been reported by Eller & Willi (1977) for the indumentum of coltsfoot (*Tussilago farfara*). However, it should be noted that the predominant vertical orientation of the leaves in a night-bud reduces the radiating surface and thereby cooling of the rosette.

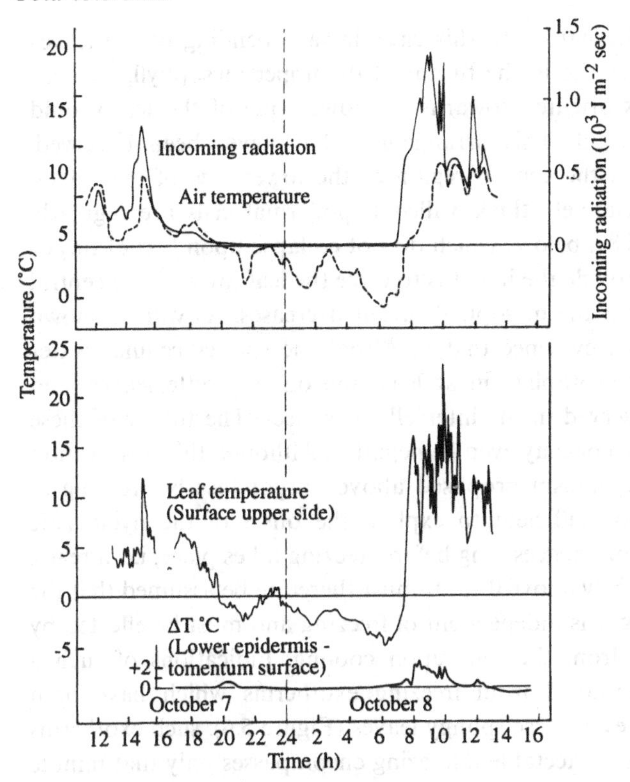

Figure 5.5. *Senecio keniensis* (Mount Kenya, 4150 m): Course of leaf temperature and of the temperature gradient between the outer surface of the felty tomentum and the epidermis beneath the indumentum (bottom panel) during 24 hours. The top panel, for purposes of reference, shows air temperature and incoming radiation (from Beck *et al.* 1982). Air temperature was recorded with a thermohygrograph sheltered by a roof of canvas, while leaf temperatures were followed using calibrated Siemens heat conductors. Radiation was measured with a Robitzsch actinograph.

Although the formation of the night-bud and its ecological effectiveness are quite obvious, the physiology of the leaf movement is not yet understood. Beck *et al.* (1980) have shown that the adult leaves of a single rosette of *Senecio keniodendron*, and of *Lobelia keniensis* and *L. telekii*, respectively, are all of the same size. Since these are the leaves forming the night-bud, tropistic movement – which is directed by growth – can be ruled out as a mechanism. Movement via growth has been suggested for *Espeletia* (Larcher 1975), but has never been demonstrated. If growth is excluded as a mechanism, turgor-governed, nyctinastic movement remains the sole alternative. In this case, inward bending of the leaves would require a decrease in the turgor of the upper mesophyll, whereas the mesophyll cells oriented towards the lower side of the leaf should have to remain turgid. This arrangement has never been observed. However, at least in the central region of the lower side of the leaves, several layers of relatively thick-walled hypodermal cells are regularly found (see Figure 5.8*a*, below) which do not collapse upon loss of turgor and hence would provide the impetus to force the leaf towards the central axis of the bud when the mesophyll turgor decreases. As will be shown below, there is good evidence that in Afroalpine species cellular water freezes outside the protoplast in at least the outer rosette leaves, and ice-caps can be observed in the intercellular spaces. The turgor of these cells then equals zero or may even be negative. Although this observation agrees with the mechanism proposed above, turgor loss by freezing of cellular water is not sufficient to explain the onset of the nyctinastic movement, which commences long before freezing takes place, i.e. already at temperatures slightly above 0 °C. It must therefore be assumed that the decrease of leaf turgor is independent of freezing and must be effected by extrusion of water from the cells upon cooling. Indications of such a process may be gathered from freezing exotherms which have been demonstrated with even these spongy leaves (Figure 5.6): such exotherms would not have been detectable if freezing encompasses only that minute volume of extracellular water which is normally present in intercellular spaces as water vapor or lining of the cell walls. Hence the occurrence of freezing exotherms suggests considerable water soaking of the extracellular spaces before the onset of ice crystallization. A proposal concerning the attainment of such an intercellular water soaking is presented on p. 95.

Compared with closure, the opening of the rosette is a very rapid process requiring only a few minutes subsequent to the first interception of direct incident sunlight by the leaves.

Insulation in other tropical plant life-forms

Insulation as a mechanism of effecting freezing avoidance may be generally relevant for the tropical alpine life-forms of the grass-tussock, the acaulescent rosette plant and the cushion plant (Hedberg & Hedberg 1979). Thus, in the instance of *Stipa* spp., the temperature in the center of the tussock was found to be 5.5 °C above that of the free air or of the soil surface on a cold morning. Similarly, with a tussock of the Afroalpine *Festuca pilgeri* subsp. *pilgeri*, a temperature difference of 7.5 °C between

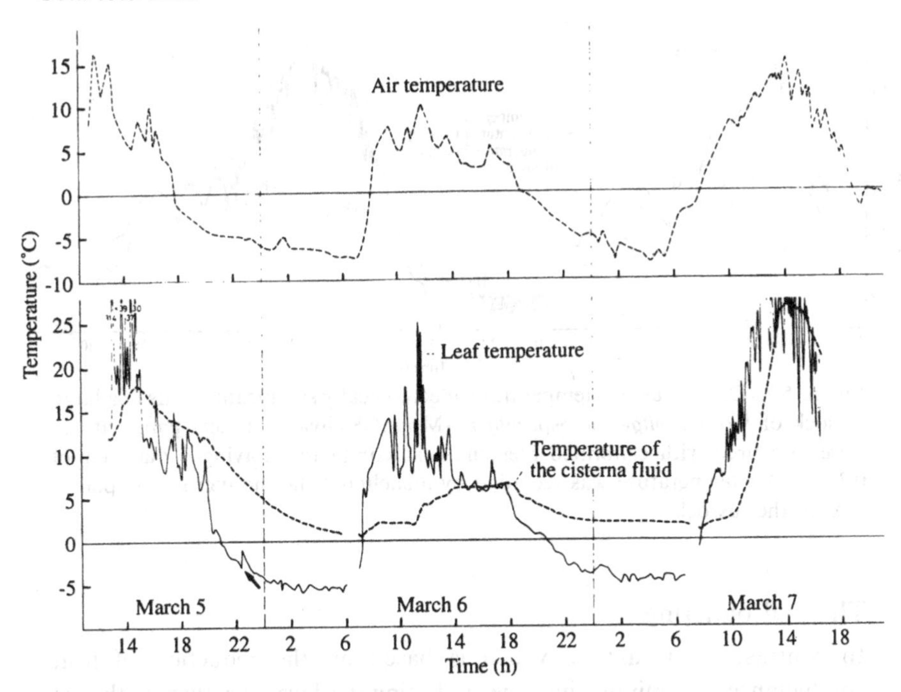

Figure 5.6. *Lobelia keniensis* (Mount Kenya, 4150 m): course of the temperature of an adult rosette leaf and of the cisterna fluid during two days. For purposes of reference the course of the air temperature is shown in the top panel. The arrow marks a freezing exotherm of leaf water (from Beck *et al.* 1982). Temperature measurement as described in the legend of Figure 5.5.

the outermost leaves and the central part of the plant where the innovation shoots originate was measured after a frosty night (Hedberg 1964). Continuous recordings of the temperatures of the tips of an adult leaf and of a shoot bud of a *F. pilgeri* tussock and of the air did not reveal such large differences as those mentioned above (Figure 5.7). However, the temperature courses show that the central portion of the tussock, embodying the innovation shoots, accumulates much more heat during the clear and sunny morning than does the outer region. This heat store then buffers, i.e. delays cooling, during the afternoon and evening. The freezing avoidance effect due to the structure of a tussock bearing many insulating dead culms and leaves may therefore be regarded as a marginal case touching upon the mechanisms of both insulation and thermal buffering. With acaulescent rosette- and cushion-plants the relations are even more complicated and not well understood.

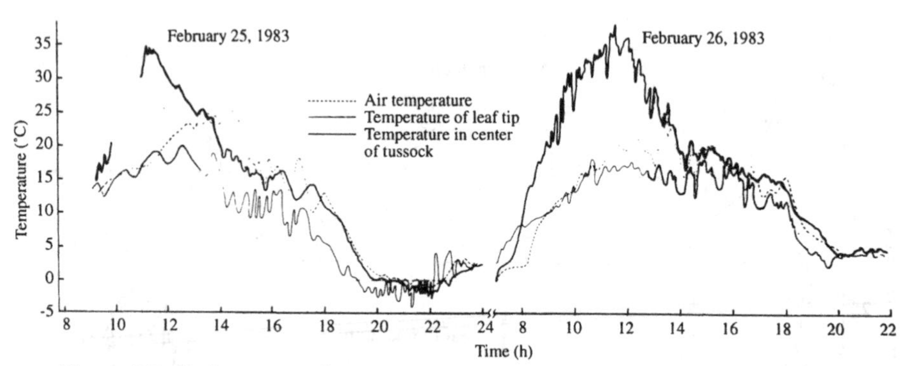

Figure 5.7. Daily course of temperature at ecologically significant points of a large tussock of *Festuca pilgeri* subsp. *pilgeri* (Mount Kenya). The leaf temperatures were measured with calibrated Siemens heat conductors having a diameter of 0.3 mm. Air temperature was recorded by a sheltered thermohygrograph placed next to the tussock.

Thermal buffering

In contrast to insulation, which is based on the reduction of heat conductance by minimizing the radiating surface together with the formation of sandwich layers of air and plant matter, the mechanism of thermal buffering requires a large mass of matter with a high heat capacity. Apart from the accumulation of solid matter in stems and leaf-bud cones, water provides a suitable thermal buffer which, in addition to its considerable heat capacity (4.22 J g^{-1} $°C^{-1}$ at 0 °C), provides 331.56 J g^{-1} of heat of fusion.

Thermal buffering by aqueous solutions is a typical feature of the Afroalpine giant lobelias. These plants produce a terminal cylindrical inflorescence of 1–3 m length after several years of vegetative growth (Young 1984). The stalk of such an inflorescence is hollow and the internal cavity, which is about 5–8 cm in diameter, contains a slightly viscous fluid reaching up to 1 m above ground level. Hence, about 3–5 litres of fluid may be collected from the inflorescence of one *Lobelia telekii*. Krog *et al.* (1979) recorded the daily course of the temperature of this fluid, starting at noon, at which +12 °C was measured. From the afternoon on, continuous decrease in the temperature was noted until midnight, at which time crystallization of water commenced. Although the air temperature had fallen to −6 °C at predawn, the temperature of the fluid and hence of the inflorescence remained at +0.1 °C throughout, until the plant again absorbed radiant heat from the sun. The proportion of ice found at the end of the cooling period was about 2% of the total fluid volume. Assuming this volume to be 3 litres, cooling from +12 to 0 °C resulted in the loss of 152 kJ, and maintenance of the temperature at the freezing point required another 20 kJ, which represents the amount of heat produced upon crystallization of 2% of 3 litres. Thus the fluid's total thermal buffer, which is equivalent to 1150 kJ, was in this case utilized by only 15%.

A similar thermal buffer, of even comparable size, is found in the non-flowering rosettes of the lobelias of the *L. deckenii* group, the imbricate leaf-bases of which form cisternae filled with an aqueous fluid. For example, 2 l of cisterna fluid have been collected from a rosette of *L. keniensis*, which was 0.5 m in diameter and consisted of 140 adult leaves (Beck *et al.* 1982). Nyctinastic inward bending of the rosette leaves raises the surface of the fluid so that the contents of the individual cisternae communicate and cover the central leaf bud. After cold nights the fluid was found to be frozen on the surface and sometimes almost completely through. Although some of the cisterna fluid may originate from rainwater (in contrast to the fluid within the inflorescence), it has been shown that an aqueous solution is excreted by the leaves: after removing the fluid and sheltering the rosette from precipitation, 0.2 l of fresh liquid could be harvested after one week (Beck *et al.* 1982). Subsequent to the removal of this thermal buffer, the rosette leaves reacted with an unusually vigorous nyctinastic bending resulting in a completely contracted night-bud. Whether the excretion of the liquid occurs during the nocturnal frost period and then could be classified as extraorgan freezing (Sakai & Larcher 1987) has not yet been examined.

Thermal buffering by aqueous solutions requires the avoidance of intense supercooling for two reasons: (i) supercooling of an extracellular or extraorgan liquid would not prevent the tissue from being subjected to subfreezing temperatures; (ii) much more energy can be withdrawn from the heat of fusion than from the heat capacity of water. Hence a thermal buffer based on a large volume of water requires nucleating agents which trigger crystallization already upon slight supercooling in order to be effective. Krog *et al.* (1979) determined a supercooling point of -4.6 ± 1.3 °C for the central fluid of the *L. telekii* inflorescence, whereas the freezing point of this liquid was close to 0 °C. The same is true for the cisterna fluid of *L. keniensis*. Both fluids contain small amounts of mucilaginous material composed of high molecular weight polysaccharides (Krog *et al.* 1979; Beck *et al.* 1982), which have been suggested to act as nucleating agents. Because of the insignificant osmolarity of these polysaccharides, the freezing point of the cistern fluid is not depressed.

Supercooling

In contrast to freezing avoidance by insulation or thermal buffering, supercooling represents a mechanism which by principle excludes simultaneous protection by freezing tolerance: whereas freezing tolerance depends on the capability of plant tissue to achieve a balanced heat and mass (water) transfer from the cell to its surroundings to enable extracellular (equilibrium) freezing, supercooling as an avoidance mechanism requires complete prevention of ice nucleation. Ice formation in highly supercooled tissues is always fatal, because the cells are not able to excrete enough water to prevent intracellular freezing during the rapid crystallization process. Hence, supercooling is beneficial only under climatic conditions where only brief periods of mild nocturnal freezing occur. In contrast to the Afroalpine regions, the typical páramo desert zone of the Venezuelan and Colombian Andes may provide such an environment (Goldstein *et al.* 1985; Sturm 1978).

Freezing of leaves in the natural environment has, accordingly, never been reported for *Espeletia* spp. (Goldstein *et al.* 1985) or for *Polylepis sericea* (Rada *et al.* 1985b; Goldstein *et al.*, Chapter 7), which as a typical crook-timber species grows at elevations of up to 4600 m. Goldstein *et al.* (1985), Rada *et al.* (1985a) and Larcher & Wagner (1976) studied the supercooling behaviour of *Espeletia* spp. and *Polylepis sericea*, respectively: as expected, ice formation and 50% tissue cryoinjury took place at the same subzero temperatures upon experimental freezing of *Espeletia* leaves. The supercooling point, i.e. the temperature required to initiate crystallization, decreased in *Espeletia* along an altitudinal gradient extending from 2800 to 4200 m from about −6.5 to −10.5 °C. The respective minimum air temperatures hardly ever fell below −4 °C, even at 4200 m. Two further observations of the above-mentioned studies may be of interest for the understanding of nucleation avoidance. First, the supercooling point was found to be linearly related to the leaf water potential, which for the particular species is more negative (and thus larger) during the dry season, when minimum air and leaf temperatures are to be observed. Part of this increment of the water potential could be due to an enhancement of the osmotic potential. The increase in supercooling capacity in the dry season could therefore perhaps be explained by the known linear relation of the supercooling point and the freezing point depression (Rasmussen & Mackenzie 1972), inasmuch as homogeneous nucleation is involved in the delay of ice formation. Second, the amount of intercellular space in specimens of *E. schultzii* growing at higher elevations tends to be reduced, which according to Levitt (1980) is a prerequisite for the efficient prevention of freezing via supercooling.

A diurnal increase in the content of soluble carbohydrates during the nocturnal cooling period has been observed with *Polylepis sericea*, which results in an enhancement of the osmotic potential and the depression of the freezing point. Since the leaves are able to supercool in the range of −6 to −8 °C in any event, a lowering of the freezing point by −2.8 °C would in itself be of little ecological significance. However, it may be regarded as a 'security valve' (Rada *et al.* 1985b) to prevent tissue from freezing in case supercooling – which on account of its high Gibbs free energy represents a labile system – might not function.

Since supercooling does not alter the water status of the cell, metabolic processes may still go on even at subzero temperatures, although at reduced velocities. Net photosynthesis thus could still be detected with the páramo plants *Eryngium humboldtii* and *Polylepis sericea* at −6 °C and with *Espeletia semiglobulata* even at −8 °C (Larcher & Wagner 1976). On the other hand, photosynthesis in leaves of Afroalpine plants was found to cease immediately upon the transition from the supercooled to the frozen state (Bodner & Beck 1987) as has been described for the moderately frost tolerant *Sempervivum montanum* of the European Alps (Larcher & Wagner 1983).

Freezing tolerance

Extracellular ice formation upon equilibrium freezing

As has been mentioned above, the outer rosette leaves of the Afroalpine groundsels and giant lobelias are usually frozen after a cold night and freezing exotherms have been repeatedly recorded (cf. Figure 5.6). Freezing of the leaf usually commences at a few spots from where it spreads to the whole leaf within a few minutes. Microscopic observation revealed that ice formation occurs outside the cells. Considerable ice-caps grow from the cell surfaces into the intercellular spaces, especially in the spongy mesophyll. Characteristic ice-lumps (cf. Figure 4A in Asahina 1978) were observed in the intercellular spaces of hypodermal strengthening tissue (Figure 5.8*b*). Using anthocyanin containing epidermal and hypodermal tissues from *Senecio keniodendron* and *Lobelia telekii*, freezing

Figure 5.8. Anatomical investigation during the freezing of leaves of *Senecio keniodendron*. Tangential sections of the hypodermal tissue of the leaf's lower surface are shown. (*a*) Non-frozen leaf; the cellular sap of the hypodermal cells is more or less stained by anthocyanin. (*b*) Frozen leaf (−6 °C) containing pronounced ice lumps in the intercellular spaces. (*c*) Frozen leaf (−6 °C) exhibiting extremely dehydrated hypodermal cells. The dark areas adjacent to the cell walls result from the concentrated anthocyanin solution in the contracted protoplasts. (*d*) Thawing leaf. Subsequently to melting of extracellularly deposited ice. The cells had rehydrated. Water vapor bubbles, created by cavitation, have substantially decreased but still can be recognized.

dehydration of the cells could unequivocally be demonstrated by the concentration of the red dye to a colored lining of the cell wall which is very thin at the flanks but very perceptible at one or both tips of the cells (Figure 5.8*c*). The central part of the cell, which is usually occupied by the (colored) vacuole, appears to be empty. Such an aspect is exactly opposite to that of a plasmolysed cell and could have resulted from cavitation, i.e. the formation of a water vapor bubble, in the vacuole. Such bubbles were clearly perceivable upon rewarming of the frozen tissue and the concomitant rehydration of the cells (Figure 5.8*d*). Cytorrhysis

(Sakai & Larcher 1987), rather than plasmolysis, may occur in the tender cells of the mesophyll; however, due to ice-cap formation and the lack of a dye in the vacuoles, this phenomenon could not be clearly demonstrated.

From leaf pieces which had just started to freeze naturally at −3 °C, about 60% of the ionic solutes of the leaves could be extracted by floating in distilled water having a temperature of 0 °C. As freezing proceeded (−6 °C), these solutes could be almost completely washed out from the leaf discs. Shortly (1 h) after natural thawing, the normal small basic rate of solute release was recorded, indicating that the cells had again resorbed the effusate into the protoplasts. A similar observation, however, with time periods of several days instead of hours, has been reported by Palta & Li (1978) for onion scale tissue upon exposure to a moderate freeze–thaw cycle.

For *Senecio*, it could be shown that the release of ionic substances from leaf discs into the surrounding water was inversely related to temperature even in the range above 0 °C (Figure 5.9), and thus, though increased by freezing, was not dependent on ice formation. It is tempting to speculate that cold inactivation of membrane-bound ion pumps may be responsible for this phenomenon. Total or partial inactivation of a plasmalemma ATPase has been reported for tissues from several plant species upon chilling or injurious freezing (Ling-Cheng *et al.* 1982).

If charged substances migrate from the protoplast into the free space upon cooling, the osmotic potential of the cellular solution increases (becomes less negative) and consequently the turgor of the tissue must drop. This consideration could provide an explanation for the commencement of nyctinastic leaf movement at non-freezing low temperatures. Since, according to conductivity measurements, the charged substances released upon exposure to such temperatures represent only a minor fraction of the total cellular solutes (about 15–30% of which are of cationic nature), complete loss of turgor and hence maximal leaf movement can only be expected by additional dehydration of the cell upon extracellular ice formation.

Leaf water potentials (Ψ_{leaf}) of frozen leaves, as determined psychrometrically in the field, were found to coincide with the chemical (water) potential of ice (Ψ_{ice}) at the respective temperatures (Beck *et al.* 1984), provided the leaf temperature did not change too rapidly. Since Ψ_{leaf} at subfreezing temperatures was identical with the osmotic potential of frozen expressed cellular sap (Ψ_{π}) at the same temperatures, the water potentials of the frozen leaves could be interpreted as representing their

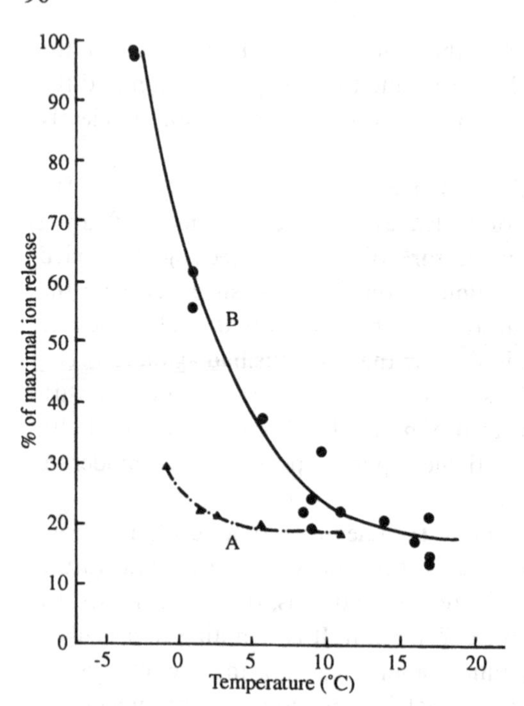

Figure 5.9. Temperature dependence of the release of ions from leaf discs of *Senecio keniodendron* floating on distilled water. Curve A was established with frost-sensitive young plants from the greenhouse. Curve B was obtained in the field (Mount Kenya) using the normal frost-hardy plants. The temperature dependence illustrated by curve B would indicate a (negative) activation energy of about 17.8 kcal mol^{-1} (ions), which is equivalent to a Q_{10} of close to -3.

osmotic potentials. Hence the water relations of the frozen leaves could be described by the equation

$$\Psi_{ice} = \Psi_{leaf} \approx \Psi_{\pi}$$

which indicates that turgor has been completely lost upon freezing and that the water potential of the leaf and that of ice are in equilibrium ('equilibrium freezing'). Under these conditions Ψ_{leaf} is governed only by the temperature and not by the original osmotic potential of the cellular sap of the unfrozen leaf. However, the proportion of the total leaf water which freezes at subfreezing temperatures depends on the concentration of the cellular solutes. Thus, in spring of 1983, the osmotic potentials of leaves of *Senecio keniodendron* and *Lobelia keniensis* were measured to be about -1 MPa. On the basis of equilibrium freezing, dehydration of the cells was calculated from this value to range from 50% (-1.5 °C) to 83%

($-4.4\,°C$). In spring of 1985 considerably more negative osmotic potentials were detected, ranging from -1.4 to -2.1 MPa. In the latter case freezing dehydration at $-4.4\,°C$ takes place to an extent of only 65%. This type of equilibrium freezing where Ψ_{ice} equals Ψ_{π} has been termed 'ideal' equilibrium freezing (Gusta *et al.* 1975; Zhu & Beck 1991). In a strongly dehydrated rigid cell it is only understandable under the premise of cavitation.

With respect to extracellular freezing, ice nucleation must be triggered in the extracellular space, while at the same time being prevented inside the protoplast. Such a requirement could simply be met by different freezing point depressions of the liquid in each of these two compartments. In addition supercooling of the extraprotoplastic solution must be minimized, in order to enable sufficient water export from the cell to the ice-crystal. In this respect leaves tolerant of freezing, in contrast to the non-tolerant ones of *Espeletia*, require only slight supercooling to initiate ice formation (Table 5.2). The actual trigger of extracellular ice nucleation, however, is as yet unknown.

Cryoprotectants

Freezing injury to freezing-tolerant plant tissues is ascribed primarily to a damaging effect of cellular solutes on the biomembranes upon concentration by freezing plasmolysis or dehydration following cavitation. The effect of extracellular freezing has been mimicked to some extent by hypertonic treatment of cells and organelles, and the release of peripheral membrane proteins, e.g. the coupling factor CF_1, from thylakoids has been shown to result from both frost and osmotic treatment (Hincha *et al.* 1984). Since the detrimental effect of concentrated inorganic solutes progresses according to the Hofmeister Series (Heber *et al.* 1981), it has been concluded that Coulomb interaction between ions and membranes is responsible for the damage. Uncharged organic compounds, such as carbohydrates or amino acids (at the isoelectric point), are known to protect membranes from freezing damage, due to their capability of hydrogen bond formation with the hydrophilic components of the membrane, thus dislodging the inorganic ions from their site of attack. In addition, the association of protective compounds bearing functional groups similar to those of water with the membrane alleviates the destructive effect of dehydration on the compounds and the architecture of the biomembrane (Chen *et al.* 1981; Crowe *et al.* 1983).

Cryprotective compounds should be present at elevated concentrations

Table 5.2 *Supercooling and freezing-point depressions of tropical alpine species*

n = number of determinations or species, respectively.

Species	Date/Reference	Minimum freezing-point of leaf (°C)	Supercooling necessary for leaf water to freeze (°C)
Senecio keniodendron	24 Feb 1983	-0.17 ± 0.1 ($n = 5$)	-7
Senecio keniensis	4 Mar 1983	-0.37 ± 0.3 ($n = 3$)	-5.5
Lobelia keniensis	5 Mar 1983	-0.23 ± 0.1 ($n = 3$)	-5.0
Lobelia telekii	22 Feb 1983	-0.17 ± 0.1 ($n = 9$)	-3.7
Lobelia telekii (fluid of the inflorescence)	Krog *et al.* (1979)	-0.3	-4.6 ± 1.3
Polylepis sericea	Rada *et al.* (1985b)	-2.8	-6 to -8 (-11 °C)
Espeletia spp.	Goldstein *et al.* (1985)		-5 to -10.8 ($n = 12$)
Draba chionophila	Azócar *et al.* (1988)		-4.0 to -5.0 ($n = 5$)

in order to be effective in displacing inorganic ions from membranes. A search for potential cryoprotectants in Afroalpine plants such as *Senecio* spp., *Lobelia* spp. and *Alchemilla argyrophylla* revealed the presence of large amounts of sucrose – up to 38% of the leaf's dry weight (Beck *et al.* 1982; Beck 1987). Notwithstanding aspects of cellular compartmentalization, a sucrose content such as this is equivalent to an intercellular concentration of about 0.1 M which, upon freezing dehydration, would approach a value of 1 . Hence, sucrose is an appropriate candidate as a cryoprotectant in Afroalpine plants. In this respect it is of particular interest that such plants in their natural habitat produce only minor amounts of assimilatory starch, whereas 80–90% of the photosynthetic carbon gain is accumulated as sucrose (see Figure 5.10). However, when specimens of these same species are grown in a greenhouse under non-freezing conditions, they accumulate starch at the expense of sucrose (E. Beck, unpublished data). This selective response to freezing and non-freezing temperatures underlines the concept of sucrose functioning as a cryoprotective substance. A daily fluctuation of the osmotic potential as a consequence of the concentration of soluble carbohydrates has also been reported for the freezing tolerant páramo species *Draba chionophila* (Azócar *et al.* 1988).

Leaf lipids

Apart from their requirement for protection by compatible solutes, plasma membranes play a crucial role in the avoidance of cryoinjury with respect to the composition of their lipids. Not only the maintenance of the lipids' liquid-crystalline state (see Wolfe 1978; Bishop 1983) during the course of substantial daily temperature fluctuation, but also the capacity to accomplish large changes in membrane surface area (Steponkus 1984), are a prerequisite of freezing tolerance. Unfortunately, little attention has been paid to the leaf lipids of tropical alpine plants until recently. The only species studied that far is *Lobelia telekii* from the alpine region of Mount Kenya (E. Beck, unpublished data). Table 5.3 shows typical results of the analysis of the polar membrane lipids of the leaves of this giant rosette plant. As expected, lipids characteristic of chloroplast thylakoids, i.e. mono- and digalactosyldiacylglycerol (MGDG, DGDG: Barber 1984), were prevalent and the ratio of these two uncharged polar lipids (1.93:1, with a preponderance of the monogalactosyl form) is similar to that measured for spinach chloroplasts (Barber *et al.* 1984) and leaves (Senser & Beck 1984). Phosphatidyl-choline and phosphatidyl-glycerol are

Table 5.3 *Polar lipids of leaves of* Lobelia telekii. Mount Kenya, 4200 m, 21.2.82, noon

Lipid	PC	PI	PE	SL	PG	DGDG	GL1	GL2	GL3	MGDG	Total
nM/g d.w.	1019	97	156	91	600	1452	255	83	204	2811	6768
mol %	15.1	1.4	2.3	1.3	8.9	21.5	3.8	1.2	3.0	41.5	
Fatty acid composition (mol %)											
14:0										0.7	0.3
16:0	25.3	31.1	26.6	20.9	17.4	24.8	18.7	27.4	25.3	5.8	16.0
16:1	0.4	3.5	2.4	4.3	0.9	1.0	3.3	7.4	4.4	0.6	1.0
16:2	0.6	5.6	2.7	1.8	30.4	0.6	3.8			2.4	4.2
16:3	1.7			2.3		1.6	3.5	6.2	7.4	30.8	13.8
17:0	1.9	2.7	2.4	2.2	0.8	0.6	0.8			0.3	0.8
17:2	6.0	7.4	4.5	7.3	1.7	3.4	21.3	7.6	2.1		3.1
18:0	1.0	8.3	5.5	6.3	1.6	1.6	3.2	6.3	5.8	0.7	1.6
18:1	3.1	3.3	3.0	4.6	2.4	1.0	3.7	6.0	3.0	1.3	1.9
18:2	38.3	19.9	38.2	28.1	26.6	10.0	13.9	14.4	5.7	5.0	14.8
18:3	21.6	13.4	12.1	18.4	17.2	54.8	25.7	19.2	43.5	50.0	40.7
20:1		4.8	2.6	3.7	1.0	0.4	2.0	5.5	2.8	0.2	0.7
Not identified										2.3	1.0

PC, phosphatidylcholine; PG, phosphatidylglycerol; PE, phosphatidylethanolamine; PI, phosphatidylinositol; SL, sulfoquinovosyl diglyceride; MGDG, DGDG, mono- and digalactosyl diglyceride, respectively; GL, unidentified glycolipid.

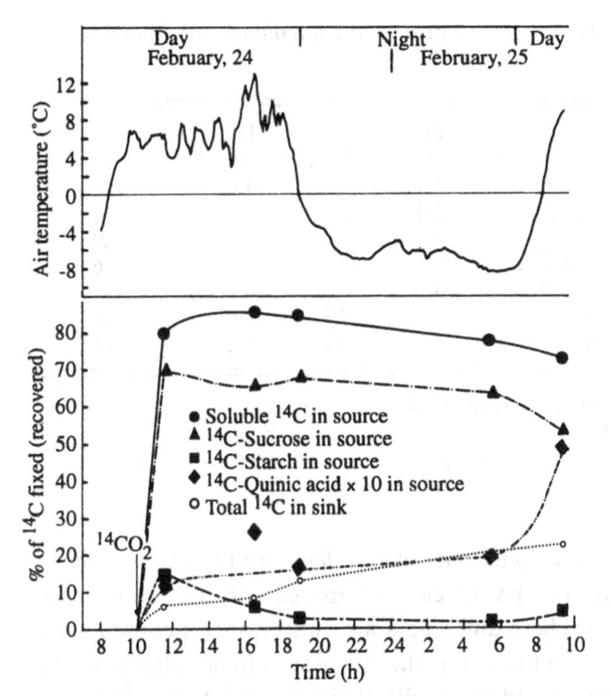

Figure 5.10. *Senecio keniodendron*: assimilation and partitioning of ^{14}C during a day–night cycle with substantial nocturnal frost (as shown by the course of the air temperature in the upper panel). $^{14}CO_2$ was administered at a concentration of 0.5% (v/v) in plastic bags to an area of about 100 cm^2 on the top part of the leaves, which, owing to its higher chlorophyll content and less shadowing by other leaves, was termed the 'source'. The lower part of the leaf, comprising about 300 cm^2, was consequently designated as the sink region. After exposure to $^{14}CO_2$ for the time indicated in the abscissa, the leaves were divided in several portions, cut into small pieces and exhaustively extracted with boiling 70% ethanol and subsequently with 1 N HCl at 100 °C for 3 hours. The latter extract contained ^{14}C-glucose as virtually the sole labelled compound, and thus represented ^{14}C-starch. The alcohol extract was analysed by paper chromatography and radioautography as described by Beck *et al.* (1982).

prominent among the polar phospholipids, a situation which appears also to be typical of leaf lipids (Senser & Beck 1984). The degree of unsaturation (i.e. the average number of double bonds per lipid molecule), which in addition to the protein/lipid ratio is a factor controlling membrane fluidity, differs from those calculated for the winter and summer states of spinach, barley or spruce, inasmuch as the galactolipids are clearly more saturated. Since this holds true in particular for the bilayer-forming DGDG, this observation may be of physiological relevance.

Table 5.4 *Degree of fatty acid unsaturation (average number of double bonds per lipid molecule)*

Species	State	PC	PG	DGDG	MGDG
Lobelia telekii	—	3.3	3.5	4.0	5.2
Spinach[a]	Summer	2.0	3.1	5.1	5.8
	Winter	2.6	3.2	5.4	6.0
Barley[b]	Frost-sensitive	3.1	3.5	4.5	5.6
	Frost-resistant	3.1	3.5	4.8	5.7
Spruce[a]	Summer	3.4	2.9	4.6	4.8
	Winter	4.0	2.2	5.2	6.1

[a] Senser & Beck 1984; [b] Havaux *et al.* 1984.
Abbreviations as for Table 5.3.

On the contrary, the phospholipids of *L. telekii* are unsaturated to an extent similar to that of the species compared, or are even more unsaturated (Table 5.4). These differences are conceivably a consequence of the demand of the climate on the tropical alpine plants to be (moderately) frost resistant and metabolically fully competent at the same time.

With respect to the pattern of fatty acids, three features deserve special reference.

1 *Lobelia telekii*, like spinach, belongs to the so-called '16:3 species' (Harwood 1980; Williams *et al.* 1983) which are characterized by a large proportion of this type of fatty acid in the chloroplast membranes and especially in MGDG. From the proportion of Δ:3 fatty acids present (Table 5.3), it must be concluded that at least 60% (maximally 80%) of the MGDG molecules must be esterified exclusively with three-fold unsaturated fatty acids. Even the value of 80% would represent an extraordinarily small proportion of Δ:3 fatty acids in MGDG compared with other '16:3 species' (Williams *et al.* 1983).
2 In contrast to the galactolipids, the phospholipids are characterized by two-fold unsaturated fatty acids, a fact which, however, appears not to be unusual (Senser & Beck 1984). Nevertheless, the large proportion of 16:2 fatty acids (especially in phosphatidylglycerol) should be noted, as this type has up to now been identified in only a

few cases (Harwood 1980) and may be considered to be an intermediate in the desaturation reaction leading to 16:3 lipids.

3 The fatty acid 17:2, which appears to be typical of an as yet insufficiently identified glycolipid, is a novelty in the context of green plant lipids (see Harwood 1980).

Since the particular functions of the various membrane polar lipids are not yet sufficiently understood (Barber 1984; Barber *et al.* 1984), these findings must assume primarily descriptive character for the present.

Metabolism of frozen leaves

As mentioned above (p. 92), frost resistance based on freezing avoidance due to supercooling brings with it the advantage that cellular water relations remain unaffected. Hence, metabolic processes may be expected to operate at subfreezing temperatures as long as activation energy is sufficient. The observation that net photosynthesis proceeds in *Espeletia* and other páramo species at temperatures of as low as −8 °C, provided that ice formation is avoided (Larcher *et al.* 1973; Larcher & Wagner 1976), is thus plausible and indicative of the advantage of freezing avoidance, as against freezing tolerance which has to put up with a decrease in metabolic activity such as photosynthesis (Larcher 1982). Admittedly, the combination of sunlight and subfreezing temperatures is not likely to occur in the natural habitat of *Espeletia.* Therefore, the temperature dependence of dark respiration and carbon transport is presumably more important for the nocturnal metabolism of tropical alpine plants than is photosynthesis. These processes have recently been studied with Afroalpine plants in which, in addition to the effects of low temperatures as such, considerable cell dehydration occurs upon freezing. Dark respiration was greatly reduced and the capacity for photosynthesis was completely lost in frozen leaves. The latter phenomenon may be ascribed to the dehydration effect of extracellular ice formation (Bodner & Beck 1987). A respiration overshoot, which is commonly observed with European plants after freezing and lasts for about 4–12 hours subsequent to thawing (Larcher *et al.* 1973), could not be observed. Carbon transport could still be observed to take place in frozen leaves, the photosynthates of which had been labelled with ^{14}C (Figure 5.10).

In order to understand the experiment detailed in Figure 5.10, it should be noted that the leaves of *Senecio keniodendron*, owing to their large size (50 × 15 cm) and their prevalent upright orientation, do not receive equal photon flux densities over all parts of their surface. The lower part of the leaf obtains much less direct radiation, and in addition contains less chlorophyll. Hence, both a source and a sink component can be assumed to exist within the same leaf. $^{14}CO_2$ was applied to a region of the source area and the incorporation of radiocarbon into source and sink segments was followed subsequently. The data suggest that the export of ^{14}C-sucrose from the source area continued at a reduced rate in the frozen leaf. About two thirds of the radioactivity found in the sink area were embodied in the midrib, and about 85% of this amount was constituted by ^{14}C-sucrose. Degradation of assimilatory starch during the dark period was not observed, since the level of this polysaccharide had already reached a minimum value at dusk. As has been shown by Schulze *et al.* (1985), those leaves of the rosette which are inserted on the west side of the plant and therefore face east usually cease photosynthetic activity at noon due to a decrease of irradiation in the afternoon, in the course of which chloroplastic starch is then degraded. Another indication of metabolic processes continuing during the frozen state is the increase of ^{14}C incorporation into a secondary plant constituent, identified as quinic acid (Figure 5.10). Although frozen leaves are at a disadvantage when compared with supercooled leaves due to the effects of cell dehydration, there is evidence that dark metabolism proceeds in both types at subfreezing temperatures.

Ecological conclusions: is temperature the limiting factor for plant life in tropical alpine habitats?

It is still a matter of debate as to which ecological factors are limiting for plant growth in the tropical alpine and nival zones. Low temperature – especially low soil temperature – has repeatedly been considered as the critical environmental element in this respect (Weberbauer 1930; Walter & Medina 1969; Larcher 1975). On the whole, however, frost resistance appears to be sufficient to enable all tropical alpine plants to withstand nocturnal subfreezing temperatures (see Table 5.1). This holds true even for locations where certain features of the terrain may favor the accumulation of cold air, thus resulting in minimum temperatures considerably lower than those quoted by the records of climatological stations. For instance, whereas the lowest temperature recorded on Mount Kenya at 4770 m elevation was −8.3 °C, Coe (1967) measured −9 °C at 4175 m at Mackinders Valley; and the author of this chapter registered values as low as −12 °C and −13 °C in a depression of the upper Teleki Valley (4200 m) in March 1983 and 1985, respectively. This hollow is covered predominantly by the cabbage groundsel and no indication of cryoinjury was observed subsequent to such cold nights. Interestingly, only a moderate gradient of increasing frost resistance could be detected with respect to increasing altitude from the valley floor to the ridges above (Table 5.1), an observation which illustrates the relative importance of the microclimatic situation.

Of course, plant species relying on supercooling to avoid freezing are much more endangered by the sporadic occurrence of particularly severe frosts, and low temperature in itself may hence be a limiting factor for plant life at high altitudes in the páramo regions. Presumably, the altitudinal distribution of the Andean genus *Espeletia* which is frost- but not freezing-tolerant may reflect the severity and duration of the nocturnal frost period (Azócar *et al.* 1988). However, the minimum air temperature cannot be held responsible for suppressing plant growth in the tropical alpine regions in general since the Afroalpine species are known to tolerate freezing of their tissues. This is indicated by the presence of flowering specimens of *Senecio keniophytum* and *Helichrysum brownei* at the top of Mount Kenya, 5200 m a.s.l. (E. Beck, unpublished data; see also Mackinder 1900), and of *Senecio telekii*, *S. meyeri-johannis*, *Helichrysum newii* and *Arabis alpina* in the Reusch Crater (5800 m) on Mount Kilimanjaro (Sampson 1953; Beck 1988). Freezing tolerance has also been shown for the 'miniature' caulescent rosette species *Draba chionophila* which is the vascular plant reaching the highest altitudes (4700 m) in the Venezuelan desert páramo (Azócar *et al.* 1988).

In contrast to the direct effects of low temperatures, indirect effects may have a considerable impact on Afroalpine plants and thus be responsible for the thinning out of the vegetation in the nival zone. Solifluction, due to repeated freezing and thawing of the water of the uppermost soil layer, must be considered in this regard. Thus, apart from globular mosses (Hedberg 1964; Beck *et al.* 1986b) and some algae and lichens which ride the crest of the moving soil, higher plants in the alpine desert are consistently found at the edge of boulders where nocturnal freezing of soil water, and hence solifluction, is reduced. From such refuges, decumbent or prostrate stems may easily spread over areas of soil disturbed by solifluction (Hedberg 1964).

Another indirect effect of altitude and low temperature is the decrease of air humidity. On Mount Kenya rainfall increases to a sharp maximum between 2500 and 3000 m and rapidly falls off once more at higher elevations (Hedberg 1964; Coe 1967). On the east side of Mount Kilimanjaro the zone of maximal precipitation is found at altitudes between 2600 and 2900 m, where 1780 mm per annum falls on the average, whereas the respective amounts of precipitation at 4900 and 5800 m are 74 and 15 mm per annum, respectively (Salt 1951). Although relative air humidity levels rapidly rise to values near 100% upon clouding, the absolute amount of water vapor, which is decisive for plant life, is, with respect to the altitude of 4000 m, 2.9 g (ml) m^{-3} at 0 °C and not more than 2.0 g (ml) m^{-3} at −5 °C. Upon rapid warming in the morning, these absolute amounts result in relative air humidity levels (at +15 °C) of 37% and 23%, respectively and vapor pressure deficits of up to 40 mPa Pa^{-1} can arise (Schulze *et al.* 1985). Since the annual mean temperature of the nival zone is about 0 °C (Larcher 1975), these tropical alpine regions appear to be considerably drier than, for example, the nearest lowland deserts in northwest Kenya, the absolute air humidity of which amounts to 20 g (ml) m^{-3} throughout the year (Knapp 1973). Thus humidity, as governed by macroclimate and temperature, is more likely to be a limiting factor than is frost. This conclusion is convincingly illustrated by the observation that the alpine desert on Mount Kilimanjaro extends to considerably lower altitudes on the drier western side of the mountain than it does on the moist southeastern slopes (Klötzli 1958; Beck *et al.* 1983). In addition, patches of wet meadows are found scattered within the alpine desert in headwater areas. Finally, flowering plants have been observed even at 5800 m at the edge of boulders, where water vapor, exhausted from fumaroles, condenses and thus provides supplementary moisture to the soil (Beck 1988).

Acknowledgements

The author is very much indebted to Prof. Dr W. Larcher (Innsbruck) and to Prof. Dr E.-D. Schulze (Bayreuth), Prof. Dr R. Scheibe (Osnabrück) and Dr P. Ziegler (Bayreuth), for critical reading of the manuscript, several useful suggestions and linguistic help. Financial support of the author's field studies by the Deutsche Forschungsgemeinschaft and the Ministry of Technical Cooperation (GTZ) is gratefully acknowledged.

References

Asahina, E. (1978). Freezing processes and injury in plant cells. In *Plant Cold Hardiness and Freezing Stress*, Vol. I, ed. P. H. Li and A. Sahai, pp. 17–36. New York: Academic Press.

Azócar, A., Rada, F. & Goldstein, G. (1988). Freezing tolerance in *Draba chionophila*, a 'miniature' caulescent rosette species. *Oecologia* **75**, 156–60.

Barber, J. (1984). Lateral heterogeneity of proteins and lipids in the thylakoid membrane and implications for electron transport. In *Advances in Photosynthesis Research*, Vol. III/2, ed. C. Sybesma, pp. 91–8.

Barber, J., Gounaris, K., Sundby, C. & Andersson, B. (1984). Lateral heterogeneity of polar lipids in the thylakoid membranes of spinach chloroplasts. In *Advances in Photosynthesis Research*. Vol. III/2, ed. C. Sybesma, pp. 159–62.

Beck, E. (1983). Frost- und Feuerresistenz tropisch-alpiner Pflanzen. *Naturwissenschaftliche Rundschau* 36, 105–9.

Beck, E. (1987). Die Frostresistenz der tropisch-alpinen Schopfbäume. *Naturwissenschaften* **74**, 355–61.

Beck, E. (1988). Plant life on top of Mt. Kilimanjaro (Tanzania). *Flora* **181**, 379–81.

Beck, E., Mägdefrau, K. & Senser, M. (1986b). Globular mosses. *Flora* **178**, 73–83.

Beck, E., Scheibe, R., Schlütter, I. & Sauer, W. (1992). *Senecio* × *saundersii* Sauer & Beck (Asteraceae), an intermediate hybrid between *S. keniodendron* and *S. keniensis* of Mt. Kenya. *Phyton* (*Horn*) **32**, 9–37.

Beck, E., Scheibe, R. & Schulze, E.-D. (1986a). Recovery from fire: observations in the alpine vegetation of western Mt. Kilimanjaro (Tanzania). *Phytocoenologia* **14**, 55–77.

Beck, E., Scheibe, R. & Senser, M. (1983). The vegetation of the Shira Plateau and the western slopes of Kibo (Mt. Kilimanjaro, Tanzania). *Phytocoenologia* **11**, 1–30.

Beck, E., Scheibe, R., Senser, M. & Müller, W. (1980). Estimation of leaf and stem growth of unbranched *Senecio keniodendron* trees. *Flora* **170**, 68–76.

Beck, E., Schulze, E.-D., Senser, M. & Scheibe, R. (1984). Equilibrium freezing of leaf water and extracellular ice formation in Afroalpine 'giant rosette' plants. *Planta* **162**, 276–82.

Beck, E., Senser, M., Scheibe, R., Steiger, H.-M. & Pongratz, P. (1982). Frost avoidance and freezing tolerance in Afroalpine 'giant rosette' plants. *Plant, Cell and Environment* **5**, 215–22.

Bishop, D G. (1983). Functional role of plant membrane lipids. In *Biosynthesis and Function of Plant Lipids*, ed. W. W. Thomson, J. B. Mudd and M. Gibbs, pp. 81–103. Baltimore, Maryland: Waverley Press.

Bodner, M. & Beck, E. (1987). Effect of supercooling and freezing on photosynthesis in freezing tolerant leaves of Afroalpine 'giant rosette' plants. *Oecologia* **72**, 366–71.

Chen, C.-H., Berns, D. S. & Berns, A. S. (1981). Thermodynamics of carbohydrate–lipid interactions. *Biophysical Journal* **36**, 359–67.

Coe, M. J. (1967). The ecology of the alpine zone of Mount Kenya. *Monographiae Biologicae 17*. The Hague: Junk.

Crowe, J. H., Crowe, L. M. & Mouradian, R. (1983). Stabilization of biological membranes at low water activities. *Cryobiology* **20**, 346–56.

Eller, B. M. & Willi, P. (1977). The significance of leaf pubescence for the absorption of global radiation by *Tussilago farfara* L. *Oecologia* **29**, 179–87.

Goldstein, G. & Meinzer, F. (1983). Influence of insulating dead leaves and low temperatures on water balance in an Andean giant rosette plant. *Plant, Cell and Environment* **6**, 649–56.

Goldstein, G., Meinzer, F. & Monasterio, M. (1984). The role of capacitance in the water balance of Andean giant rosette species. *Plant, Cell and Environment* **7**, 179–86.

Goldstein, G., Rada, F. & Azócar, A. (1985). Cold hardiness and supercooling along an altitudinal gradient in Andean giant rosette species. *Oecologia* **68**, 147–52.

Gusta, L. V., Burke, M. J. & Kapoor, A. C. (1975). Determination of unfrozen water in winter cereals at subfreezing temperatures. *Plant Physiology* **56**, 707–9.

Harwood, J. L. (1980). Plant acyl lipids: structure, distribution and analysis. In *The Biochemistry of Plants*, ed. P. K. Stumpf and E. F. Conn, Vol. 4, *Lipids*, ed. P. K. Stumpf, pp. 1–55. New York: Academic Press.

Hauman, L. (1955). La région afroalpine en phytogeographie centro-africaine. *Webbia* **11**, 467–9.

Havaux, M., Lannoye, R., Chapman, D. J. & Barber, J. (1984). In *Advances in Photosynthesis Research*, Vol. IV/4, ed. C. Sybesma, pp. 459–62.

Heber, U., Schmitt, J. M., Krause, G. H., Klosson, R. J. & Santarius, K. A. (1981). Freezing damage to thylakoid membranes *in vitro* and *in vivo*. In *Effects of Low Temperatures on Biological Membranes*, ed. G. J. Morris and A. Clarke, pp. 263–88. London: Academic Press.

Hedberg, O. (1964). Features of afroalpine plant ecology. *Acta Phytogeographica Suecica* **48**, 1–144.

Hedberg, I. & Hedberg, O. (1979). Tropical-alpine life-forms of vascular plants. *Oikos* **33**, 297–307.

Hincha, D. H., Schmidt, J. E., Heber, U. & Schmitt, J. M. (1984). Colligative and non-colligative freezing damage to thylakoid membranes. *Biochimica et Biophysica Acta* **769**, 8–14.

Jeffrey, C. (1986). The Senecioneae in East Tropical Africa. *Kew Bulletin* **41**, 873–943.

Kandler, O., Dover, C. & Ziegler, P. (1979). Kälteresistenz der Fichte. I. Steuerung von Kälteresistenz, Kohlehydrat- und Proteinstoffwechsel durch Photoperiode und Temperatur. *Berichte der Deutschen Botanischen Gesellschaft* **92**, 225–41.

Klötzli, F. (1958). Zur Pflanzensoziologie des Südhanges der Alpinen Stufe des Kilimandscharo. *Berichte des Geobotanischen Forschungsinstituts Rübel 1957*, 33–59.

Knapp, R. (1973). *Die Vegetation von Afrika*, p. 377. Stuttgart: G. Fischer Verlag.

Krog, J. O., Zachariassen, K. E., Larsen, B. & Smidsrod, O. (1979). Thermal buffering in Afro-alpine plants due to nucleating agent-induced water freezing. *Nature* **282**, 300–1.

Larcher, W. (1975). Pflanzenökologische Beobachtungen in der Paramostufe der venezolanischen Anden. *Anzeiger der mathematisch–naturwissenschaftlichen Klasse der Österreichischen Akademie der Wissenschaften* **11**, 194–213.

Larcher, W. (1982). Typology of freezing phenomena among vascular plants and evolutionary trends in frost acclimatization. In *Plant Cold Hardiness and Freezing Stress*, Vol. II, ed. P. H. Li and A. Sakai, pp. 417–26. New York: Academic Press.

Larcher, W., Heber, U. & Santarius, K. A. (1973). Limiting temperatures for life functions. In *Temperature and Life*, ed. H. Precht, J. Christophersen, H. Hensel and W. Larcher, pp. 199 and 202. Berlin: Springer-Verlag.

Larcher, W. & Wagner, J. (1983). Ökologischer Zeigerwert und physiologische

Konstitution von *Sempervivum montanum. Verhandlungen der Gesellschaft für Ökologie* (Festschrift für Heinz Ellenberg) Vol. XI, pp. 253–264.
Larcher, W. & Wagner, J. (1976). Temperaturgrenzen der CO_2-Aufnahme und Temperaturresistenz der Gebirgspflanzen im vegetationsaktiven Zustand. *Oecologia Plantarum* **11**, 361–74.
Levitt, J. (1980). *Responses of Plants to Environmental Stresses*, Vol. I, pp. 116–62. New York: Academic Press.
Ling-Cheng, J., Long-hua, S., He-zhu, D. & De-lan, S. (1982). Changes in ATPase activity during freezing injury and cold hardening. In *Plant Cold Hardness and Freezing Stress*, Vol. II, ed. P. H. Li and A. Sakai, pp. 243–59. New York: Academic Press.
Mackinder, H. K. (1900). A Journey to the Summit of Mt. Kenya, British East Africa. *Geographical Journal* **15**, 453–86.
Meinzer, F. & Goldstein, G. (1985). Some consequences of leaf pubescence in the Andean giant rosette plant *Espeletia timotensis. Ecology* **66**, 512–20.
Monasterio, M. (1986). Adaptive strategies of *Espeletia* in the Andean desert páramo. In *High Altitude Tropical Biogeography*, ed. F. Vuilleumier and M. Monasterio, pp. 49–80. New York: Oxford University Press.
Palta, J. P. & Li, P. H. (1978). Cell membrane properties in relation to freezing injury. In *Plant Cold Hardiness and Freezing Stress*, ed. P. H. Li and A. Sakai, pp. 93–115. New York: Academic Press.
Rada, F., Goldstein, G., Azócar, A. & Meinzer, F. (1985a). Freezing avoidance in Andean giant rosette plants. *Plant, Cell and Environment* **8**, 501–7.
Rada, F., Goldstein, G., Azócar, A. & Meinzer, F. (1985b). Daily and seasonal osmotic changes in a tropical treeline species. *Journal of Experimental Botany* **36**/167, 989–1000.
Rasmussen, D. H. & MacKenzie, A. P. (1972). Effect of solute on the ice-solution interfacial free energy; calculation from measured homogeneous nucleation temperatures. In *Water Structure at the Water–Polymer Interface*, ed. H. H. G. Jellinek, pp. 126–45. New York: Plenum.
Sakai, A. & Larcher, W. (1987). *Frost Survival of Plants.* Berlin, Springer-Verlag.
Salt, G. (1951). The Shira Plateau of Kilimanjaro. *Geographical Journal* **117**, 150–64.
Sampson, D. N. (1953). Notes on the Flora of Kilimanjaro. *Tanganyika Notes and Records* **34**, 68–9.
Schulze, E.-D., Beck, E., Scheibe, R. & Ziegler, P. (1985). Carbon dioxide assimilation and stomatal response of afroalpine giant rosette plants. *Oecologia* **65**, 207–13.
Senser, M. (1982). Frost resistance in spruce *Picea abies* (L.) Karst.: III. Seasonal changes in the phospho- and galactolipids of spruce needles. *Zeitschrift für Pflanzenphysiologie* **195**, 229–39.
Senser, M. & Beck, E. (1979). Kälteresistenz der Fichte. II. Einfluß von Photoperiode und Temperatur auf die Struktur und photochemischen Reaktionen von Chloroplasten. *Berichte der Deutschen Botanischen Gesellschaft* **92**, 243–59.
Senser, M. & Beck, E. (1984). Correlation of chloropast ultrastructure and membrane lipid composition to the different degrees of frost resistance achieved in leaves of spinach, ivy, and spruce. *Journal of Plant Physiology* **117**, 41–55.
Smith, A. P. (1974). Bud temperature in relation to nyctinastic leaf movement in an Andean giant rosette plant. *Biotropica* **6**, 263–6.

Smith, A. P. (1979). Function of dead leaves in *Espeletia schultzii* (Compositae), an Andean caulescent rosette species. *Biotropica* **11**, 43–7.
Steponkus, P. L. (1984). Role of the membrane in freezing injury and cold acclimation. *Annual Review of Plant Physiology* **35**, 543–84.
Sturm, H. (1978). Zur Ökologie der andinen Paramoregion. In *Biogeographica*, Vol. 14, ed. J. Schmithüsen. The Hague: Dr W. Junk.
Troll, C. (1943). Die Frostwechselhäufigkeit in den Luft- und Bodenklimaten der Erde. *Metereologische Zeitschrift* **60**, 161–71.
Walter, H. & Medina, E. (1969). Die Bodentemperatur als ausschlaggebender Faktor für die Gliederung der subalpinen und alpinen Stufe in den Andean Venezuelas. *Berichte der Deutschen Botanische Gesellschaft* **82**, 275–81.
Weberbauer, A. (1930). Untersuchungen über die Temperaturverhältnisse des Bodens im hochandinen Gebiet Perus und ihre Bedeutung für das Pflanzenleben. *Englers Botanische Jahrbücher* **63**, 330–49.
Williams, J. P., Khan, M. U. & Mitchell, K. (1983). Galactolipid biosynthesis in leaves of 16:3- and 18:3-plants. In *Advances in Photosynthesis Research*, Vol. III/2, ed. C. Sybesma, pp. 28–39.
Wolfe, J. (1978). Chilling injury in plants – the role of membrane lipid fluidity. *Plant, Cell and Environment* **1**, 241–7.
Young, T. P. (1984). The comparative demography of semelparous *Lobelia telekii* and iteroparous *Lobelia keniensis* on Mt. Kenya. *Journal of Ecology* **72**, 637–50.

6

Anatomy of tropical alpine plants

SHERWIN CARLQUIST

Introduction

Despite recent interest in alpine ecology, we have little information on anatomy of alpine plants. This is as true of tropical alpine plants as it is of those from temperate mountain areas. The reasons for the lack of studies in anatomy of tropical alpine plants are perhaps surprising.

First, one can cite the tendency for plant anatomists to work little on comparative problems and, when they do, to work in terms of particular taxonomic groups and to express their data in systematic terms rather than in ecological ones. To be sure, interest in ecological plant anatomy has increased in the latter half of the 20th century, and one can cite more studies concerning tropical alpine plants in recent years. The curious climatic regimes of high equatorial mountains make anatomical adaptations of especial interest, as will be seen from the relatively few examples cited in this chapter. Adaptations to frost and to drought are central in tropical plants, but these adaptations are different from those in plants of extremely cold or extremely dry regions.

A second reason for lack of studies on anatomy of tropical alpine species is a by-product of the working habits characteristic of biologists. Ecologists, unless highly theoretical in orientation, tend to use the outdoors as a laboratory. For the plant anatomist, laboratory work must be done indoors, and the habit of working both in the field and in the laboratory apparently does not come easily to most plant anatomists. Indeed, few plant anatomists who do comparative work collect the material they study. Tropical alpine areas are often difficult of access, so that plant anatomists are quite unlikely to visit these areas. Perhaps more significant, plant anatomists tend not to be well trained in plant ecology and physiology, so that kinds of data needed for structure/function correlations are often not collected. Plant physiologists tend not to initiate studies in plant anatomy very often. A collaboration among workers seems a desirable way of proceeding with studies on anatomy of tropical alpine plants. Such collaborations have been very few, so that the present chapter will have the purpose of calling attention to possible findings rather than reporting them.

A third circumstance that has delayed anatomical studies on tropical alpine plants is the tendency for anatomical studies to favor woody species and species of economic importance. Woody plants are often thought to contain more anatomical diversity. In fact, herbaceous plants offer just as many features of anatomical interest. The studies cited below indicate the inherent interest of tropical alpine plants, and should promote further studies by enterprising workers.

Anatomy of rosette trees

The chief genera of rosette trees in tropical alpine areas are *Espeletia* (Asteraceae) in the northern Andes and *Senecio* (Asteraceae) on the high African volcanoes. *Lobelia* (Campanulaceae, subfamily Lobelioideae; or Lobeliaceae) is often cited, but note should be taken that the high elevation species of *Lobelia* are not arborescent; the woodier species of *Lobelia* tend to be at the bottom of the alpine zone or below. *Lobelia* in the alpine zone tends to be what some authors term 'rosulan' (with a very condensed stem), 'caulorosulan' (with a short stem, usually under 1 m). The term 'megaherb' is sometimes applied to plants like the tropical alpine *Lobelia* species, although it has been used more commonly for large-leaved herbs such as *Gunnera* or large rosette plants of subantarctic islands (e.g. *Pleurophyllum, Stilbocarpa*). The woodiest species of *Lobelia* in Africa can be found in the mossy forest, as in the case of *L. gibberoa* Hemsl. or *L. lanuriensis* De Wild. The chief references with respect to habit, habitat and taxonomy are: *Senecio* (Hauman 1935; Hedberg 1957, 1964; Nordenstam 1978); *Espeletia* (Smith & Koch 1935; Rock 1972); and *Lobelia* (Hauman 1933; Hedberg 1957, 1964).

Anatomy of all three genera is certainly not the same but certain main features are similar. One can say that all three genera have a wide pith and a thick cortex or bark (Hauman 1935; Rock 1972). The pith width results from the action of a primary thickening meristem in *Espeletia* (Rock 1972) and the other genera should be investigated in this respect. There is evidence for a water storage function of the wide pith (Goldstein & Meinzer 1983; Goldstein *et al.* 1984). Another possible function of the pith is structural, providing a device for achieving a cylinder wide in diameter (and thus structurally superior to a narrow cylinder for resistance to shear) and capable of bearing larger numbers of leaves than could a narrower stem. There seems little reason to doubt the water storage function of the stem in these genera, because in these genera the stems are succulent by any definition, whether studied in terms of anatomical preparations (Hauman 1935; Rock 1972) or by macroscopic dissections (Hedberg 1964, p. 51).

The xylem characteristics of *Senecio*, *Espeletia* and *Lobelia* show little or no indication of xeromorphy (Carlquist 1958, 1962a, 1966, 1969; Hauman 1935; Rock 1972). Although the high-alpine species of *Lobelia* (*L. telekii* Schweinf.: Carlquist 1969) and *Espeletia* (*E. timotensis* Cuatr.: Rock 1972) have short vessel elements, the shortness of vessel elements in these species might be related indirectly to the rosette habit (rosulan) in these species rather than to xeromorphy.

Features that validate mesomorphy in wood anatomy of the three genera cited above include wide diameter of vessels, low number of vessels per group (many vessels solitary), and low number of vessels per mm^2 of transection (Carlquist 1958, 1962a, 1969; Rock 1972). These features are shown for *Lobelia gibberoa* in Figure 6.1. A comparison between wood of rosette trees and wood of shrubs and other kinds of trees has been offered earlier (Carlquist 1966, 1975). The latter study shows that rosette trees stand next to stem succulents in their quantitative wood characteristics. One might, in fact, call *Senecio*, *Espeletia* and *Lobelia* (montane species) stem succulents. The broad leaf surfaces would be atypical for succulents. The vessels in the rosette trees certainly do not seem to be narrow enough or dense enough to provide redundancy (or possibly other) characteristics that are common in xeromorphic woods (for a review, see Carlquist 1988).

Areas of rays, as seen in tangential sections of woods, are rather great for the rosulan species of *Espeletia* and *Lobelia*, and the lack of secondary walls on ray cells in those species (as revealed by examination with polarized light: Rock 1972) has led Rock to claim a water storage function for these rays. The rays of the more arborescent species in these genera are tall, but relatively narrow (Figure 6.2), and do not suggest a marked water storage function. Absence of growth rings (Figure 6.1) is definitely characteristic of all three genera, although A. P. Smith (personal communication) has found distinct growth rings in the stem of *S. keniodendron* at 4200 m on Mount Kenya and in *E. humbertii* Cuatr., an arborescent species, at treeline in the Venezuelan Andes. The presence of growth rings in these species probably reflects the strongly seasonal rainfall in both of

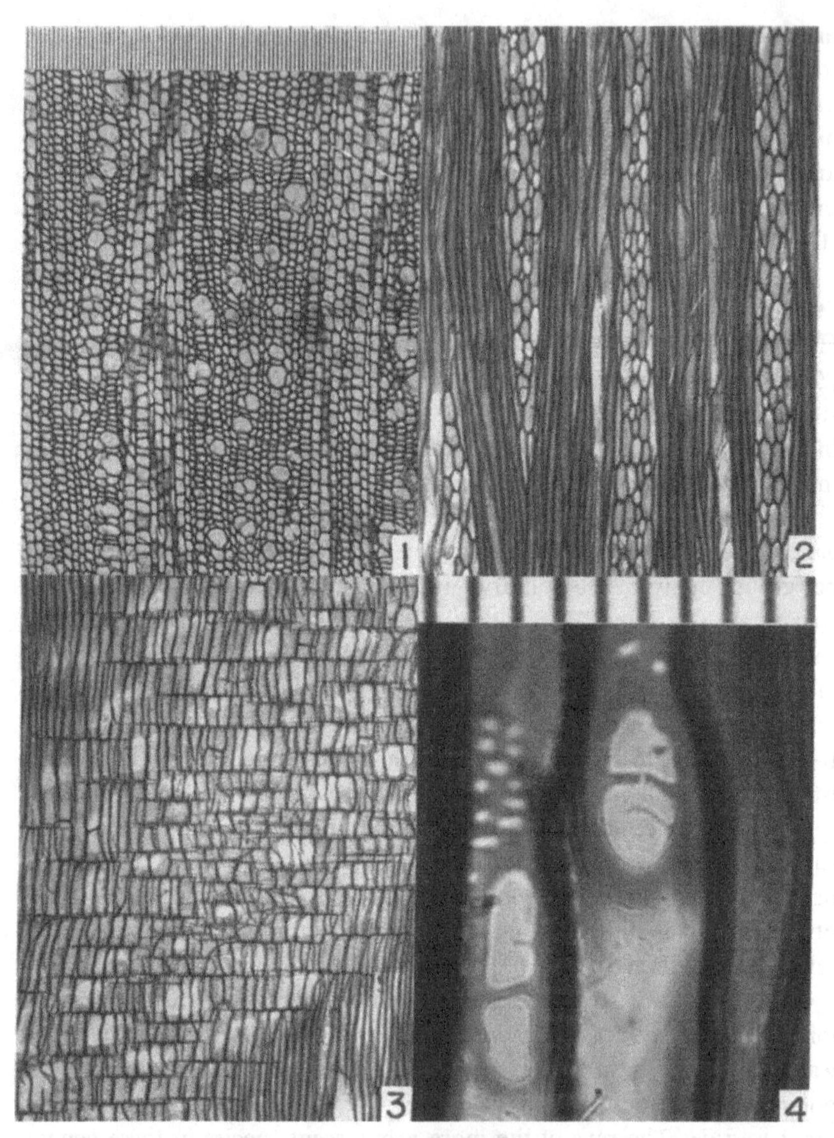

Figures 6.1–6.4. Wood sections of *Lobelia* (Campanulaceae). Figs. 6.1–6.3. *L. gibberoa* (Carlquist 2829). Fig. 6.1. Transection; note relative paucity of vessels. Fig. 6.2. Tangential section. Rays are tall, moderately wide. Fig. 6.3. Radial section; most ray cells are upright. Fig. 6.4. *Lobelia shaferi* (Yw-40053). Perforation plates from radial section, showing modified scalariform condition. Figures 6.1–6.3, magnification scale above Figure 6.1 (finest divisions = 10 μm). Figure 6.4, scale above Figure (divisions = 10 μm).

these exceptional regions. On the whole, most tropical alpine sites where rosette trees grow are not so strongly seasonal that strongly marked growth rings characteristically occur.

The slow and moderate accumulation of secondary xylem, correlated with succulence, is characteristic of the three genera cited above. In turn, this has the effect of preserving a juvenilistic secondary xylem conformation, which is being termed paedomorphosis in wood (Carlquist 1962b). Paedomorphosis has multiple expressions in wood anatomy. Horizontal subdivision of ray cells is lessened, so that erect ray cells are more abundant than would be expected in comparison to procumbent cells (Figure 6.3). Procumbent cells are found in some primary rays and little-altered primary rays, and may be related to leaf traces (Rock 1972). Lateral wall pitting of vessels in the secondary xylem often resembles that of metaxylem vessels in the three genera, in contrast to the alternate pitting that quickly succeeds metaxylem patterns in typically woody species. Simple perforation plates characterize all of the three genera, as one might expect in these families, on the basis of their numerous specialized features (Carlquist 1958, 1962a, 1969). However, occasional perforation plates that are not scalariform in the ordinary sense but much altered versions best termed multiperforate or irregularly scalariform, do occur in a very small proportion of vessel elements (Figure 6.4). The occurrence of this small proportion of multiperforate plates is very likely a manifestation of paedomorphosis, since metaxylem vessels do have scalariform perforation plates in some Campanulaceae and Asteraceae, whereas secondary xylem vessels in these families typically have simple perforation plates (Bierhorst & Zamora 1965). Mabberley (1974) confused scalariform *lateral wall pitting* of vessels (common in rosette trees) with scalariform perforation plates (rare in rosette trees). He also confused the theory of paedomorphosis with Bailey's (1944) idea that the primary xylem is a refugium of primitive xylem features. Mabberley's misunderstanding and consequent theorizing in terms of the Durian theory have been criticized (Carlquist 1980, 1988), and ideas of paedomorphosis in wood have continued to be accepted in the various taxonomic groups where this phenomenon is characteristic (Carlquist 1988). Predominantly herbaceous dicotyledon families that have become secondarily woody in the relatively non-seasonal climates of tropical montane areas are among the categories of dicotyledons that show paedomorphosis in wood (Cumbie 1983).

The leaf anatomy of the three genera of rosette trees cited above suggests forms of xeromorphy in species of higher elevations: leaves of the lower elevation species are thinner, broader, and lacking in the various features that connote xeromorphy. This has been noted by Rock (1972) for *Espeletia*, and can be readily observed in the leaves of *Senecio* and *Lobelia*; both of these genera have leaves less complex than those of *Espeletia. Espeletia* has pockets on the abaxial surfaces of leaves ('areolar cavities' of Rock). Water storage undoubtedly does occur in the thicker leaves of the high elevation species of all three genera, although identifying exactly which cells in the leaf store water and to what degree they do so is not easy (Rock 1972). Hypodermis occurs in leaves both in *Espeletia* (Rock 1972) and *Senecio* (Hauman 1935), and may serve a water storage function.

Distinctive functions performed by the leaves in the high elevation species of all three genera can be described in anatomical terms. Old leaves clothe the stems in *Senecio keniodendron* (Hedberg 1964) and *Espeletia* (Cuatrecasas 1934), and have the effect of preventing freezing of the stems (Goldstein & Meinzer 1983). The immediate anatomical cause of this behavior is failure of formation of an abscission layer in these genera; that has not been established by observations, but can be supposed with reasonable assurance.

All three genera of rosette trees cited above have nyctinastic leaf movements (Hedberg 1964; Smith 1974). The anatomical structure underlying this behavior has not been established. One would expect that parenchyma cells near the petiole base and on its abaxial side would fill with water during the night. Nyctinastic movements in rosulan and caulorosulan plants can be seen in some other plants as well as the tropical alpine rosette trees; beet and chard (*Beta vulgaris* L.) show such daily cycles, for example.

Hairs on leaves of some species of *Senecio*, *Espeletia* and *Lobelia* are sufficiently dense so as to be significant in the insulating qualities of the leaves (Hedberg 1964). These hairs are, in all three genera, elongate uniseriate hairs, sometimes densely borne (Hauman 1935; Rock 1972). Some species of *Senecio* have hairs only on petioles of leaves (Hauman 1935), but this is a maturation effect. On primordia of these *Senecio* leaves, hairs occur over the entire surface, but differential loss occurs as leaves in particular species mature (original observation).

Sclerophyllous shrubs

Hedberg & Hedberg (1979) find that tropical alpine plants fall into the categories of rosette plants, tussock grasses, acaulescent rosette plants, cushion plants, and sclerophyllous shrubs. Acaulescent rosette plants of tropical alpine regions have been little studied anatomically, but superficial examination suggests no profound differences in leaf texture or stem nature from related rosette plants of other geographical regions. Likewise, tussock grasses, although a distinctive element in tropical alpine zones, seem unlikely to differ modally as a group from tussock grasses in other regions. As individual species, however, species comparisons would doubtless prove interesting.

Haleakala and other Hawaiian volcanoes

Leaf anatomy has been studied synoptically for very few tropical alpine sclerophyllous shrubs, but the Hawaiian species of *Geranium*, all of which are high elevation plants, have distinctive modes of leaf anatomy (Carlquist & Bissing 1976). In addition to features of gross morphology (leaf area, number of leaf teeth) that distinguish the species, such anatomical characters as outer epidermal wall thickness and leaf thickness can be combined to form an index to the relative mesomorphy or xeromorphy of particular species. Hypodermis prominence increases with xeromorphy of species locality, lending credence to the idea that hypodermis is often a water storage tissue. *Geranium tridens* Hillebrand has isolateral leaves, whereas the other species have bifacial leaves. The felty covering of hairs on at least some surfaces of the leaves of the high elevation species *G. cuneatum* Hillebrand and *G. tridens* is highly reflective, and although trichomes on leaves ordinarily do not lower ultraviolet light absorption by leaves very much, in extreme situations this might be one possible function. This and other potential functions of trichomes in these species are worthy of investigation.

Sclerophyllous shrubs of Haleakala and the other Hawaiian volcanoes experience climatic regimes sufficiently extreme that adaptations by wood are of significance and, in fact, characterize all of the species in the alpine zones of these mountains. To be sure, alpine conditions are equalled only by desert areas in transpiration potential (Smith & Geller 1979), but freezing also occurs. Many Hawaiian alpine genera are also represented in dry lowland areas (Carlquist 1970).

Wood of dryland evergreen shrubs of Mediterranean-type climates is characterized, to a surprisingly large extent, by the occurrence of vasicentric tracheids and, to a lesser degree, true tracheids (Carlquist 1985). Vasicentric tracheids and true tracheids offer subsidiary conductive systems that could maintain water columns to leaves even if all vessels embolize. In addition, they offer redundancy in the conductive system by virtue of being

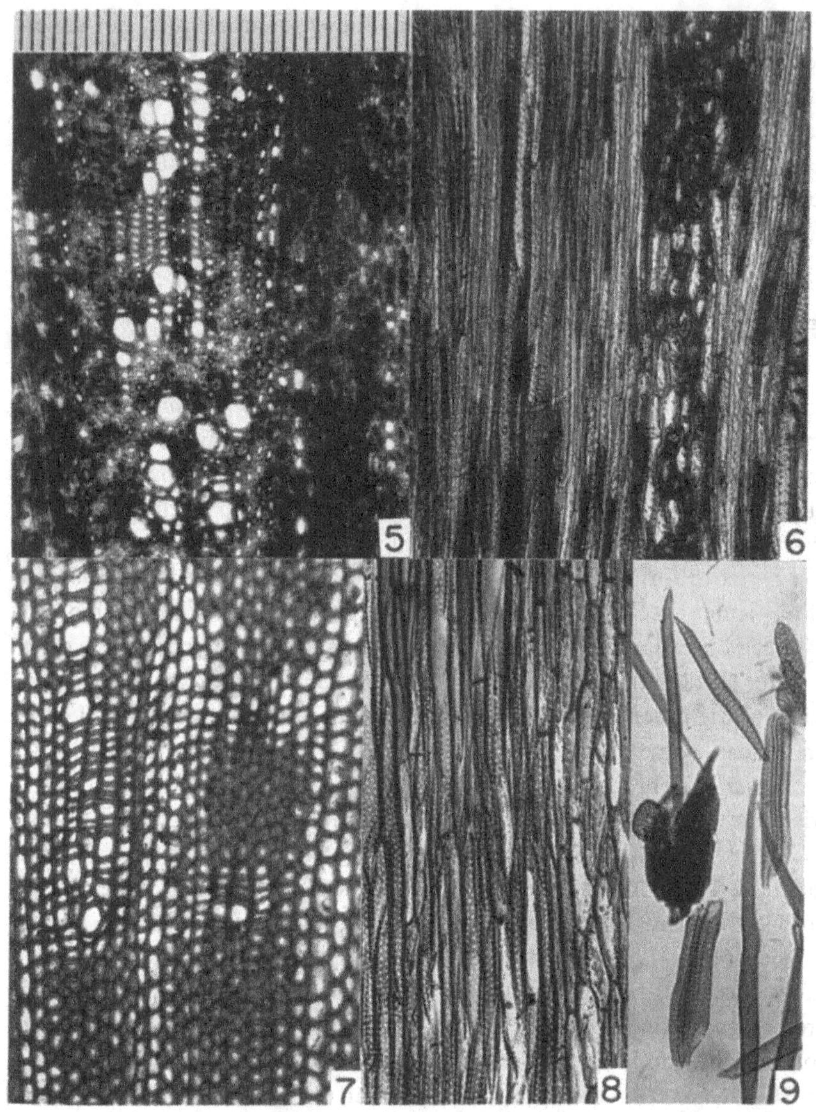

Figures 6.5–6.9. Wood preparations of Hawaiian alpine plants. Figs. 6.5, 6.6. *Geranium tridens* (Carlquist 546). Fig. 6.5. Transection of wood; a fascicular area is in the center; the rays, at either side, are filled with dark-staining compounds. Fig. 6.6. Tangential section of wood, with fascicular area at left (vessels and vasicentric tracheids in centre of that area, ray at right). Figs. 6.7–6.9. *Dubautia menziesii* (Carlquist H17). Fig. 6.7. Transection of wood, ray in center and at right; vessels are narrow and grade into vasicentric tracheids. Fig. 6.8. Tangential section, ray at right; most cells are narrow vessels or vasicentric tracheids. Fig. 6.9. Cells from a wood maceration; a vasicentric tracheid is visible, above (diagonally

conductive cells additional to vessels, and merely redundancy alone could explain the effectiveness of this cell type in resisting the effects of drought or freezing.

Vasicentric tracheids, despite their rarity in the world flora, prove to be common in the Haleakala alpine flora. *G. tridens* (Figures 6.5, 6.6) and the other Hawaiian species of *Geranium* have vasicentric tracheids (original data). In *Geranium*, the central portions of each fascicular area (zones in wood between rays) consists of vessels and vasicentric tracheids, whereas the margins of the fascicular areas adjacent to rays consist of libriform fibers. Vasicentric tracheids are abundant in wood of the high-alpine species *G. tridens*, but are less abundant in wood of the more mesomorphic species from lower on Haleakala, *G. arboreum* A. Gray.

Vasicentric tracheids are relatively uncommon in the large family Asteraceae (Carlquist 1985), but are abundant in the Haleakala species of this family. Wood of *Dubautia menziesii* (A. Gray) Keck, which may be found at the summit of Haleakala, features patches of what appear to be variously narrow vessels (Figures 6.7–6.9). In macerations, many of these prove to be vasicentric tracheids. *Tetramolopium humile* Hillebrand also occurs at the summit of Haleakala, and is a dramatic example of vasicentric tracheid presence. In the wood of this species (Figures 6.10, 6.11), vasicentric tracheids outnumber narrow vessels; only a few libriform fibers are present (Carlquist 1960). Vasicentric tracheids appear to increase in abundance in various phylads in proportion to the dryness experienced by any given species (Carlquist 1985). This may explain why vasicentric tracheids are more abundant in *T. humile*, which has roots shallower than those of *D. menziesii*. In addition, *D. menziesii* has thick and markedly succulent leaves (Carlquist 1959a), in contrast to those of *T. humile*, which appear to have little water storage capacity although their upright orientation may reduce transpiration somewhat. Physiological measures show that leaves of Hawaiian species of *Dubautia* in dry localities have mechanisms for dealing with limited water availability compared to species from more mesic localities (Robichaux & Canfield 1985).

Other species with vasicentric tracheids that occur near the summit of Haleakala include *Santalum haleakalae* Hillebrand (Santalaceae) and *Argyroxiphium sandwicense* DC. (Asteraceae). The latter is considered in detail later.

Caption for Figs. 6.5–6.9 (*cont.*).

placed), its lower tip touches a pair of vasicentric tracheids and a parenchyma cell. Figures 6.5–6.9, magnification scale above Figure 6.5 (divisions = 10 μm).

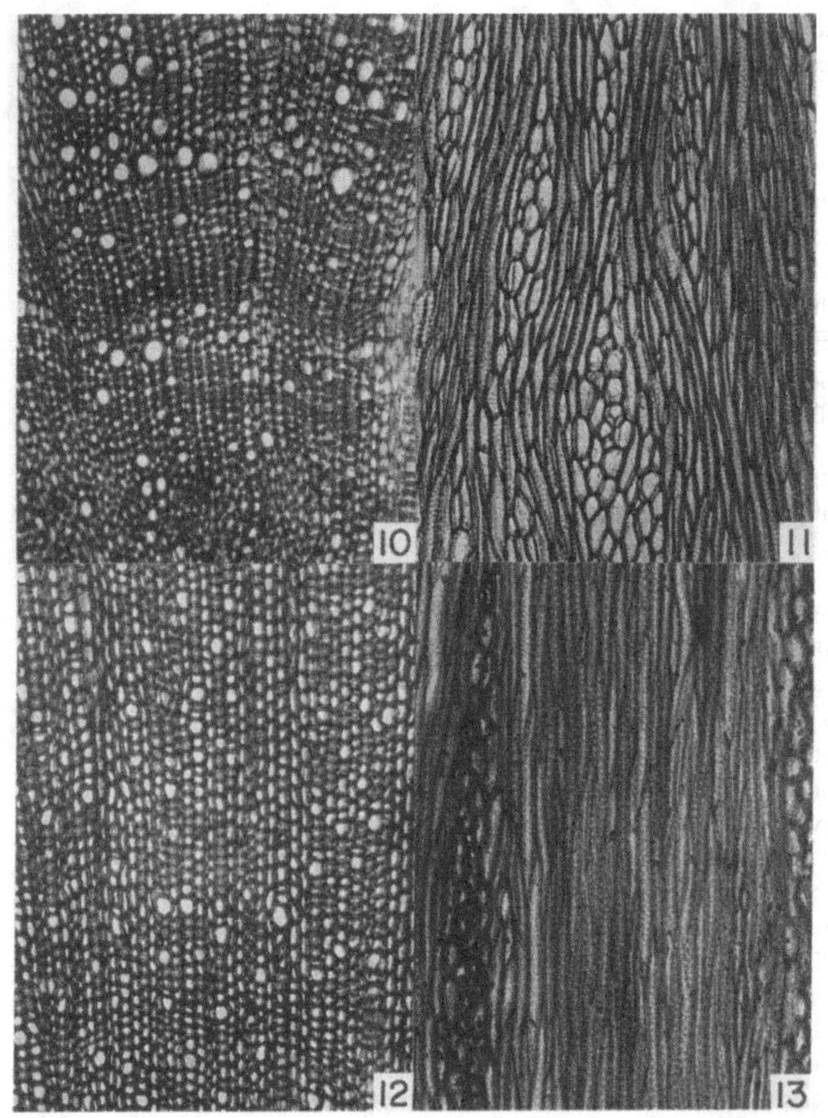

Figures 6.10–6.13. Wood sections of Hawaiian alpine and Andean superpáramo plants. Figs. 6.10, 6.11. *Tetramolopium humile* (Carlquist H18). Fig. 6.10. Transection. Vague growth rings evident; cells that appear to be libriform fibers are mostly vasicentric tracheids. Fig. 6.11. Tangential section, showing rays and narrow vessels (very few libriform fibers are present in this particular section). Figs. 6.12, 6.13. *Loricaria thuyoides* (Yw-20733). Fig. 6.12. Transection. No growth rings are evident; narrow vessels are scattered among vasicentric tracheids and a few parenchyma cells. Fig. 6.13. Tangential section; rays at left and right; axial parenchyma portion is mostly vasicentric tracheids. Figures 6.10–6.13, magnification scale above Figure 6.5.

A wood in which the entire background is composed of tracheids (thereby termed 'true tracheids' for purposes of clarity in my 1985 paper) offers a maximum of potential safety in terms of water-column failure. Woods with true tracheids comprise a minority of angiosperms, and many of these species appear to be restricted to wet habitats, in accordance with the ideas, seemingly supported by work in wood evolution, that claim angiosperms to have originated under moist conditions (Carlquist 1975, 1988). However, these phylads, by virtue of greater conductive safety of woods, would be ideal for dry areas if vessels can have simplified perforation plates and other vegetative structures can be modified for dryland water economy. Such species do appear to occur in dryland areas quite commonly (Carlquist 1985). On Haleakala and other Hawaiian alpine areas, this type of wood anatomy is represented in *Styphelia tameiameiae* (Cham.) F. Muell. (Epacridaceae) and *Vaccinium* spp.

Páramos

Although the anatomy of particular species listed from the Colombian páramos by Cuatrecasas (1934) is mostly unknown, one can cite types of wood anatomy reported for the genera to which they belong. There is usually little qualitative difference among congeners with respect to wood anatomy, so this indirect form of reportage can be used as a first approximation. Cuatrecasas lists as main shrubby associates of *Espeletia* the following: *Alchemilla* spp. (Rosaceae); *Berberis quiundiensis* Kunth (Berberidaceae); *Diplostephium* spp. (Asteraceae); *Gaultheria* spp. (Ericaceae); *Hypericum* spp.–especially *H. laricifolium* Juss. (Clusiaceae); *Miconia salicifolia* Naud. (Melastomataceae); *Myrteola vassinioides* (HBK.) Berg (Myrtaceae); *Senecio rigidifolius* Badillo (Asteraceae); and *Vaccinium* spp. (Ericaceae). Of these, *Hypericum* and *Berberis* were cited as having vasicentric tracheids in my 1985 survey. *Alchemilla, Gaultheria, Myrteola* and *Vaccinium* all have true tracheids (Metcalfe & Chalk 1950; Carlquist 1985). Rock (1972) reports that *Diplostephium venezuelense* Cuatr. and *Senecio rigidifolius* have vascular tracheids – tracheids formed in latewood (although often confused with vasicentric tracheids by various authors). Thus, only *Miconia salicifolia* belongs to a genus for which one of these unusual wood adaptations is not reported; very likely, it has other distinctive forms of wood xeromorphy.

The superpáramo areas, higher than the zone where *Espeletia* occurs, are drier and colder than the páramos. The genus *Loricaria* occurs in the superpáramo. *Loricaria thuyoides* (Lam.) Sch. Bip. (Asteraceae) belongs to a family in which libriform fibers are to be expected in wood. However, *Loricaria* has tracheids so abundant that libriform fibers have vanished altogether (Figures 6.12, 6.13). Termed vascular tracheids for this species earlier (Carlquist 1961), these tracheids now must be termed vasicentric tracheids (Carlquist 1985). The wood of *Loricaria* offers optimal safety, rather like a conifer wood in high redundancy of conducting cells, most of which are tracheids. Like a conifer, *Loricaria* is microphyllous.

Afroalpine

The shrubby flora of alpine East Africa contains the following genera of shrubs according to the flora of Hedberg (1957): *Alchemilla* spp. (Rosaceae); *Anthospermum usambarense* K. Schum. and *Galium* spp. (Rubiaceae); *Blaeria* spp., *Erica arborea* L. and *Philippia* spp. (Ericaceae); *Crassocephalum* spp. and *Helichrysum* spp. (Asteraceae); *Hypericum* spp. (Clusiaceae); *Pelargonium whytei* Bask. (Geraniaceae); *Protea kilimanscharica* Engl. (Proteaceae); and *Thesium kilimanscharicum* Engl. (Santalaceae). Of these, the genera *Hypericum* and *Protea*, and the families Geraniaceae and Santalaceae have been shown to possess vasicentric tracheids. True tracheids are known in *Anthospermum* (Koek-Noorman & Puff 1983) and *Galium* (Carlquist 1985) of the Rubiaceae. *Erica* has true tracheids (Carlquist 1985), and very likely the other ericaceous genera cited, *Blaeria* and *Philippia*, do also. *Alchemilla* belongs to a tribe of Rosaceae characterized by true tracheids, according to my unpublished observations. Thus, the majority of shrubs in the African alpine flora have either vasicentric tracheids or true tracheids, and very likely have other wood features adapted for these extreme sites. The species of *Helichrysum* should be studied for possible wood xeromorphy, as should *Adenocarpus mannii* (Hook f.) Hook f., a legume shrub of the region. *Crassocephalum* tends to occur in moist areas at the bottom of the alpine zone.

Argyroxiphium, a unique tropical Alpine genus

Argyroxiphium (Asteraceae, tribe Heliantheae, subtribe Madinnae), endemic to the Hawaiian Islands, is quite distinctive in its adaptations to tropical (or subtropical) alpine zones, despite the fact that it hybridizes (very infrequently in nature) with *Dubautia*. As a rosette plant, *A. sandwicense* DC. (Figure 6.14) shows no cessation in growth, and that is demonstrated in wood anatomy of all species as well, since growth rings are absent (Figure 6.17). The rosette of *A. sandwicense* undoubtedly protects the apical meristem from frost by virtue of its hairs and the dense crowding of leaves. In *A. kauense* (Rock & Neal) Deg. & Deg., short trunks are formed below the rosettes, suggesting, in comparison to *A. sandwicense*, that *A. sandwicense* is constrained by frost and that *A. kauense* has been released from it by virtue of its lower elevation habitat.

The brilliant appearance of leaves in *A. sandwicense* is produced by non-glandular trichomes that are flattened in their distal portions (Figure 6.16) and highly reflective, giving the plant the name silversword. This suggests that protection from UV may be a selective factor, although trichome covers of leaves often function in reducing transpiration by serving as a windbreak or offer a moderate degree of insulation against frost. Separation of these factors presents an interesting challenge for future investigators.

The leaves of *A. sandwicense* as seen in transection (Figure 6.15) contain some of the most remarkable foliar adaptations known. *Argyroxiphium* leaves are thick, related to the innovation of three sets of bundles as opposed to the single set of bundles (as is typical in angiosperm leaves) in the related genera *Dubautia* and *Wilkesia* (Carlquist 1957, 1959a). In terms of form alone, the reduction of surface to volume ratio represents a xeromorphic adaptation. What is much more surprising, however, is the development of massive water-retaining gels in the intercellular spaces of the leaf (Carlquist 1957). This may be observed on living plants: the gels can be squeezed from a broken leaf. These gels are not without parallel in Asteraceae, for they occur in other genera of subtribe Madiinae: *Blepharizonia*, *Hemizonia* and *Madia* (Carlquist 1959b). The leaves of the 'greenswords' – *A. grayanum* (Hillebr.) Degener and *A. virescens* Hillebr. – have gel storage also, as does the bog species *A. caliginis* Forbes (formerly spelt *A. caligini*), which is evidently a recent entrant into the bog habitat from a dryland ancestry. The leaves of *A. caliginis* do show some adaptation to the bog habitat in their development of more massive hydathodes near the leaf margins than occurs in *A. sandwicense*. The gels in *Argyroxiphium* represent a form of water storage, and the distribution within Madiinae and within the plants in the annual species dramatizes this function (Carlquist 1959b). The occurrence of the gels in annual species shows that these accumulations are probably not related to frost resistance. The effectiveness of the gels in the water economy of *Argyroxiphium* leaves has been quantified by the study of Robichaux & Morse (1990) and physiological data on the functioning of leaves are compared for *A. sandwicense* and *Dubautia menziesii* by Robichaux *et al.* (1990).

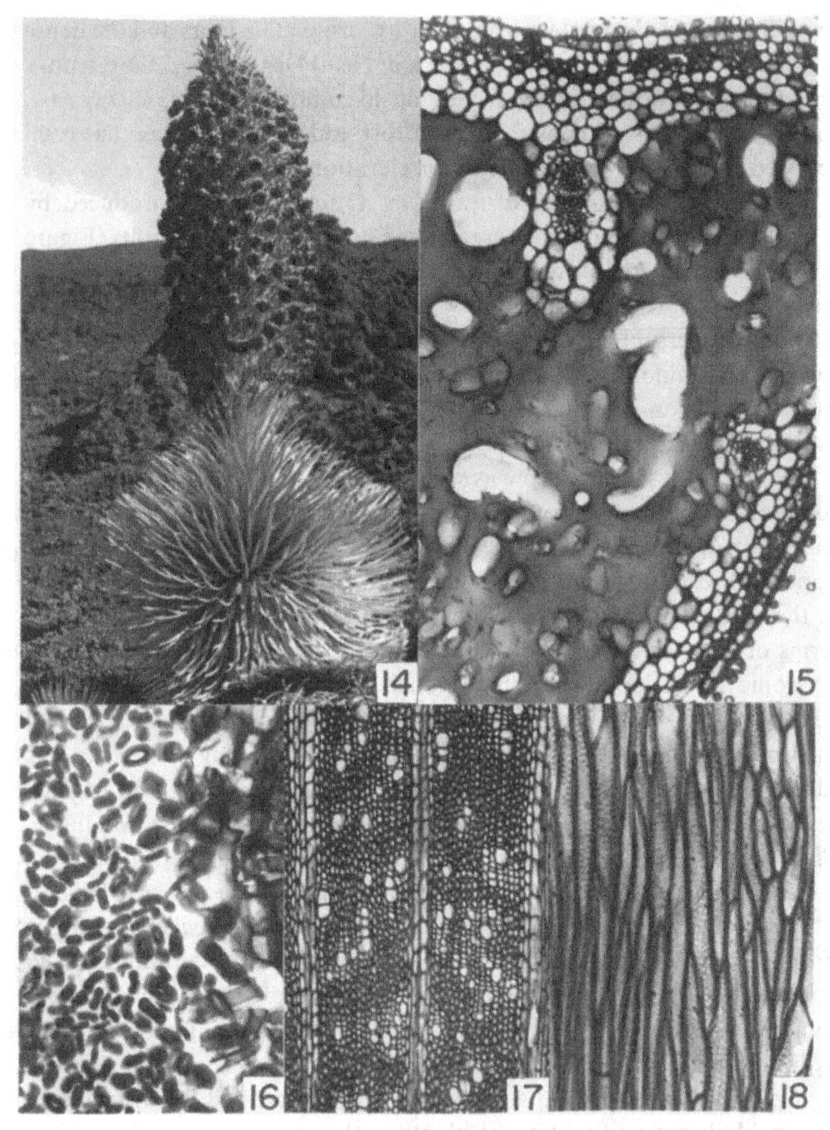

Figures 6.14–6.18. *Argyroxiphium* habit and anatomical details. Fig. 6.14. Flowering individual of *Argyroxiphium sandwicense*, Haleakala Crater, 7 July 1966. Fig. 6.15. Leaf from *A. sandwicense*, transection, showing that most of the central portion of the leaf is filled with a gel. Fig. 6.16. Transection of hairs from leaf surface of *A. sandwicense*; distal portion of trichomes is flat in transection. Figs. 6.17, 6.18. *A. kauense* (Carlquist 2110). Fig. 6.17. Wood transection, showing that vessels are narrow and tend to grade into vasicentric tracheids. Fig. 6.18. Tangential section; bordered pits can be seen in many fibriform cells; most of these are vasicentric tracheids, but some narrow vessels are present (ray at right).

The wood of *Argyroxiphium* (Figures 6.17, 6.18) features the presence of a few vasicentric tracheids. Because these occur in other alpine Asteraceae of the Hawaiian Islands cited above, *Dubautia menziesii* and *Tetramolopium humile*, this may not seem very remarkable until one notes that vasicentric tracheids are rare in woods of Asteraceae (Carlquist 1988). The multiseriate rays of *Argyroxiphium* wood represent relatively little-altered primary rays (pith rays), as is expected in a rosulan (*A. sandwicense*) or caulorosulan (*A. grayanum, A. kauense*) plant. Vessel diameter is less in *A. sandwicense* than in the bog species *A. caliginis* and *A. grayanum* (Carlquist 1958), indicating that even though the stem of *A. sandwicense* is thick and could be regarded as potentially capable of water storage, the wood is xeromorphic. The vessels of *A. kauense* (Fig. 6.17) are intermediate in diameter between these two extremes.

The glandular trichomes on the inflorescences of *Argyroxiphium* secrete a resin-like substance. This is probably not a mechanism for reduction of transpiration, but rather a means for deterring insect predation; it is very viscid and bitter, and often coats the epidermis of most inflorescence structures. Abundance of these secretions in the annual tarweeds shows that their presence is not related to frost resistance.

The sum of anatomical features of *Argyroxiphium* shows close and intricate adaptation to the dryland habitats of alpine Hawaii. The two bog species are held to be recent derivatives in which time has been insufficient for much change in that adaptation – either loss of xeromorphic features or introduction of mesomorphic ones. However, even in the bog species, some degree of xeromorphic adaptation may still be of positive value. In fact, on clear days in the Puu Kukui bog where these species grow, leaf heating and transpiration are very likely pronounced.

Cushion plants

One would like to be able to describe the nature of leaf anatomy and other anatomical characters of the cushion plants so characteristic of high elevations in the equatorial Andes (notably *Azorella* of the Apiaceae). Cushion plants are less characteristic of other tropical alpine areas, but one may see mat-like growth forms not dissimilar to cushion plants in such tropical areas as alpine New Guinea, and the adaptations of mat-like

Caption for figs. 6.14–6.18 (*cont.*).

Figure 6.14, plant is 1.6 m tall. Figures 6.15, 6.16, magnification scale above Figure 6.5. Figure 6.17, scale above Figure 6.1.

plants in tropical alpine areas may not be far removed from those of cushion plants.

Napp-Zinn (1984, pp. 171–88) reviews the literature on leaf anatomy of Andean plants, and that literature proves to be rather considerable. However, Napp-Zinn's tables group plants of particular Andean habitats, despite the fact that within each habitat a wide range of habits is represented. That may account for the fact that, while some generalizations concerning leaf anatomy of the Andean alpine plants are entertained by Napp-Zinn (1984), he seems to feel these generalizations do not hold. What one wants is a comparison of leaves of cushion plants of different genera similar in habit and leaf size. Alternatively, one could develop valuable understanding of ecological leaf anatomy by comparing for a particular genus, such as *Azorella*, species from a range of habits (including cushion plants) and habitats. When the ranges in leaf anatomy for several such genera represented in tropical alpine areas are compared, we may have an idea of anatomical modalities related to the cushion plant habitat and their significance. Ideally, one would like to see development of physiological data concomitant with study of anatomy of these plants.

References

Bailey, I. W. (1944). The development of vessels in angiosperms and its significance in morphological research. *American Journal of Botany* **31**, 421–8.
Bierhorst, D. W. & Zamora, P. M. (1965). Primary xylem elements and element associations of angiosperms. *American Journal of Botany* **52**, 657–710.
Carlquist, S. (1957). Leaf anatomy and ontogeny in *Argyroxiphium* and *Wilkesia* (Compositae). *American Journal of Botany* **44**, 696–705.
Carlquist, S. (1958). Wood anatomy of Heliantheae (Compositae). *Tropical Woods* **108**, 1–30.
Carlquist, S. (1959a). Vegetative anatomy of *Dubautia*, *Argyroxiphium*, and *Wilkesia* (Compositae). *Pacific Science* **13**, 197–210.
Carlquist, S. (1959b). Studies in Madinae: anatomy, cytology, and evolutionary relationships. *Aliso* **4**, 171–236.
Carlquist, S. (1960). Wood anatomy of Astereae (Compositae). *Tropical Woods* **113**, 54–84.
Carlquist, S. (1961). Wood anatomy of Inuleae (Compositae). *Aliso* **5**, 21–37.
Carlquist, S. (1962a). Wood anatomy of Senecioneae (Compositae). *Aliso* **5**, 123–46.
Carlquist, S. (1962b). A theory of paedomorphosis in dicotyledonous woods. *Phytomorphology* **12**, 30–45.
Carlquist, S. (1966). Wood anatomy of Compositae: a summary, with comments on factors controlling wood evolution. *Aliso* **6**(2), 25–44.
Carlquist, S. (1969). Wood anatomy of Lobelioideae (Campanulaceae). *Biotropica* **1**, 47–72.

Carlquist, S. (1970). *Hawaii, A Natural History.*Garden City, New York: Natural History Press.
Carlquist, S. (1975). *Ecological Strategies of Xylem Evolution.* Berkeley: University of California Press.
Carlquist, S. (1980). Further concepts in ecological wood anatomy, with comments on recent work in wood anatomy and evolution. *Aliso* **9**, 499–553.
Carlquist, S. (1985). Vasicentric tracheids as a drought survival mechanism in the woody flora of southern California and similar regions; review of vasicentric tracheids. *Aliso* **11**, 37–68.
Carlquist, S. (1988). *Comparative Wood Anatomy.* Berlin: Springer-Verlag.
Carlquist, S. & Bissing, D. R. (1976). Leaf anatomy of Hawaiian geraniums in relation to ecology. *Biotropica* **8**, 248–59.
Cuatrecasas, J. (1934). Observaciones geobotanicas en Colombia. *Trabajos del Museo Nacional de Ciencias Naturales, ser. Botanica* **27**, 1–137.
Cumbie, B. G. (1983). Developmental changes in the wood of *Bocconia vulcanica* Donn. Smith. *International Association of Wood Anatomists Bulletin, n.s.*, **4**, 131–40.
Goldstein, G. & Meinzer, F. (1983). Influence of insulating dead leaves and low temperature on water balance in an Andean giant rosette plant. *Plant, Cell and Environment* **6**, 649–56.
Goldstein, G., Meinzer, F. & Monasterio, M. (1984). The role of capacitance in the water balance of Andean giant rosette species. *Plant, Cell and Environment* **7**, 179–86.
Hauman, L. (1933). *Les* 'Lobelia' *géants des montagnes du Congo Belge.* Brussels: Marcel Hayex, Imprimeur.
Hauman, L. (1935). Les '*Senecio*' arborescents du Congo. *Revue de Zoologie et de Botanique Africaines* **28**, 1–76.
Hedberg, O. (1957). Afroalpine vascular plants. A taxonomic revision. *Symbolae Botanicae Upsalienses* **15**(1), 1–411.
Hedberg, O. (1964). Features of Afroalpine plant ecology. *Acta Phytogeographica Suecica* **49**, 1–144.
Hedberg, O. & Hedberg, I. (1979). Tropical-alpine life-forms of vascular plants. *Oikos* **33**, 197–307.
Koek-Noorman, J. & Puff, C. (1983). The wood anatomy of Rubiaceae tribes Anthospermeae and Paederiae. *Plant Systematics and Evolution* **143**, 17–45.
Mabberley, D. J. (1974). Pachycauly, vessel elements, islands, and the evolution of arborescence in 'herbaceous' families. *New Phytologist* **73**, 977–84.
Metcalfe, C. R. & Chalk, L. (1950). *Anatomy of the Dicotyledons.* Oxford: Clarendon Press.
Napp-Zinn, K. (1984). Anatomie des Blattes, II. Blattanatomie der Angiospermen. B. Experimentelle und ökologische Anatomie des Angiospermenblattes. *Handbuch der Pflanzenanatomie VIII(2B)*, pp. 1–1431. Berlin: Gebrüder Borntraeger.
Nordenstam, B. (1978). Taxonomic studies in the tribe Senecioneae (Compositae). *Opera Botanica* **44**, 1–84.
Robichaux, R. & Canfield, J. E. (1985). Tissue elastic properties of eight *Dubautia* species with 13 pairs of chromosomes. *Pacific Science* **39**, 191–4.
Robichaux, R., Carr, G. D., Liebman, M. & Pearcy, R. W. (1990). Adaptive radiation of the Hawaiian silversword alliance (Compositae – Madiinae): ecological, morphological, and physiological diversity. *Annals of the Missouri Botanical Garden* **77**, 674–72.

Robichaux, R. & Morse, S. R. (1990). Extracellular polysaccharide and leaf capacitance in a Hawaiian bog species, *Argyroxiphium grayanum* (Compositae – Madiinae). *American Journal of Botany* **77**, 134–8.
Rock, B. N. (1972). Vegetative anatomy of *Espeletia* (Compositae). Thesis, University of Maryland.
Smith, A. C. & Koch, M. (1935). The genus *Espeletia*: a study in phylogenetic taxonomy. *Brittonia* **1**, 479–530.
Smith, A. P. (1974). Bud temperature in relation to nyctinastic leaf movement in an Andean giant rosette plant. *Biotropica* **6**, 263–6.
Smith, W. K. & Geller, G. N. (1979). Plant transpiration at high elevations: theory, field measurements, and comparisons with desert plants. *Oecologia* **41**, 109–22.

7

Environmental biology of a tropical treeline species, *Polylepis sericea*

G. GOLDSTEIN, F. C. MEINZER and F. RADA

Introduction

It is likely that *Polylepis* (Rosaceae) occurs naturally at higher elevations than any other arborescent angiosperm genus in the world. The 15 species (Simpson 1979) are confined to the South American Andes where they occur primarily in tropical alpine environments. Some *Polylepis* species tend to form discrete forest stands reaching elevations over 5000 m, well above the upper continuous forest limit (timberline). Throughout their high altitude distribution most members of this genus are exposed to rigorous climatic conditions in which diurnal temperature variations by far exceed seasonal ones and night frosts are frequent.

The genus is exclusively arborescent (trees or shrubs), with individuals ranging in height from 1 m to no more than 30 m. The trees tend to have twisted, crooked stems and branches, particularly in open, exposed habitats. The form and branching pattern of some individuals resembles those of *krummholz* trees found in temperate alpine regions. The bark is deep red in color and consists of several layers of thin, exfoliating sheets. Although the exfoliating bark is particularly thick at the base of the stem or large branches, the insulating effect is by no means comparable to that of the marcescent leaves that surround the stem of the adjacent giant rosette plants (Smith 1979; Goldstein & Meinzer 1983). The leaves are compound and alternate but often appear whorled owing to the compression of internodes at the branch tips. The leaflets are small, dark green above and are covered with dense, silvery trichomes on the underside in several species. An extensive documentation of leaf anatomy and several important species-specific characters of potential adaptive value such as the degree of deciduousness is to be found in Simpson's (1979) revision.

Polylepis trees seldom grow as isolated individuals. They tend to form small forest 'islands' above the continuous forest limit. Several hypotheses have been proposed to explain the distributiom of *Polylepis* and its success at high elevations. Except for a few works suggesting that the patchy type of distribution is the result of past human activities (Ellenberg 1958a, b; Vareschi 1970), most studies attribute the success of *Polylepis* to the special microclimatic conditions associated with the rocky, protected habitats in which it occurs (Rauh 1956; Troll 1959; Walter & Medina 1969). Even though the species growing above 3300–3500 m are usually restricted to microenvironments that produce 'lower elevation' climatic conditions, they are still exposed to below-freezing temperatures at night. Such low temperatures would cause tissue injury and metabolic disturbances such as depressed photosynthesis and decreased stomatal reactivity (e.g. Fahey 1979; Takahashi 1981; Teskey *et al.* 1984) in plants not adapted to these conditions. After a critical review of all the published works up to 1978, Simpson (1979) concluded that the question of the ecological distribution of the genus will remain unsettled until both the physiology of *Polylepis* is better understood and the microclimatic conditions of the areas in which it occurs are more thoroughly documented.

Morphological adaptations in *Polylepis* are not as conspicuous as in the other arborescent growth forms of the high altitude American Tropics. For example, *Polylepis* trees do not possess an insulating layer of dead leaves around the stem, or a pith water reservoir, or a parabolic terminal rosette of leaves with nyctinastic movements as found in giant rosette plants (Meinzer & Goldstein 1986). This poses the question of how *Polylepis* trees prevent frost damage of the stem and leaf tissues and the formation of freezing-induced gas embolisms in the xylem, given that the trees are functionally active throughout the year.

The purpose of this chapter is to summarize and draw together recent studies on microclimatic conditions, frost resistance mechanisms, water relations and carbon economy in *Polylepis sericea* Wedd, the most abundant *Polylepis* species in the Venezuelan Andes. It is our belief that the success of *P. sericea* at high elevations is not solely a consequence of the microclimatic and physical conditions of the areas in which it occurs, but is also dependent on the special features of its carbon economy and frost resistance.

Microclimate and physical characteristics of *P. sericea* forest stands

The altitudinal air temperature lapse rate in the northern part of the Andes chain where *Polylepis sericea* grows is approximately -0.6 °C per 100 m

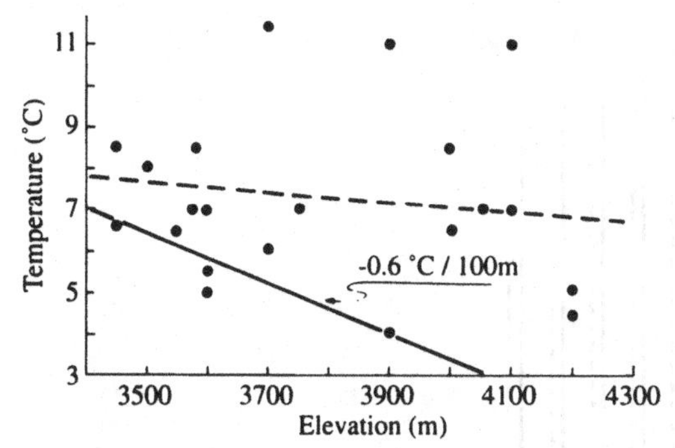

Figure 7.1. Mean air temperature as a function of elevation in *Polylepis sericea* forest stands above 3400 m. The solid line represents the air temperature lapse rate (−0.6 °C per 100 m elevation) for the Venezuelan Andes (adapted from Arnal 1983).

elevation. Mean daily air temperatures in *P. sericea* forest islands (above 3000 m), however, do not follow this temperature gradient (Figure 7.1). The lack of correlation between air temperature and elevation in these forest stands suggests that special microclimatic conditions exist there. As expected, air temperature extremes are far less pronounced inside forest islands, than in the adjacent, open páramo (Figure 7.2). Minimum air temperatures inside *P. sericea* stands were usually more than 1.5 °C higher and maximum temperatures were more than 3 °C lower than in páramo sites dominated by caulescent giant rosette plants. These forest islands of *P. sericea*, which are generally found near massive rock outcroppings and talus slopes, can be considered as true thermal refugia where freezing temperatures are much less frequent than in the open páramo, even at altitudes up to 4500 m. The microclimatic information available suggests that both the shelter and the thermal stability provided by the rocks help to dampen diurnal air temperature variations even more than the presence of the tree canopy cover alone (Rada 1983).

It has been suggested that by allowing warm air to reach considerable depth through the crevices between the rocks, talus slopes provide better soil conditions than those of the surrounding tropical alpine habitats (páramo) (Walter & Medina 1969). Warmer soils should create better conditions for water and nutrient uptake and thereby improve establishment and growth of tree seedlings. Smith (1977) tested this hypothesis by transplanting seedlings of *P. sericea* in open páramo vegetation and sheltered rocky rites. Seedling mortality during the first year was 100% on bare soil. Seven transplanted individuals (14%) in rocky areas were still

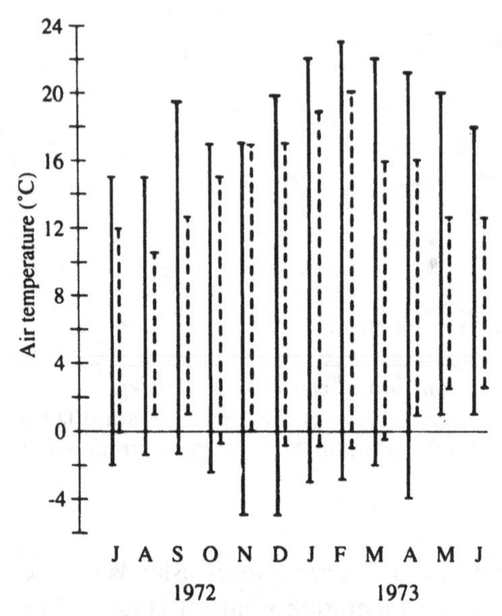

Figure 7.2. Absolute minimum and maximum monthly air temperature ranges in the Mucubají Páramo at 3550 m (solid lines) and inside adjacent *P. sericea* stands (broken lines) from July 1972 to June 1973 (adapted from Azócar & Monasterio 1980).

alive one year after planting. With few exceptions, mortality was restricted to the dry period of December to April, when soil surface temperatures and soil moisture levels were at their lowest.

The temperature of *P. sericea* leaves also tends to exhibit reduced variation compared with leaves of giant rosette plants growing in the surrounding páramo (Figure 7.3). The leaves of *P. sericea* are small and the upper leaf surface is glabrous. In contrast, leaves of giant rosette plants are typically more than 19 times wider and both leaf surfaces are covered by a 2–3.5 m thick pubescent layer. Energy balance models and experimental data indicate that leaf pubescence in the giant rosette species *Espeletia timotensis* exerts its principal influence on leaf temperature through increased boundary layer resistance to heat transfer rather than through reduced absorptance to solar radiation (Meinzer & Goldstein 1985). Under clear conditions at 4200 m elevation the temperature of a pubescent leaf would be higher than that of a non-pubescent one during the day. At night, reduced convective heat transfer from air to a pubescent giant rosette leaf would cause it to be cooler than a glabrous leaf. Figure 7.3 also shows that at night, temperatures of *E. timotensis* leaves are indeed

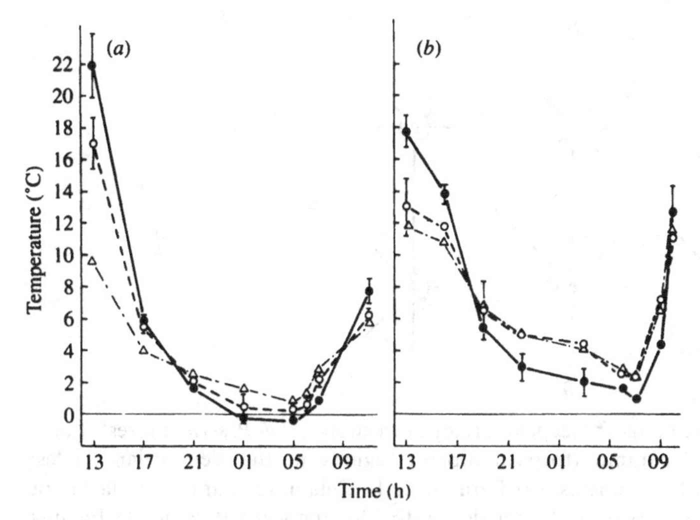

Figure 7.3. Daily courses of air temperature (△) and leaf temperature for *P. sericea* (○) and the giant rosette plant *Espeletia timotensis* (●), during (*a*) a dry season day (December 1982) and (*b*) a wet season day (June 1982). Night-time temperatures were lowest during the dry season due to unimpeded radiational cooling.

consistently more than 1 °C lower than those of glabrous *P. sericea* leaves. During the day this temperature pattern is reversed, particularly during periods of high incoming solar radiation.

Arnal (1983) described the physical characteristics of 256 forest islands in previously glaciated areas of the Venezuelan Andes where *P. sericea* occurs. His data failed to show a predominant slope orientation of the *P. sericea* sites (Figure 7.4) as had been suggested earlier by Walter & Medina (1969). More importantly, Arnal (1983) obtained a highly significant multiple linear relationship between tree density and increasing slope angle and number of rocks on the soil surface. This and the results of Smith's (1977) transplant experiment indicate that rocky substrates on relatively steep slopes provide good quality sites for seedling establishment and tree growth. This type of site not only dampens temperature fluctuations on a daily basis, but also reduces water loss from the soil surface, resulting in higher soil moisture levels during the dry season.

Carbon metabolism

Temperature, with its effects on carbon balance, is one of the most important factors determining the upper altitude limit of trees in temperate

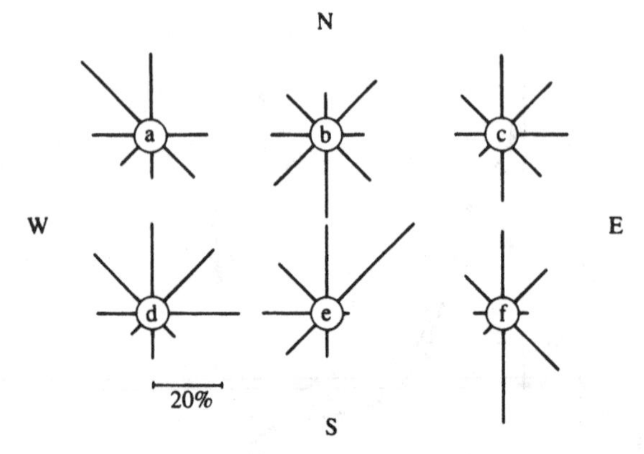

Figure 7.4. Percentage frequencies of slope direction of 256 *P. sericea* forest stands in six different páramo (tropical alpine) regions in the Venezuelan Andes: (a) Páramo de Los Conejos, (b) Páramo de la Culata, (c) Páramo de la Sierra Nevada, (d) Páramo de la Sierra de Santo Domingo, (e) Páramo de Piedras Blancas, and (f) Páramo el Escorial (adapted from Arnal 1983).

zone mountains (Tranquillini 1979). In some respects, high elevation tropical trees are exposed to even harsher environmental conditions than treeline species in cold temperate zones. While temperate trees can become dormant in winter and in a sense escape the most severe portion of the annual environmental regime, tropical treeline species are exposed to changes from summer-like to winter-like conditions in less than 24 hours. These conditions, which require maintenance of a continuously high level of physiological activity, may impose special constraints on the carbon economy of these trees. This section deals with the effects of temperature on photosynthesis and dark respiration in *Polylepis sericea.* Carbon balance data for other higher elevation tree species in the Venezuelan Andes will be cited for comparison.

At least four aspects of its carbon balance distinguish *P. sericea* from its temperate zone treeline counterparts and many other evergreen trees: (i) maximum photosynthetic capacity, (ii) upper temperature limit for net photosynthesis, (iii) temperature span between the upper temperature limit and the optimum and (iv) dark respiration rate. The photosynthetic capacity of *P. sericea* (9 μmol m^{-2} s^{-1}) is roughly twice that of temperate treeline species (Larcher 1969; Pisek *et al.* 1973; Tranquillini 1979). This maximum photosynthetic rate is attained at approximately 13 °C (Figure 7.5), with the rate of photosynthesis falling off rapidly above and below this optimum temperature. The upper temperature limit for net photosynthesis in temperate treeline species is typically 35–45 °C with a span

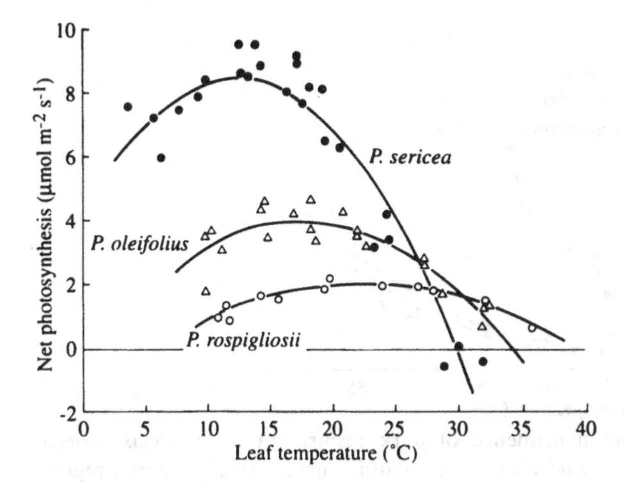

Figure 7.5. Temperature dependence of net photosynthesis in *Polylepis sericea* (●), *Podocarpus oleifolius* (△) and *Podocarpus rospigliosii* (○) as determined in an open gas exchange system. The quantum flux density in all the experiments was maintained at 1100 μmol m^{-2} s^{-1} (adapted from Jaimes 1985).

of 20–25 °C between the upper limit and the optimum (Pisek *et al.* 1973). This contrasts with an upper limit of only 30 °C in *P. sericea* (Figure 7.5) which is one of the lowest reported for any vascular plant species, and a span of only 17 °C between the upper limit and the optimum. Larcher (1975) reports a low temperature limit for net photosynthesis of −5 °C for *P. sericea* transported from Venezuela to Austria. If the second degree polynomial regression curve fitted to the points in Fig. 7.5 is extrapolated to zero net assimilation, the corresponding leaf temperature is −5 °C. Such cold daytime temperatures are seldom, if ever, observed in *P. sericea* forest islands, even at its absolute altitudinal limit of 4500 m in Venezuela. Thus, if photosynthesis is not inhibited by short night-time periods of subzero temperatures, positive carbon balance could always be maintained when photosynthetically active radiation levels are above the light compensation point.

The temperature dependence of photosynthesis has also been determined for *Podocarpus oleifolius* and *P. rospigliosii* (Figure 7.5), two dominant coniferous species growing in the Venezuelan Andes (Jaimes 1985). The elevational range of *P. oleifolius* is 2200–3300 m, while *P. rospigliosii* occurs at lower altitudes between 1700 and 2500 m. As could be expected, the temperature optimum for photosynthesis increases with decreasing altitudinal range as does the upper temperature compensation point;

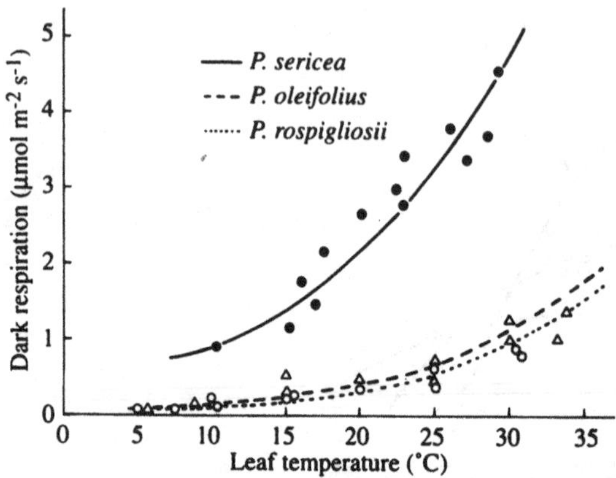

Figure 7.6. Temperature dependence of dark respiration in *Polylepis sericea*, *Podocarpus oleifolius* and *Podocarpus rospigliosii*. Curves are exponential regressions fitted to the data (obtained from F. C. Meinzer and G. Goldstein, unpublished data, and from Jaimes 1985).

however, the temperature span between the optimum and the upper compensation point remains small. The temperature optima for photosynthesis of the three species tend to coincide with the mean maximum temperatures at the sites where the material was collected, and this gradient follows approximately the same lapse rate as that of air temperature (−0.6 °C per 100 m elevation in the Venezuelan Andes).

The temperature dependence of dark respiration in these three species follows a typical positive exponential function over the temperature range in which no thermal injury occurs (Figure 7.6). The respiration rate at 20 °C, a standard temperature often used for comparison among species, is 2.3 μmol m^{-2} s^{-1} in *P. sericea*, several times higher than that of the two lower elevation coniferous tree species. Temperate high elevation tree species are noted for having higher dark respiration rates than lowland species. However, dark respiration on a leaf area basis is several times higher in *Polylepis* than in its temperate high elevation counterparts, even at the low night-time temperatures that prevail in its habitat. These respiration rates resemble those for winter deciduous trees rather than evergreens (Larcher 1969; Pisek *et al.* 1973; Goldstein 1981). The respiration rates of the two *Podocarpus* species are similar to those of temperate montane species.

These features of carbon balance in *P. sericea* probably reflect a response to the selective pressures operating in cold tropical environments with frequent freezing temperatures. For example, a high rate of dark respiration in plants from cold climates has been cited as an adaptation permitting adequate release of chemical energy at low temperatures for repair of cellular damage, particularly membrane damage due to chilling and freezing (Tranquillini 1979; Levitt 1980). Such injury may be cumulative over time and its severity increased if exposure to low temperatures is followed by exposure to high light intensity (Berry & Bjorkman 1980). In the high Venezuelan Andes, frequent subfreezing night temperatures are usually followed by clear skies and, therefore, high irradiance conditions in the morning hours, particularly during the dry season. The especially high respiration rate of *P. sericea* may be in part a response to daily exposure to periods of chilling or freezing temperatures followed by exposure to high light intensities.

P. sericea leaves exhibit large diurnal changes in osmotic potential with maximum osmotic concentrations occurring during the early morning at the time of minimum temperatures (see below, p. 140). This increase in osmotically active solutes at night implies an investment of energy and, therefore, would require an adequate dark respiration rate at low night temperatures. Although one of the species used for comparison (*Podocarpus oleifolius*) occurs at the upper limit of continuous forests (3300 m) its respiration rate is much lower than that of *P. sericea.* The continuous forest limit in Venezuela also coincides with the altitude at which night frosts begin to occur (Monasterio & Reyes 1980). Thus, the seemingly abrupt change in respiration rates with increasing elevation may reflect the rapid increase in risk of night frosts and consequent need for repair of injured tissues. A high respiration rate coupled with a relatively high photosynthetic capacity may be a prerequisite for tree survival at high elevations in the tropics.

It is not known to what extent photosynthesis and respiration in *P. sericea* show temperature acclimation. Temperate treeline and montane species usually show a considerable capacity for photosynthetic and respiratory temperature acclimation (Pisek *et al.* 1973; Berry & Bjorkman 1980; Black & Bliss 1980; Mooney & West 1964). As *P. sericea* trees are exposed only to diurnal and not seasonal temperature fluctuations, there would be little selective advantage in possessing mechanisms for acclimation to long-term temperature changes. The 17 °C difference between the photosynthetic optimum and the upper temperature limit of 30 °C suggests that *P. sericea* may indeed possess little flexibility in this regard. Furthermore, the range of maximum temperatures along the entire altitudinal range of *P. sericea* in Venezuela (11 °C) is much smaller than seasonal fluctuations in maximum temperature at a constant altitude in

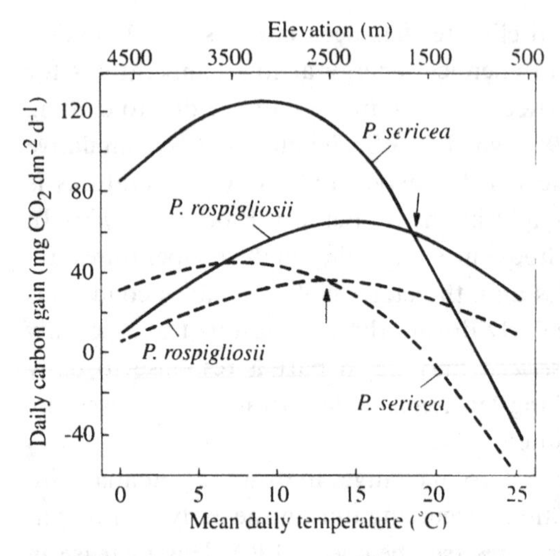

Figure 7.7. Calculated daily carbon balance as a function of mean daily temperature (and elevation) in *Polylepis sericea* and *Podocarpus rospigliosii*. Simulations were obtained for clear days (——), and days with afternoon clouds (- - -) typical of the Venezuelan Andes. Arrows indicate mean daily temperatures and elevations that produce the same simulated carbon gain in both species under clear and cloudy days.

temperate zone mountains. The two tropical coniferous species used here for comparison, and tropical savanna trees (Sarmiento *et al.* 1985) also have small temperature spans between their temperature optima and upper limits for positive net photosynthesis. Furthermore, preliminary transplant experiments indicate that seedlings of the two coniferous species show almost no short-term photosynthetic temperature acclimation responses (G. Goldstein and F. Meinzer, unpublished data). All of these results suggest that the lack of photosynthetic temperature acclimation potential may be a general phenomenon in trees of tropical mountains.

The lack of seasonal temperature variation and acclimation potential simplify carbon budget calculation. When a mathematical model of carbon balance of *P. sericea* and *Podocarpus oleifolius* is simulated for days with different mean air temperatures and afternoon overcast conditions (Figure 7.7), typical of the high Venezuelan Andes, it is found that below 2600 m *P. oleifolius*, in spite of its lower photosynthetic capacity, has a more positive carbon balance than *P. sericea.* This is approximately the lower altitudinal limit of *P. sericea* and the point in which *Podocarpus oleifolius* should outcompete *P. sericea* for carbon resource gain. Predicted carbon balance in *P. sericea* deteriorates rapidly with decreasing altitude due to the high dark respiration rate and steep drop in photosynthetic rate above the optimum temperature. On the other hand, net assimilation is still very positive at the upper distribution limit and, therefore, other factors may be limiting *P. sericea* growth above this elevation. Thus, in the absence of significant temperature acclimation as suggested above, some of the features of carbon metabolism that may explain the success of *P. sericea* at high elevations may also partially explain its lower altitude limit.

Frost resistance mechanisms

Plants in high tropical mountains exhibit a wide range of frost resistance mechanisms (Sakai & Larcher 1987; Beck, Chapter 5). For example, leaves of *Draba chionophila*, a small rosette plant found at the highest altitudes reported for vascular plants in the Venezuelan Andes (*c.* 4700 m), as well as leaves of Afroalpine *Senecio* and *Lobelia* plants, are freezing-tolerant, with freezing injury appearing only at temperatures lower than the temperature at which extracellular ice formation begins (Azócar *et al.* 1988; Beck *et al.* 1982, 1984). In Andean giant rosette plants, on the other hand, leaf, bud and internal stem tissues are protected solely by freezing avoidance mechanisms (Goldstein *et al.* 1985; Rada *et al.* 1985a). In many cases freezing avoidance offers adequate protection against the slight frosts that occur in tropical high mountains throughout the year (Larcher 1982). Of the five possible freezing avoidance mechanisms described by Levitt (1980), (i) freezing point depression through decrease in osmotic potential, and (ii) supercooling are found in *P. sericea.*

Leaves of *P. sericea* exhibit diurnal osmotic adjustment of the order of 0.4–1.0 MPa, depending on the minimum night temperature. Minimum leaf osmotic potential (Ψ_π) and therefore lowest tissue freezing point is generally attained at 0500–0600 h, coinciding with minimum leaf temperatures rather than minimum leaf water potential (Ψ_L, Figure 7.8). Changes in levels of soluble carbohydrates seem to correspond to the pattern of changing Ψ_π (Figure 7.8), suggesting that the decline in Ψ_π throughout the night results, at least in part, from the accumulation of soluble sugars in the leaves. The diurnal variation in freezing point of the cell sap corresponding to the osmotic fluctuations shown in Figure 7.8 would be 0.5–1.0 °C. This small enhancement of freezing avoidance may in itself be important from an ecological standpoint because in the sites where *P.*

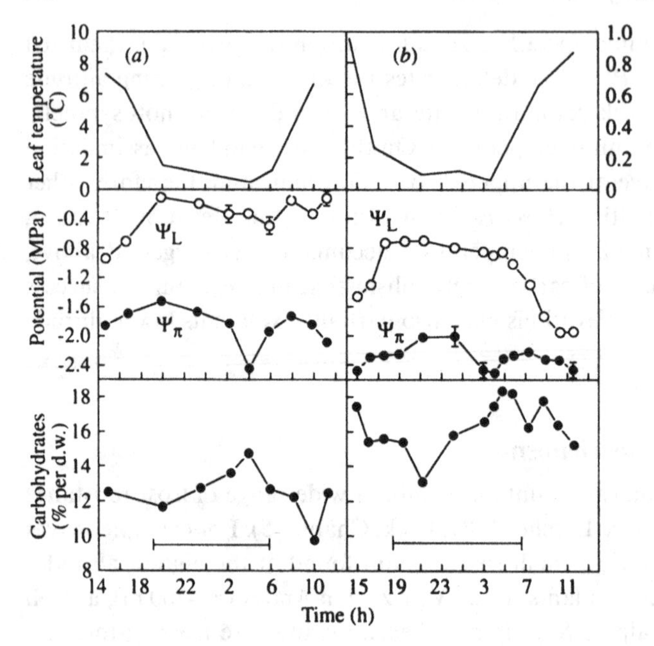

Figure 7.8. Daily courses of leaf temperature, leaf water potential (Ψ_L), leaf osmotic potential (Ψ_π) and soluble carbohydrate levels for *P. sericea* during (*a*) a wet season day (20 December 1983) and (*b*) a dry season day (13 February 1984). The horizontal segments in the lower panel indicate the night-time period (from Rada *et al.* 1985b).

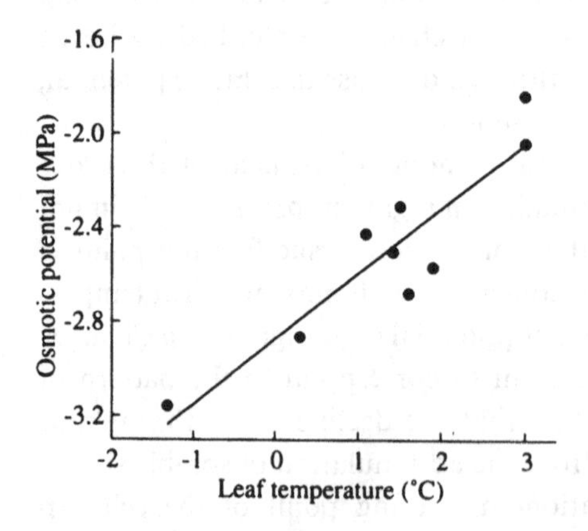

Figure 7.9. Minimum night-time osmotic potential in relation to minimum leaf temperature of *P. sericea*. The solid line represents a linear regression fitted to the data ($r^2 = 0.82$) (from Rada *et al.* 1985b).

sericea grows daily temperature fluctuations are not very pronounced and minimum temperatures do not fall much below 0 °C. The magnitude of the nocturnal decline in Ψ_π of *P. sericea* leaves appears to be directly related to minimum leaf temperature rather than to changes in Ψ_L (Figure 7.9). This contrasts with water stress-induced osmotic adjustment reported for many other species. Tyree *et al.* (1978) also observed a relationship between Ψ_π and temperature in which Ψ_π of *Tsuga canadiensis* leaves showed a marked seasonal decrease as temperatures fell below 0 °C with the onset of winter.

P. sericea leaves do not exhibit freezing tolerance. When leaf temperature is experimentally lowered at a constant rate, the points of 50% tissue injury and release of heat of fusion by freezing of supercooled water approximately coincide (−6.0 to −8.0 °C). This poses a question about the adaptive significance of the freezing point depression observed in this species, since supercooling appears to be effective well below the lowest calculated freezing point of the leaf tissue (−3.8 °C, corresponding to −3.2 MPa osmotic potential). The transient supercooled state, however, is extremely labile and can seldom be maintained for more than a few hours under natural conditions (Larcher 1982). Thus, the temperature-induced changes in the freezing point of *P. sericea* leaves may serve as a safety valve to prevent tissue freezing and damage when supercooling is no longer effective (Rada *et al.* 1985b). It is interesting to note that leaves of caulescent giant rosette species have a greater supercooling capacity than *P. sericea,* while *Podocarpus oleifolius* growing near the upper continuous forest limit has a much smaller supercooling capacity (Table 7.1). Giant rosette plants are usually exposed to much lower nightly temperatures than *P. sericea* while *Podocarpus oleifolius* is seldom exposed in its habitat to freezing temperatures. Of these species, only *P. sericea* exhibits temperature-induced changes in osmotically active solutes (Rada 1983).

Results from laboratory experiments suggest an alternative interpretation concerning the ecological significance of the variable freezing point depression response described above for *P. sericea.* Plants which were preconditioned at different temperatures showed three clear responses to decreasing temperature (Figure 7.10): (i) a decrease in osmotic potential, (ii) increase in soluble carbohydrates, and (iii) a decrease in supercooling point (increase in supercooling capacity). The approximately 0.5 MPa decrease in Ψ_π over the temperature range shown in Fig. 7.10 would result in only a 0.41 °C lowering of the freezing point as derived from the van't Hoff and Raoult equations. If the relationship between the lowering of

Table 7.1 *Supercooling capacity and freezing injury temperature in several tissues of two tree species* (Podocarpus oleifolius *and* Polylepis sericea) *and two giant rosette species* (Espeletia spicata *and* E. timotensis) *from the Venezuelan Andes*

Species	Organ or tissue	Supercooling capacity (C°)	Freezing injury (°C)	Lowest temperature recorded (°C)
Podocarpus oleifolius	leaf	−4.0	−1.8	3.8
	bud	−2.7		
	stem	−3.8		
Polylepis sericea	leaf	−7.3	−8.2	−0.3
	bud	−11.4		
	stem	−7.6	−8.9	−0.8
Espeletia spicata	leaf	−15.8	−13.8	−2.8
	bud	−5.4	−5.0	2.0
	stem pith	−4.8	−5.0	3.0
	root	−4.1	−4.5	4.0
Espeletia timotensis	leaf	−13.7	−12.8	−1.5
	bud	−6.0	−4.0	2.0
	stem pith	−5.0	−4.8	2.0
	root	−5.4		7.0

the supercooling point and the decrease in the freezing point observed in the laboratory holds under natural conditions, then low temperature preconditioning will produce a larger decrease in the supercooling point than in the freezing point (Figure 7.10). Zachariassen (1982), working with insect hemolymph, found that supercooling point depression was associated with the accumulation of low molecular weight cryoprotective substances, such as glycerol, sorbitol, mannitol and others, but could also be related to sugars such as glucose. The increase in the soluble carbohydrate levels observed at night in *P. sericea* may indicate that these sugars are acting as a cryoprotectant to help decrease the supercooling point or to make supercooling more effective under natural conditions.

In summary, in an environment with frequent, slight frosts, the large temperature-driven fluctuations in osmotic potential observed in *P. sericea* leaves, combined with their relatively high supercooling capacity, appear to be the principal means of avoiding freezing, and therefore tissue injury. At 4500 m, the upper altitudinal limit of *P. sericea*, the minimum

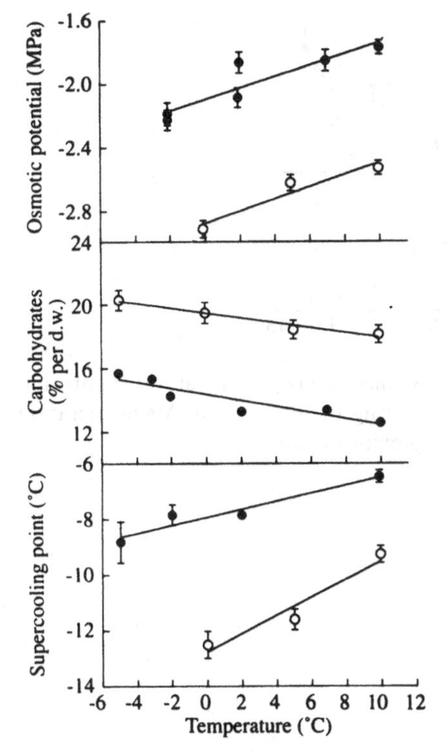

Figure 7.10. Leaf osmotic potential, soluble carbohydrate content and supercooling points obtained for *P. sericea* as preconditioning temperature was decreased in approximately 5 °C steps under controlled conditions. Plant material was obtained 20 February 1984 (•), and 9 April 1984 (o). The latter sampling day coincided with the end of an unusually long dry season (from Rada *et al.* 1985b).

absolute air temperature is approximately −6 °C. At this temperature the combined effects of osmotically induced freezing point depression and enhanced supercooling capacity may no longer be effective as a freezing avoidance mechanism under natural conditions, thus allowing leaf tissue injury and metabolic disturbance to occur.

Water relations

Stomatal conductance in *P. sericea* is typically high in the morning hours, sometimes even higher than at midday (Figure 7.11), suggesting that cold night-time temperatures do not reduce stomatal opening capacity during the subsequent day. In temperate zone tree species, on the other hand, exposure to relatively brief periods of chilling or freezing temperatures

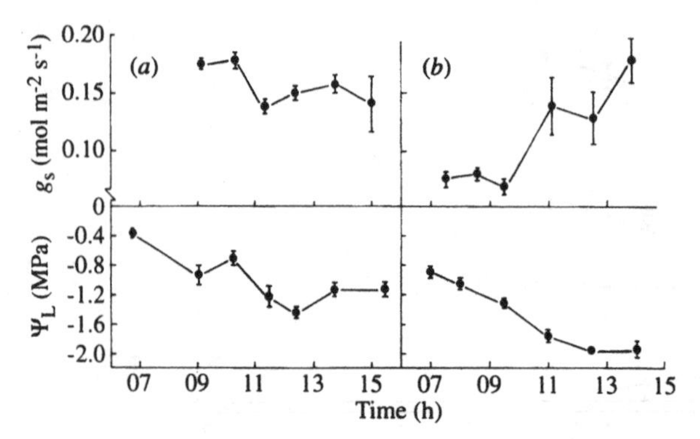

Figure 7.11. Daily courses of stomatal conductance (g_s) and leaf water potential (Ψ_L) of *P. sericea* on two different days during the dry season. Air temperature fell slightly below 0 °C in (*b*) during the previous night.

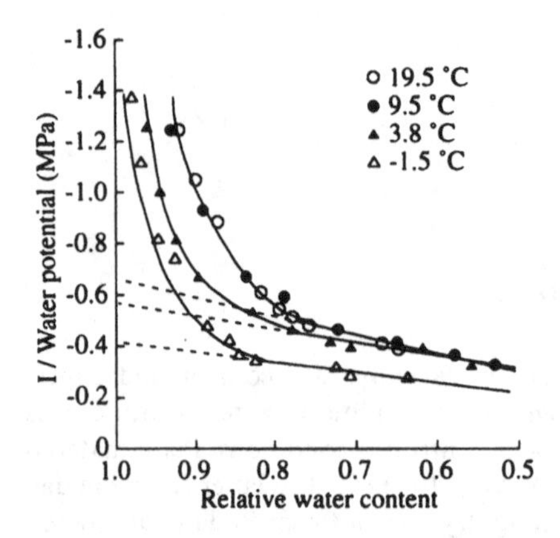

Figure 7.12. Representative pressure–volume curves for *P. sericea* leaves preconditioned at various temperatures during 24 hours. Leaves were kept at their preconditioning temperatures during pressure–volume measurements.

can cause stomatal closure (Teskey *et al.* 1984), and decreased enzymatic activity (Bauer *et al.* 1975). In these species decreased transpiration as winter approaches appears to be a mechanism by which desiccation is avoided during the period when the soil may be frozen, rendering the soil water unavailable. Such a mechanism in tropical treeline species, however, could be detrimental for total annual carbon balance, as prolonged stomatal closure following exposure to night-time chilling or freezing temperatures would not only reduce water vapor loss but also CO_2 uptake.

As described earlier, cold air temperatures induce diurnal fluctuations

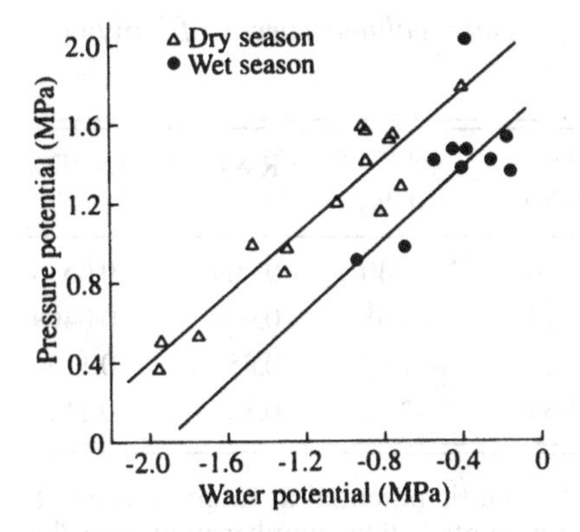

Figure 7.13. Turgor in relation to leaf water potential for a dry season day (13 February 1984), and a wet season day (20 December 1983). Turgor was calculated by subtracting osmotic potential determined psychrometrically from leaf water potential measured with a pressure chamber. The lines are regressions fitted to the data ($r = 0.94$, $p \leq 0.01$ for △; $r = 0.84$, $p \leq 0.01$ for ●) (from Rada *et al.* 1985b).

in osmotic potential of *P. sericea* (Figures 7.8, 7.9). Figure 7.12 shows representative pressure–volume curves determined for leaves from plants preconditioned in the laboratory at one of four different temperature levels for several hours. In plants preconditioned at temperatures above 9 °C the osmotic potential at full turgor was −1.5 MPa compared with −2.4 MPa in plants preconditioned at −1.5 °C. This 0.9 MPa decrease in osmotic potential after preconditioning at below-zero temperatures provides additional evidence that short-term osmotic changes under field conditions are driven by temperature.

Transient diurnal water stress, on the other hand, does not appear to induce short-term osmotic adjustment and consequent turgor maintenance in *P. sericea* (Figure 7.13). On a seasonal basis, however, maintenance of higher levels of turgor in *P. sericea* via osmotic adjustment does seem to be significant (Figure 7.13). Rada *et al.* (1985b) described a decrease in osmotic potential of almost 1.0 MPa between the end of the wet season in November and middle of the dry season in February (Table 7.2). This adjustment would reduce the likelihood of turgor loss during extraction of water from drying soils. Neither diurnal nor seasonal osmotic adjustment has been found in high elevation *Espeletia* species subjected to similar annual patterns of soil water availability.

The effects of a severe seasonal drought could also be reduced by special characteristics of sites in which *P. sericea* is predominantly found. The

Table 7.2 *Parameters obtained from pressure–volume curves of* Polylepis sericea

	Ppt (mm)	Et (mm)	Ψ_π^{100} (MPa)	Ψ_π^{0} (MPa)	RWC°	N_S/Dw
Nov 1982	59.1	71.7	−1.16	−1.40	0.909	0.0575
Dec 1982	19.0	69.7	−1.13	−1.38	0.909	0.0469
Jan 1983	6.4	84.4	−1.55	−1.85	0.851	0.1156
Feb 1983	9.3	108.5	−1.63	−2.13	0.820	0.1123

Ψ_π^{100}, osmotic potential at saturation; Ψ_π^{0}, osmotic potential at turgor loss point, RWC°, relative water content at turgor loss point; N_S/Dw, number of osmoles/kg dry weight. Climatic information is also included: Ppt, mean monthly precipitation; Et, mean monthly evapotranspiration (for October 1982, Ppt = 80.4 mm and Et = 67.3 mm). From Rada *et al.* 1985b.

presence of talus slopes and large boulders is likely to decrease the rate of water loss from the soil surface. Deep roots have been observed growing through the crevices between the rocks (Arnal 1983). Although seasonal changes in soil water potentials were not monitored, predawn Ψ_L in *P. sericea* was not as negative during the dry season as in other woody plant species growing in the open páramo.

Conclusions

Special microclimatic regimes and physiological features permit *Polylepis sericea* trees to grow at very high elevations in the tropical Andes. The potential for low temperature limitation on water availability, CO_2 uptake and growth appears to have led to a series of physiological adaptations: (i) high photosynthetic capacity and high dark respiration rates to permit an adequate supply of chemical energy for metabolic work such as repair of cellular damage, stomatal opening in the early morning, and production of cryoprotective substances, (ii) daily osmotic adjustment to enhance the supercooling capacity of the leaves and lower the freezing point of the cell sap at night, preventing tissue injury, and (iii) seasonal changes in osmotic potential that may help maintain water uptake and turgor, and thus a positive carbon balance during the dry season. These features allow *P. sericea* a certain degree of homeostasis in the face of an external environment characterized by strong daily fluctuations in temperature and water availability and seasonal shortage of soil water.

To increase our understanding of the ecological distribution of *P. sericea* and the factors responsible for its success in cold tropical climates, further information is required on the mechanisms underlying nocturnal accumulation of osmotically active solutes and on the full adaptive significance of the high dark respiration rates observed under controlled environmental conditions.

Acknowledgements

We are grateful to Noel M. Holbrook for comments on an earlier version of this chapter.

References

Arnal, H. (1983). Estudio ecologico del Bosque alti-Andino de *Polylepis sericea* Wedd. en la Cordillera de Mérida. Unpubl. Licentiature Dissertation, Universidad Central de Venezuela, Caracas.

Azócar, A. & Monasterio, M. (1980). Estudio de la variabilidad Meso y Microclimatica en el Páramo de Mucubají. In *Estudios Ecologicos en los Páramos Andinos*, ed. M. Monasterio. Mérida, Venezuela: Ediciones de la Universidad de los Andes.

Azócar, A., Rada, F. & Goldstein, G. (1988). Freezing tolerance in *Draba chionophila*, a 'miniature' caulescent rosette species. *Oecologia* **75**, 156–60.

Bauer, H. J., Larcher, W. & Walker, R. B. (1975). Influence of temperature stress on CO_2-gas exchange. In *Photosynthesis and Productivity in Different Environments*, ed. J. P. Cooper. Cambridge: Cambridge University Press.

Beck, E., Senser, M., Scheibe, R., Steiger, H. M. & Pongratz, P. (1982). Frost avoidance and freezing tolerance in Afroalpine giant rosette plants. *Plant, Cell and Environment* **5**, 215–22.

Beck, E., Schulze, E.-D., Senser, M. & Scheibe, R. (1984). Equilibrium freezing of leaf water and extracellular ice formation in Afroalpine 'giant rosette' plants. *Planta* **162**, 276–82.

Berry, J. A. & Bjorkman, O. (1980). Photosynthetic response and adaptation to temperature in higher plants. *Annual Review of Plant Physiology* **31**, 491–543.

Black, R. A. & Bliss, L. C. (1980). Reproduction ecology of *Picea mariana* (Mill.) Bsp. at treeline near Inuvik, North-west Territories, Canada. *Ecological Monographs* **50**, 331–54.

Ellenberg, H. (1958a). Wald oder Steppe. Die natürliche Pflanzendecke der Andean Perus. I. *Die Umschau* **21**, 465–8.

Ellenberg, H. (1958b). Wald oder Steppe. Die natürliche Pflanzendecke der Andean Perus. III. *Die Umschau* **22**, 679–81.

Fahey, J. H. (1979). The effect of night frost on the transpiration of *Pinus contorta* spp. *latifolia*. *Oecologia Plantarum* **14**, 483–90.

Goldstein, G. (1981). Ecophysiological and demographic studies of white spruce (*Picea glauca* (Moench) Voss) at treeline in the central Brooks Range of Alaska. PhD dissertation, University of Washington, Seattle.

Goldstein, G. & Meinzer, F. (1983). Influence of insulating dead leaves and low temperatures on water balance in an Andean giant rosette species. *Plant, Cell and Environment* **6**, 649–56.

Goldstein, G., Rada, F. & Azócar, A. (1985). Cold hardiness and supercooling along an altitudinal gradient in Andean giant rosette species. *Oecologia* **68**, 147–52.

Jaimes, M. (1985). Mecanismos de regulación del intercambio de gases en dos especies de la selva nublada. Unpubl. Licentiature Dissertation, Universidad de los Andes, Mérida, Venezuela.

Larcher, W. (1969). The effect of environmental and physiological variables on the carbon dioxide gas exchange of trees. *Photosynthetica* 3, 167–98.

Larcher, W. (1975). Pflanzenökologische Beobachtungen in der Paramostufe der Venezolanischen Andean. *Anzeiger der mathematisch–naturwissenschaftliche Klasse Österreichischen Akademie der Wissenschaften* **11**, 194–213.

Larcher, W. (1982). Typology of freezing phenomena among vascular plants and evolutionary trends in frost acclimation. In *Plant Cold Hardiness and Freezing Stress*, ed. P. H. Li and L. Sakai, pp. 417–26. New York: Academic Press.

Levitt, J. (1980). *Responses of Plants to Environmental Stresses.* Vol. 1. *Chilling, Freezing and High Temperature Stresses.* New York: Academic Press.

Meinzer, F. & Goldstein, G. (1985). Some consequences of leaf pubescence in the Andean giant rosette plant *Espeletia timotensis. Ecology* **66**, 512–20.

Meinzer, F. C. & Goldstein, G. (1986). Adaptations of water and thermal balance in Andean giant rosette plants. In *On the Economy of Plant Form and Function*, ed. T. J. Givnish, pp. 381–411. New York: Cambridge University Press.

Monasterio, M. & Reyes, S. (1980). Diversidad ambiental y variacion de la vegetacion en los Páramos de los Andes Venezolanos. In *Estudios Ecologicos en los Páramos Andinos*, ed. M, Monasterio, pp. 47–91. Mérida, Venezuela: Ediciones de la Universidad de los Andes.

Mooney, H. A. & West, M. (1964). Photosynthetic acclimation of plants of diverse origin. *American Journal of Botany* **51**, 825–7.

Pisek, A., Larcher, W., Vegis, A. & Napp-Zinn, K. (1973). The normal temperature range. In *Temperature and Life*, ed. H. Precht, J. Christophersen, H. Hensel and W. Larcher, pp. 102–4. New York: Springer-Verlag.

Rada, F. (1983). Mecanismos de resistencia a temperaturas congelantes en *Espeletia spicata* y *Polylepis sericea.* Unpublished Masters Disseration. Universidad de los Andes, Mérida, Venezuela.

Rada, F., Goldstein, G., Azócar, A. & Meinzer, F. (1985a). Freezing avoidance in Andean giant rosette plants. *Plant, Cell and Environment* **8**, 501–7.

Rada, F., Goldstein, G., Azócar, A. & Meinzer. F. (1985b). Daily and seasonal osmotic changes in a tropical treeline species. *Journal of Experimental Botany* **36**, 989–1000.

Rauh, W. (1956). Peruanische Vegetationsbilder, I: Die grossen Gegensatze: Die Kustenwuste und due Walder des Osten. *Die Umschau* **56**, 140–3.

Sakai, A. & Larcher, W. (1987). *Frost Survival of Plants. Responses and Adaptation to Freezing Stress.* Berlin: Springer-Verlag.

Sarmiento, G., Goldstein, G. & Meinzer, F. (1985). Adaptive strategies of woody species in neotropical savannas. *Biological Reviews* **60**, 315–55.

Simpson, B. (1979). A review of the genus *Polylepis* (Rosaceae: Sanguisorbeae). *Smithsonian Contributions to Botany* 43.

Smith, A. P. (1977). Establishment of seedlings of *Polylepis sericea* in the Paramo (Alpine) zone of the Venezuelan Andes. *Bartonia* **45**, 11–14.
Smith, A. P. (1979). Function of dead leaves in *Espeletia schultzii* (Compositae), an Andean caulescent rosette plant. *Biotropica* **11**, 43–7.
Takahashi, K. (1981). Changes in xylem pressure potential in *Abies sachalinesis* Mast. seedlings caused by subfreezing temperature and cold wind. *XVII IUFRO World Congress.*
Teskey, R. O., Hinckley, T. M. & Grier, C. C. (1984). Temperature induced changes in the water relations of *Abies amabilis* (Dougl.) Forbes. *Plant Physiology* **74**, 77–80.
Tranquillini, W. (1979). *Physiological Ecology of the Alpine Timberline.* New York: Springer-Verlag.
Troll, C. (1959). Die tropische Gebirge. *Bonner geographische Abhandlungen* **25**, 1–93.
Tyree, M. T., Cheung, N. S., MacGregor, M. E. & Talbot, A. J. B. (1978). The characteristics of seasonal and ontogenic changes in the tissue–water relations of *Acer, Populus, Tsuga* and *Picea. Journal of Botany* **56**, 635–47.
Vareschi, W. (1970). *Flora de los Páramos de Venezuela.* Mérida, Venezula: Ediciones del Rectorado, Universidad de los Andes.
Walter, H. & Medina, E. (1969). La temperatura del suelo como factor determinante para la caracterizacion de los pisos Subalpino y Alpino de los Andes de Venezuela. *Boletín Venezolano de Ciencias Naturales* **115/116**, 201–10.
Zachariassen, K. E. (1982). Nucleating agents in cold-hardy insects. *Comparative Biochemistry and Physiology* **73**, 557–62.

8

Morphological and physiological radiation in páramo *Draba*

WILLIAM A. PFITSCH

Introduction

Draba is one of several genera that are distributed throughout the north temperate latitudes and at high elevations in Central and South America from Mexico to Tierra del Fuego (Good 1974). Individuals of *Draba* typically are among the last vascular plants to drop out at the upper elevational limit to plant growth in North and South America. In the northern Andes, species of *Draba* occur in continuous páramo vegetation at 3500 m and are typical of rocky habitats and discontinuous vegetation up to 4800 m elevation (Cuatrecasas & Cleef 1978; Monasterio 1981a).

The basic vegetative form of the genus *Draba* is a rosette. The 'typical' form of north-temperate alpine drabas is a perennial, loose mat to tight cushion of small-leaved rosettes (Hitchcock 1941). The variety of morphologies present within the *Draba* in the highest elevations of the Venezuelan Andes is remarkable in contrast to their morphologically uniform north-temperate alpine congeners, although in both areas species of *Draba* occur in what might be regarded as similar habitats: protected rocky cliffs to exposed rock outcrops, ridges and scree slopes at the limit of plant growth. In the tropical Andes two quite distinct growth forms occur: upright-branching shrubs (Sections *Calodraba* and *Dolichostylis*), and thick-stemmed rosettes with relatively large leaves (Section *Chamaegongyle*: Schulz 1927) (Figure 8.1).

The year-round growing season of the páramo may have represented a release from the strong selective limitations of the temperate alpine environment, consequently allowing the expression of a diversity of relatively unusual forms. The large leaf surface area of the stemmed rosettes (up to 400 cm^2 per rosette vs 2–5 cm^2 for alpine species) would be difficult to produce during the short alpine growing season, and the

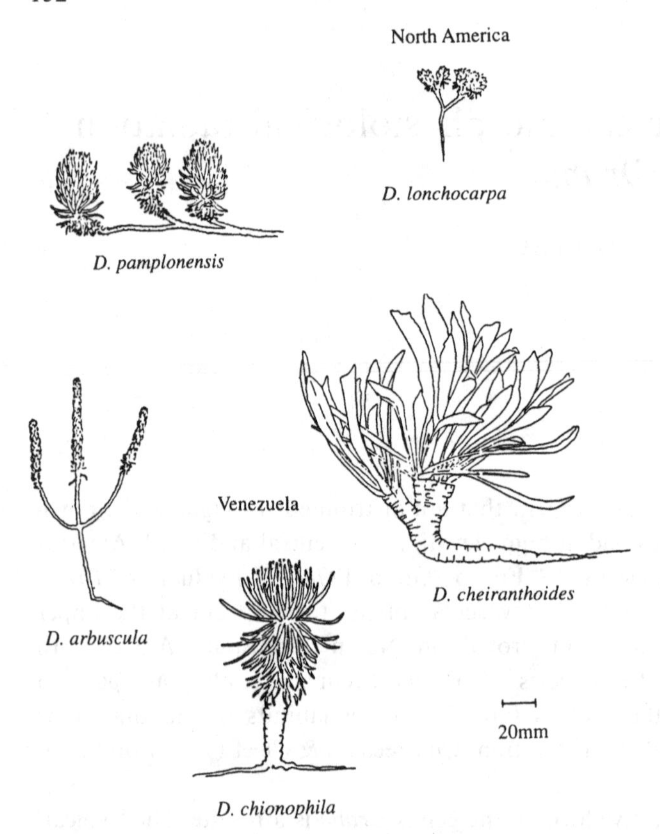

Figure 8.1. Line drawings of a typical North American alpine *Draba*, *D. lonchocarpa*, and four species of *Draba* from the Venezuelan páramo showing the diversity of life-forms to be found in the páramo.

maintenance of large, evergreen leaves through the alpine winter would require a substantial carbon investment. Although large-leaved drabas occur in extremely protected habitats in alpine and boreal areas, the relative frequency of occurrence is much lower than in the páramo (Pfitsch 1986). The vertical growth of the shrub form would be vulnerable to wind-blast, abrasion by ice, and desiccation during the alpine winter, especially in the exposed rock outcrop habitats where these species are typically found in the páramo.

The phylogenetic relationships of the neotropical species of *Draba* are not known. It is possible that the three taxonomic groupings of Schulz (1927) represent the descendants of different immigrant taxa. However, the uniform chromosome number of the Venezuelan drabas ($1n = 24$),

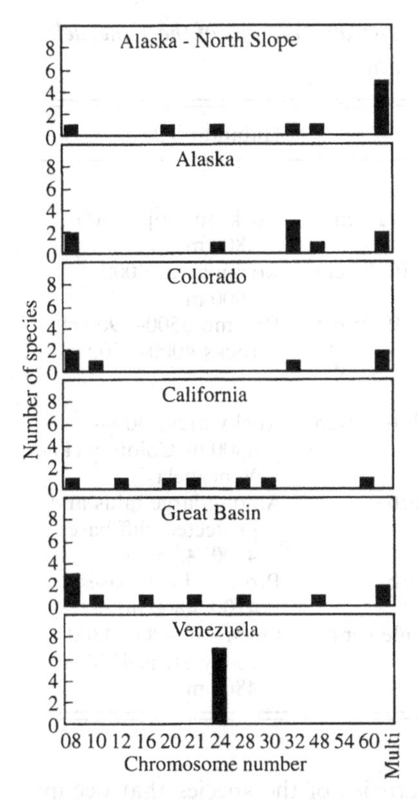

Figure 8.2. Frequency distributions of chromosome number (1*n*) for *Draba* species in regional floras from Alaska to Venezuela. Only perennial species are included. Chromosome counts for Venezuelan material were performed by N. Galland, other numbers are from Bolkhovskikh *et al.* (1969). Flora references: Colorado (Harrington 1954); Great Basin (Holmgren & Reveal 1966); California (Munz & Keck 1970); Alaska (Welsh 1974); Alaska North Slope (Wiggins & Thomas 1962).

compared to the diversity of chromosome numbers found in other alpine floras ($1n = 8$ to >60: Figure 8.2) is one piece of evidence that supports the notion of a monophyletic origin. Such uniformity in chromosome number is similar to other cases involving immigrant taxa in island floras (Raven 1973). A complete analysis of the shared and derived traits of these species, and of putative ancestors, would be necessary before a monophyletic origin could be asserted.

This chapter will describe some of the microenvironmental differences between habitats occupied by páramo drabas, and the morphological,

Table 8.1 *Seven species of* Draba *from the páramo regions of the Venezuelan Andes, their morphology and typical habitats*

Species	Morphology	Habitat
Section *Dolicostylis*		
D. arbuscula Hooker f.	Upright shrub 8–20 cm	Rock outcrops 3500–4800 m
D. empetroides Brandt, Gilg, and D. E. Schulz	Upright shrub 10–25 cm	Rocky areas 4000–4800 m
D. lindenii Hooker f.	Upright shrub 10–35 cm	Páramo 3500–3900 m, rocks 4000–4300 m
Section *Chamaegongyle*		
D. pamplonensis	Prostrate shrub 6–10 cm	Rocky areas 4000–4800 m, Colombia and Venezuela
D. cheiranthoides Hooker f. var *leiocarpa* O. E. Schulz	Large-leafed herb	Among large talus and protected cliff bases 4200–4300 m
D. bellardii Blake	Large-leafed herb	Protected cliff bases 4200–4800 m
D. chionophila Blake	Monocarpic caulescent rosette	Open soil 4200–4300 m, rocky areas 4500–4800 m

physiological, and life history characteristics of the species that occupy the different habitats. The seven taxa that will be considered include representatives of the major páramo growth forms, and they span the range of habitats occupied by *Draba* in the upper elevations of the Venezuelan páramo (3600–4800 m) (Table 8.1).

Microenvironment

Extreme daily temperature fluctuations and low mean temperatures are the basis for climatic limitations to plant growth in the páramo (Hedberg 1964). In Venezuela there is an annual, 4–5 month long dry season, a time that is particularly stressful for plants because minimum and maximum temperatures are more extreme and there is a potential for soil moisture deficit (Monasterio 1981a; Smith 1981; Pfitsch 1988).

Open soil habitats in the páramo of Piedras Blancas have perhaps the most severe diurnal stress of any páramo habitat (Pfitsch 1988). Surface temperatures in the dry season typically fluctuate from below freezing before dawn to 25–30 °C in the afternoon. Without large rocks or significant plant cover (<1%) to provide stability, the soil is extremely susceptible to movement by frost action. Surface and air temperatures rise rapidly in the morning, but soil temperatures in the rooting zone at −5 cm frequently remain close to freezing for 2 hours after dawn. Plants therefore are exposed to high evaporative demand during a time when low soil and root temperatures may restrict water uptake, a situation with potential for the development of severe short-term plant water deficits. Soil water potential in the rooting zone remained high (> −0.3 MPa) through the 1983 and 1984 dry seasons due to the low plant cover and the mulching effect of a crust of dry soil (Pfitsch 1988). Increased stress during the dry season in these habitats therefore results from the greater severity of diurnal stress rather than the development of seasonal water deficits.

The presence of massive rock outcrops can alter the microenvironment of the surroundings in a number of ways depending on aspect and exposure. An important effect of rock outcrops is their high heat capacity that serves as a thermal reservoir resulting in a decreased rate of nocturnal cooling (Pfitsch 1988). The decline in air temperature during the night is moderated as a result, and minimum soil temperatures remain well above freezing (3–5 °C: Pfitsch 1988). The potential for problems with water uptake induced by low soil and root temperatures is therefore reduced.

The surfaces of exposed outcrops heat rapidly and can reach very high temperatures (> 30 °C: Pfitsch 1988). Boundary layer effects may contribute to the heating of air near the surface depending on wind speed and topography. Sandy soils augmented by windblown silts and organic matter collect in crevices, so roots crowd into limited soil areas. Soil moisture can become depleted as prolonged periods of high evaporative demand and reduced precipitation continue through the dry season. Soil water potentials of < −1.5 MPa were measured in exposed rock outcrop habitats during both the 1983 and 1984 dry seasons (Pfitsch 1988). Stress during the dry season therefore can arise from a combination of high diurnal evaporative demand because of high surface temperatures, and a seasonal decline in soil moisture availability.

Sheltered areas of large rock outcrops and protected cliff bases have a considerably more moderate microclimate than the more exposed habitats described above. Precipitation runoff from large surface areas collects in these frequently moss-covered areas, resulting in high soil moisture availability through the dry season. The shelter of the rocks provides thermal buffering due to their high heat capacity. Protected cliff bases and crevices receive reduced direct insolation during much of the day and have large boundary layer effects with resulting high humidities and reduced evaporative demand. As a result, plants in the sheltered outcrop habitats are protected from high and low temperatures and experience little danger of either diurnal or seasonal water deficit.

Morphology

Individual leaf area and per-stem leaf area differ by more than two orders of magnitude among species of *Draba* from the Venezuelan páramo (Table 8.2). All the species sampled had relatively thick leaves, with specific weights in the range reported for sclerophyllous Mediterranean species (Mooney & Miller 1985). The species that occupy exposed habitats have narrow leaves, an attribute that minimizes the danger of overheating if transpiration is restricted. Unlike the leaf temperatures of species of *Espeletia*, which can be relatively uncoupled from the ambient temperature (Meinzer & Goldstein 1985), the leaf temperatures of the narrow-leaved species of páramo *Draba* track air temperature very closely (Pfitsch 1988).

The upright-branching imbricate-leaved shrubs found on exposed rock outcrops, *D. arbuscula* and *D. empetroides*, have a number of characteristics that can reduce transpirational water loss and contribute to the avoidance of water deficits. They have the smallest leaves and the least total leaf area of the páramo drabas. Their pubescent and steeply inclined leaves minimize the interception of radiation at midday when atmospheric vapor pressure deficits are greatest, thereby reducing overheating and increased transpirational demand. The inclination of the leaves of *D. arbuscula* gradually became steeper (angle from vertical decreased from 25° to 15°) and the total leaf area per stem declined as the 5-month dry season progressed at Piedras Blancas (Pfitsch 1986). The upright shrub growth form removes the transpirational surface from the extremely high temperatures of the rock outcrop surfaces, but entails a cost in terms of additional carbon investment in stem tissue compared to the large leafy rosette species *D. cheiranthoides* and *D. bellardii*.

Other shrubby species that have larger, horizontally arranged leaves and greater total leaf areas per stem, are found from the continuous páramo vegetation of lower elevations (e.g. *D. lindenii*) to prostrate forms in protected sites at high elevations (e.g. *D. pamplonensis*).

The monocarpic rosettes of *D. chionophila* are found in two distinct habitats: large rosettes (leaves 5–10 cm in length) occur in protected rocky ravines at high elevation (4700 m – Pico Espejo); smaller individuals (leaves 2–4 cm) are found in the frost-heaved soils at Piedras Blancas. The

Table 8.2 *Size relations for stems or rosettes of seven species of* Draba *from the páramo region of the Venezuelan Andes.* (*Means* $\pm$ *SE*)

Species	*n*	Number of leaves	Leaf area (cm^2)	Leaf dry weight (mg)	Leaf specific weight ($g\ m^{-2}$)	Leafed stem weight (mg)	Stem: leaf ratio (g/g)
D. arbuscula	127	70	2.8	35	140	19	0.48
		(2)	(0.2)	(2)	(10)	(1)	
D. empetroides	5	69	2.8	36	120	11	0.29
		(18)	(0.6)	(9)	(0)	(3)	
D. lindenii	11	51	13.3	165	130	22	0.12
		(5)	(1.2)	(14)	(10)	(4)	
D. pamplonensis	5	78	16.5	208	140	71	0.30
		(9)	(5.1)	(51)	(10)	(26)	
D. cheiranthoides	3	[a]	434.3	7994	180	1409	0.14
			(74.5)	(1363)	(0)		
D. bellardii	5	39	110.8	1160	100	80	0.08
		(4)	(16.0)	(208)	(0)	(11)	
D. chionophila	40	75	31.0	614	210	74	0.26
from Piedras Blancas		(3)	(5.8)	(103)	(0)	(11)	
D. chionophila	6	184	134.1	3155	230	693	0.21
from Pico Espejo		(8)	(7.4)	(270)	(10)	(82)	

[a] Leaves not counted, stem measured on one plant only.

vertically oriented leaves of this species are somewhat succulent, with a flattened-conical cross section, resulting in the highest specific leaf weight of these páramo drabas (Table 8.2). The proportion of total aboveground biomass composed of stem material is nearly as high for *D. chionophila* as it is for the imbricate-leaved shrubs.

The largest leaves (up to 2 cm wide and 25 cm long) and greatest total leaf area per rosette are found on the branching herbaceous species, *D. bellardii* and *D. cheiranthoides* var. *leiocarpa*, that occur in protected areas at cliff bases or among talus boulders (Table 8.2). The abundant moisture availability and low exposure to direct insolation minimize the danger of drought stress in these sheltered sites, making large leaf surfaces a viable option. These species have the least investment in stem biomass of the páramo drabas (Table 8.2).

Osmotic properties and water storage capacity

Pressure–volume curves (Tyree & Jarvis 1982) of fully hydrated tissue collected in February 1984 were analysed using the method of Hinckley *et al.* (1980). Osmotic potential at full hydration (Ψ_{sat}) was estimated by extrapolation of the *y* intercept of the linear portion of the relationship between tissue relative water content (*x*) and the inverse of water potential (*y*). The point of turgor loss (Ψ_{tlp}) was determined by repeated linear regression, adding points of increasing relative water content until there was a significant (>5%) deviation from previous estimates of the slope.

The seven species of *Draba* can be separated into two groups in terms of their osmotic potential at saturation and turgor loss (Table 8.3). The majority of the species have relatively high values for saturated osmotic potentials ($\Psi_{sat} = -0.8$ to -1.3 MPa) and turgor loss ($\Psi_{tlp} = -1.1$ to -1.6 MPa); the two imbricate-leaved shrubs had more negative values ($\Psi_{sat} = -1.5$ to -1.9 MPa; $\Psi_{tlp} = -2.0$ to -2.4 MPa). A more negative osmotic potential enables a plant to remain metabolically active and continue expansive growth to lower water potentials. A greater total volume of soil moisture may be exploited in habitats where soil moisture declines on a seasonal basis. The shrub *Draba arbuscula* was shown to be capable of limited seasonal osmotic adjustment; Ψ_{tlp} decreased by about 0.4 MPa during the 1983–4 dry season at Piedras Blancas (Pfitsch 1986). The Ψ_{tlp} of the monocarpic rosette, *D. chionophila*, did not decline, even for individuals that had been transplanted into the rocky habitat of *D. arbuscula* (Pfitsch 1986). This difference among species in ability to adjust

Table 8.3 *Pressure–volume curve characteristics (mean ± SE) for seven species of* Draba *from Venezuelan páramos*

Samples were collected and rehydrated for at least 24 hours in February 1984 (dry season). For *D. bellardii*, *D. cheiranthoides* and *D. chionophila* from Pico Espejo, curves were done on individual leaves and whole-plant capacitance estimates were conducted separately. Analyses for other species were conducted on entire stems or rosettes. Means differing by $p < 0.05$ are indicated by different letters (Student–Newman–Keuls test). All ANOVA were significant at $p < 0.0001$. ANOVA for water storage capacity included both leaf and whole-plant estimates.

Species	*n*	Saturated osmotic potential (MPa)	Turgor loss point (TLP) MPa	Relative water (RWC) at TLP (%)	Plant water storage capacity (mm/MPa)	Leaf water storage capacity (mm/MPa)
D. arbuscula	4	−1.51 b (0.07)	−2.09 a (0.07)	78 ed (2)	−25.45 b (4.05)	
D. empetroides	4	−1.86 a (0.17)	−2.38 a (0.05)	71 e (3)	−12.61 cb (2.33)	
D. lindenii	4	−1.27 bc (0.09)	−1.59 b (0.12)	81 dc (2)	−22.05 b (5.82)	
D. pamplonensis	4	−1.01 dc (0.16)	−1.34 b (0.19)	86 cb (2)	−20.54 b (4.47)	
D. bellardii	4	−0.93 dc (0.12)	−1.28 b (0.15)	96 a (0)	−19.89 b (1.90)	−39.98 a (7.36)
D. cheiranthoides	3	−1.28 cb (0.05)	−1.48 b (0.05)	90 ba (1)	−19.18 b (6.47)	−46.76 a (2.67)
D. chionophila Pico Espejo	6	−1.00 dc (0.08)	−1.14 b (0.08)	92 ba (1)	−5.56 c (0.53)	−11.00 cb (1.41)
D. chionophila Piedras Blancas	5	−0.72 d (0.07)	−1.37 b (0.11)	72 e (2)	−3.95 e (0.66)	

to soil drought suggests that seasonal drought has played a role in determining species distribution in the páramo.

The large difference in plant size in the populations of *D. chionophila* at Piedras Blancas (small rosettes; Table 8.2) and Pico Espejo (large rosettes) made it possible to compare the pressure–volume characteristics of entire shoots with those of individual leaves. The leaves from Pico Espejo plants and shoots from Piedras Blancas plants did not differ significantly in Ψ_{sat} or Ψ_{tlp} (Table 8.3). They did differ dramatically in the relative water content at the turgor loss point, with entire rosettes losing close to 30% of their saturated water content before turgor loss; individual leaves lost less than 10% of their saturated water content before turgor loss (Table 8.3). The rosettes had a much greater change in osmotic potential from saturation to turgor loss than did the leaves, presumably because of accumulation of solutes from the relatively much greater volume of water lost from the rosette. Assuming that the leaves on the smaller rosettes at Piedras Blancas have similar pressure volume properties as the larger leaves on the Pico Espejo plants, these results suggest that water storage in the stem tissue supplies water to maintain leaf relative water content, and hence leaf water potential, at high levels while the water content of the entire shoot declines.

The question of whether species differ in water storage capacity in aboveground tissues was addressed using the methods of Nobel & Jordan (1983). Water storage capacity is defined as the slope of the relationship between accumulative transpiration (mm^3 water per mm^2 of leaf area) versus the decline in water potential to the turgor loss point (MPa). A steep slope (large negative values) indicates that the tissue had little storage capacity. The water potential of the larger leaves of *D. bellardii* and *D. cheiranthoides* declined very rapidly as transpirational water was lost (Table 8.3). The water storage capacity of entire rosettes of these species was also low, similar to the stems of the shrubby species. The rate of decline in water potential with the loss of transpirational water was least in entire rosettes of *D. chionophila*. Stem and leaf tissue in this species each contribute to the high water storage capacity; individual leaves of *D. chionophila* have higher water storage capacity than entire stems or rosettes of other species. Water storage in fully hydrated rosettes of *D. chionophila* can supply average transpiration rates for 2–3 hours before turgor loss (Pfitsch 1986), an interval that is similar to the storage capacity of the giant rosette *Espeletia timotensis* (Goldstein *et al.* 1984). This interval is equal to the time after dawn that rooting zone temperatures in the open soil habitat at Piedras Blancas remain close to freezing while leaf temperatures are rising (Pfitsch 1988). The water storage therefore provides a mechanism by which this species may avoid the loss of turgor during this period of temperature-induced water transport disequilibrium.

Life history characteristics

The vegetative axis of an individual rosette is monocarpic in all species of *Draba*; the rosette dies after a terminal inflorescence is produced. Differences among species in life history results from variation in whether new vegetative meristems become active following the production of an inflorescence. Life histories in the genus *Draba* include weedy annual species at low elevations in temperate latitudes, the typical polycarpic alpine and páramo species that branch after reproduction, and relatively long-lived monocarpic species. The latter two life histories occur among species of páramo *Draba*. In simplistic terms, the multiple reproductive events of a polycarpic life history allow an individual to spread the risk of establishing successful progeny among years; the single reproductive event of a monocarpic life history does not. A necessary condition for the success of monocarpy must be a relatively constant probability of seedling establishment among reproductive years. Without that, an individual reproducing in an unfavorable year for seedling establishment runs the risk of having no successful offspring.

Both protected and exposed outcrop habitats have environmental characteristics that favor species with a means of spreading the risk of reproduction among years, such as a polycarpic life history. Protected areas on rock outcrops are often densely vegetated. Plant interactions in these areas may reduce the probability of seedling establishment, with successful seedling establishment being restricted to the unpredictable occurrence of disturbance opening up space in the community. On exposed rock outcrops, year-to-year variation in the probability of seedling establishment could be related to differences among years in the severity of drought during the dry season. Drought stress due to soil drought and high surface temperatures would result in high seedling mortality in a severe dry season. Although soil movement due to freezing and thawing can cause seedling mortality, rocks and vegetation may decrease frost action locally and increase seedling survival in cold environments (Mark, 1965; Sohlberg & Bliss 1984). The thermally buffered environment provided by the rock mass of the outcrops reduces the potential for frost-induced seedling mortality. Therefore, in years when precipitation during the dry season maintains high levels of moisture in the thin exposed outcrop soils the seedlings of most species may be able to become established, which contributes to the relatively high levels of local plant species diversity in these páramo habitats (Monasterio 1981b; Pfitsch 1986).

The low vascular plant species diversity of the open soil habitat occupied by *D. chionophila* at Piedras Blancas (Monasterio 1981b; Pfitsch 1986) is evidence of the effectiveness of the nearly nightly occurrence of soil freezing in restricting establishment of most plant species. Given that soil heaving by frost and needle ice are extremely limiting to seedling establishment and that a monocarpic life history requires a high probability of seedling establishment, it is ironic that the only monocarpic *Draba* in the Venezuelan Andes is found in the páramo environment in which frost action is most severe. The irony can perhaps be explained by the predictability of the seedling microenvironment of the open soil habitat. Little variability in soil moisture occurred in *D. chionophila* habitat either between dry and wet seasons, or between years (Pfitsch 1988). The few seedlings of *D. chionophila* found during this study were buried in the soil below the level of frost action. Apparently seedlings persist until they have a firm rooting anchor and then develop enough leaves to appear on the surface. With an ability to establish seedlings under conditions that most species find hostile, a monocarpic species can have a high and relatively constant potential for juvenile survivorship. The monocarpic life history may represent a means to accelerate the rate of population expansion. The surprisingly short lifespans (3–10 years) estimated for *D. chionophila* at Piedras Blancas may contribute to a relatively high potential rate of population growth (Pfitsch, 1986).

The evolution of the monocarpic life history of *Draba chionophila* has apparently resulted from a telescoping of the reproductive events of a polycarpic life into a single reproductive event, rather than a reduction to an annual or biennal life history. Multiple floral axes develop that are homologous to the axes of polycarpic species that develop after a period of vegetative growth. This species has retained its ability to sprout vegetatively after minimal flowering under controlled environment conditions, although this was not observed in the field.

The stemmed-rosette growth form of *D. chionophila*, which is morphologically similar to the giant rosettes in the genus *Espeletia*, is due in part to its monocarpic life history. The central pith that remains alive throughout the life of the plant has a very high concentration of soluble sugar (>40% dry weight), and probably serves as an energy reservoir for the development of flowers and fruits. The observed contribution of this tissue to the water storage capacity of these rosettes may be a secondary effect of its presence as a carbohydrate supply, or *vice versa.*

Conclusions

Microenvironmental differences related to the geomorphological features of the páramo contribute to its ecological diversity. The diversity of habitats in the Venezuelan páramo have been exploited by species of *Draba* with differing morphological, physiological, and life history characteristics.

The open soil habitat of *D. chionophila*, with its extreme daily temperature fluctuation, has perhaps the archetypal páramo environment. High soil moisture availability makes rehydration possible, so diurnal water deficits due to low morning soil temperatures can be moderated with storage reservoirs of transpirational water in the succulent leaves and the thick aboveground stem. The central pith may serve a dual function for this monocarpic species: storage of sugars to provide energy for flowering and seed production, and of water to meet short-term transpirational demands.

Rock outcrops with varying degrees of exposure present a diversity of microenvironments, but all share the beneficial effects of the high thermal capacity of rock which contributes to higher nocturnal temperatures, and reduces the probability of water uptake problems due to low soil temperatures. The moderate environment of the protected rock outcrop habitats of *D. bellardii* and *D. cheiranthoides* imposes little thermal or water stress on a daily or seasonal basis. Both species have large, relatively mesophytic leaves with a high osmotic potential and minimal water storage capacity.

Exposed rock outcrops have high surface temperatures which contribute to high evaporative demand. The species of *Draba* that are characteristic of such habitats, *D. arbuscula* and *D. empetroides*, have an imbricate-leaved, upright shrub growth form similar to many species of páramo plants in a number of taxonomic groups. This form results in moderated diurnal stress by removing leaves from the extreme temperatures of the soil boundary layer. A gradual decline in soil moisture availability during the annual dry season can impose a seasonal drought stress. *D. arbuscula* is able to adjust osmotically, lowering osmotic potential in response to decreasing water availability. The ability of *D. arbuscula* and *D. empetroides* to maintain turgor at low tissue water potentials enables them to exploit a greater volume of the potentially limited moisture supply.

Acknowledgements

I thank J. Ammirati, L. Bliss, T. Dawson and W. DiMichele for comments on this manuscript, T. Dawson for field assistance, N. Galland for chromosome counts, and A. Alexander for field assistance and performing the line drawings. Funding for this study was provided by a Dissertation Improvement Award from the U.S. National Science Foundation.

References

Bolkhovskikh, X., Grif, V., Matvejeva, T. & Zakharyeva, O. (1969). *Chromosome Numbers of Flowering Plants*. Leningrad.

Cuatrecasas, J. & Cleef, A. M. (1978). Una nueva crucifera de la Sierra Nevada del Cocuy (Colombia). *Caldasia* **12**(57), 145–58.

Goldstein, G., Meinzer, F. & Monasterio, M. (1984). The role of capacitance in the water balance of Andean giant rosette species. *Plant, Cell and Environment* **7**, 179–86.

Good, R. (1974). *The Geography of Flowering Plants*. London: Longman.

Gould, S. J. & Vrba, E. S. (1982). Exaptation – a missing term in the science of form. *Paleobiology* **8**, 4–15.

Harrington, H. D. (1954). *Manual of the Plants of Colorado*. Denver: Sage Books.

Hedberg, O. (1964). Features of Afroalpine plant ecology. *Acta Phytogeographica Suecica* **49**, 1–144.

Hinckley, T. M., Duhme, F., Hinckley, A. R. & Richter, H. (1980). Water relations of drought hardy shrubs: osmotic potential and stomatal reactivity. *Plant, Cell and Environment* **3**, 131–40.

Hitchcock, C. L. (1941). *A Revision of the* Drabas *of Western North America*. Seattle: University of Washington Press.

Holmgren, A. H. & Reveal, J. L. (1966). *Checklist of the Vascular Plants of the Intermountain Region*. Ogden, Utah: U. S. Department of Agriculture.

Mark, A. F. (1965). Flowering, seeding, and seedling establishment of narrow-leaved tussock, *Chionchloa rigida*. *New Zealand Journal of Botany* **3**, 180–95.

Meinzer, F. & Goldstein, G. (1985). Some consequences of leaf pubescence in the Andean giant rosette plant *Espeletia timotensis*. *Ecology* **66**, 512–20.

Monasterio, M. (1981a). Los páramos Andinos como region natural caracteristicas biogeográficas generales y afinidades con otras regions andinas. In *Estudios Ecológicas en los Páramos Andinos*, ed. M. Monasterio, pp. 15–27. Mérida, Venezuela: Universidad de Los Andes.

Monasterio, M. (1981b). Las formaciones vegetales de los páramos de Venezuela. In *Estudios Ecológicas en los Paramos Andinos*, ed. M. Monasterio, pp. 93–158. Mérida, Venezuela: Universidad de los Andes.

Mooney, H. A. & Miller, P. C. (1985). Chaparral. In *Physiological Ecology of North American Plant Communities*, ed. B. F. Chabot and H. A. Mooney, pp. 213–31. New York: Chapman and Hall.

Munz, P. A. & Keck, D. D. (1970). *A California Flora*. Berkeley: University of California.

Nobel, P. S. & Jordan, P. W. (1983). Transpiration stream of desert species:

resistances and capacitances for a C_3, a C_4 and a CAM plant. *Journal of Experimental Botany* **34**, 1379–91.

Pfitsch, W. A. (1986). *Draba* in the Venezuelan paramo: microenvironment, growth, and water relations. PhD dissertation, University of Washington, Seattle.

Pfitsch, W. A. (1988). Microenvironment and the distribution of two species of *Draba* (Brassicaceae) in the Venezuelan paramo. *Arctic and Alpine Research* **20**, 333–41.

Raven, P. A. (1973). Evolution of subalpine and alpine plant groups in New Zealand. *New Zealand Journal of Botany* **11**, 177–200.

Schulz, O. E. (1927). *Cruciferae* – Draba *et* Erophila. Leipzig: Engelmann.

Smith, A. P. (1981). Growth and population dynamics of *Espeletia* (Compositae) in the Venezuelan Andes. *Smithsonian Contributions to Botany* **48**, 1–45.

Sohlberg, E. H. & Bliss, L. C. (1984). Microscale pattern of vascular plant distribution in two high arctic plant communities. *Canadian Journal of Botany* **63**, 2033–42.

Tyree, M. T. & Jarvis, P. (1982). Water in tissues and cells. In *Water Relations and Carbon Assimilation. Encyclopedia of Plant Physiology, New Series*, vol. 12B, ed. O. L. Lange, P. S. Nobel, C. S. Osmond & H. Ziegler, pp. 35–78. New York: Springer-Verlag.

Welsh, S. L. (1974). *Anderson's Flora of Alaska and Adjacent Parts of Canada.* Provo, Utah: Brigham Young University.

Wiggins, I. L. & Thomas, J. H. (1962). *A Flora of the Alaskan Arctic Slope.* Toronto, Canada: University of Toronto Press.

9

Sediment-based carbon nutrition in tropical alpine *Isoetes*

JON E. KEELEY, DARLEEN A. DEMASON, RENEE GONZALEZ and KENNETH R. MARKHAM

Introduction

Isoetes (Isoetaceae) is a genus of small herbaceous plants often aligned with *Lycopodium* and *Selaginella*. There are more than 150 species distributed worldwide, typically in aquatic habitats (Tryon & Tryon 1982). A particularly intriguing aspect of the physiology of these plants is the presence of Crassulacean Acid Metabolism (CAM) (Keeley 1981, 1982), a photosynthetic pathway commonly associated with terrestrial xerophytes. CAM was selected for in these species by the daytime carbon limitation characteristic of their oligotrophic aquatic habitats (Keeley & Busch 1984; Boston & Adams 1985).

Across its range, *Isoetes* has radiated into a variety of aquatic as well as some terrestrial habitats and these environments have selected for a number of different structural–functional syndromes (Keeley 1987). Aquatic species occur in lacustrine habitats where they are permanently submerged throughout their life cycle and in amphibious environments where they alternate seasonally between aquatic and terrestrial conditions. In general, all aquatic species so far tested possess a well-developed CAM pathway while under water but lose this pathway when grown in an aerial environment. True terrestrial species of *Isoetes* are few in number although such species are known from most parts of the world. They readily fall into one of two groups: vernally active, summer-deciduous species at relatively low elevations in temperate latitudes; and evergreen species restricted to very high elevations (> 3500 m) in tropical latitudes. The former species show no CAM activity (even if artificially submerged), possess stomata and presumably depend entirely on C_3 photosynthesis. The tropical alpine species have variable levels of CAM metabolism but are unique in the terrestrial vascular plant kingdom in their complete lack of stomata. There is some evidence (Keeley *et al.* 1984) that these taxa rely upon the sediment for their source of photosynthetic carbon.

Sediment-based carbon nutrition in the terrestrial *Isoetes* (*Stylites*) *andicola*

Dependence upon the sediment for carbon acquisition is known in aquatic plant species, particularly those with the 'isoeted' growth form (e.g. Wium-Andersen 1971; Sondergaard and Sand-Jensen 1979; Richardson *et al.* 1984; Boston *et al.* 1987). These species are restricted to oligotrophic lakes where the very low inorganic carbon levels in the water, coupled with the very high diffusive resistance of water, limit carbon availability. In these environments the much higher carbon levels in the sediment put a premium on carbon acquisition through the roots.

On land the photosynthetic organs of higher plants normally have access to atmospheric CO_2 through stomatal pores and the dependence upon the sediment for carbon acquisition was unknown until it was reported for *Isoetes andicola* (Keeley *et al.* 1984).

Field work in the Peruvian Andes by Rauh and Falk in the 1950s brought to light a relatively unusual taxon in the Isoetaceae (although the species was collected by earlier botanists, Asplund No. 11830, 1940, STC). Amstutz (1957) named this *Stylites andicola*, separating it from the well-known genus *Isoetes* by the terrestrial habitat, elongated and branching stem with unbranched roots initiated along one side of the stem and sporangia elevated above the sporophyll base. Rauh & Falk (1959) provided further reason for separation by reporting a chromosome number of $2n = c.$ 58, a number quite distinct from the $2n = 22$, or multiples thereof, typical of the genus *Isoetes.* Further collections of Andean *Isoetes*, however, have revealed taxa with many of the traits thought to be unique to *Stylites* (Kubitzki & Borchert 1964; Gomez 1980; Karrfalt & Hunter 1980; Fuchs-Eckert 1982; Hickey 1985) and recent counts of *Stylites* have given a chromosome number of $2n = 44$ (Hickey 1984; L. D. Gomez and J. Keeley, unpublished data). Thus, there are good reasons for including *Stylites* within *Isoetes* and thus the correct name for 'stylites' is *Isoetes andicola* (Amstutz) Gomez.

Isoetes andicola is endemic to several widely scattered populations mostly above 4000 m in central and southern Peru. Reports by Feuerer that the species 'is quite common' in west central Bolivia (Hickey 1985) are in error. Thorough examination of Bolivian sites described by Feuerer (personal communication) uncovered no *I. andicola* (J. Keeley, personal observations); possibly the similar appearing *Plantago rigida*, which was quite abundant at these sites, was mistaken for *I. andicola*. The report of *I. andicola* from Colombia (Cleef 1981) is also based on a misidentified specimen (J. Keeley, personal observation). The habitat of *I. andicola* is not well documented but, based on reports in the literature, personal observations and data on herbarium specimens, it appears to be largely restricted to sites around the outer edges of bogs or lakes. At most sites it is distributed at an elevation of *c.* 1 m above the surface level of the standing water. Such microhabitats would be saturated during the wet season but potentially very xeric during the dry season (e.g. Table 9.1).

This species forms small rosettes from the apices of elongated, often branching, stems (Figure 9.1). It is unusual among *Isoetes*, both terrestrial and aquatic, in the extent of biomass underground. Over half of the biomass is tied up in roots and only a fraction of the leaf material is

Figure 9.1. *Isoetes andicola*, 'stylites', collected near Junin, Peru; scale is a 15 cm ruler.

Table 9.1 *Climatic data for La Oroya, Departamento Junin, Peru* (11° 31′ S, 75° 56′ W), *a station at 3712 m within 50 km of several* Isoetes andicola *populations*

	Jan	Feb	Mar	Apr	May	Jun	Jul	Aug	Sep	Oct	Nov	Dec
Precipitation (mm)	61	71	75	31	3	1	22	9	61	36	44	4.0
Temperature (°C)												
Mean minimum	2.9	3.9	3.4	3.1	0.1	−0.7	0.9	0.9	3.3	3.7	3.7	2.0
Mean maximum	16.5	16.0	18.2	17.5	18.0	17.0	15.8	16.8	16.0	19.1	19.0	17.0

Table 9.2 *Dry weight, carbon, and nitrogen distribution in leaves, stems and roots of* Isoetes andicola *collected near Junin, Peru* ($n = 4$)

	Leaves			
	Green portions	White portions	Stems	Roots
	$\bar{X} \pm$ SD	$\bar{X} \pm$ SD	$\bar{X} \pm$ SD	$\bar{X} \pm$ SD
Oven dry weight (g)	0.055 ± 0.011	0.148 ± 0.027	0.361 ± 0.032	0.821 ± 0.086
%	4.0	10.7	26.0	59.3
C (mg g^{-1} ODW)	399.0 ± 1.9	398.0 ± 1.4	420.8 ± 4.0	413.3 ± 4.1
N (mg g^{-1} ODW)	14.8 ± 0.8	8.0 ± 0.6	16.6 ± 1.2	8.0 ± 0.5

From J. A. Raven, unpublished data.

chlorophyllous and occurs aboveground (Table 9.2). Although there is no significant difference between the different parts in terms of C per unit of dry weight, the green leaves have significantly more N than underground achlorophyllous parts of leaves or roots (Table 9.2). Only the tips of the leaves are chlorophyllous (4% of total biomass) and these barely emerge from the sediment (Figure 9.2). Roots extend to at least 50 cm depth, although 80% occur in the upper 20 cm.

Structurally, the leaves of *Isoetes andicola* are unusual among terrestrial vascular plants in the lack of stomata. Additionally, these leaves are covered by a very thick cuticle. These observations led to the initial investigation of its unique physiology (Keeley *et al.* 1984), where it was demonstrated that there is very little CO_2 conductance across the leaf epidermis; dual isotope porometer measurements gave CO_2 conductances of 0.001–0.005 mm s^{-1} (values nearly an order of magnitude lower than those of desert cacti with closed stomata: Kluge & Ting 1978). Studies with $^{14}CO_2$ confirmed this and demonstrated that the bulk of the carbon for photosynthesis is obtained through the root system (Table 9.3). Under field conditions the CO_2 captured by roots is potentially much greater than observed in Table 9.3 since, during transplanting, the original root systems died and were only partially regenerated at the time of these experiments; the root/green-leaf ratio for the plants sampled in Table 9.3 was 2.5 ± 1.1 ($n = 12$), which contrasts markedly with the much higher ratio (14.9) for the field-collected plants shown in Table 9.2. Additionally, the free CO_2 concentrations in the sediment measured in the field (Table 9.4) are significantly higher than the 0.78 mol m^{-3} used for the laboratory studies reported in Table 9.3.

Internal pathway of carbon uptake in *Isoetes andicola*

The internal route taken by carbon moving from the roots to the leaves in *I. andicola* is unknown, although the case for CO_2 diffusion through intercellular gas space is strong. The leaves have four large lacunal air canals, which occupy more than half of the cross-sectional airspace (Figure 9.3*a*), and are surrounded by mesophyll cells with abundant chloroplasts (Figure 9.3*b*). The roots are up to 3 mm in diameter, unbranched and relatively hollow (Figure 9.3*c*), due to the collapse of cortical tissue early in development (Rauh & Faulk 1959). The leaves, stems and roots were studied anatomically in order to obtain precise measures of the volume of airspace along the root–stem–leaf pathway. These organs were fixed in FAA, dehydrated through an alcohol series

Figure 9.2. *In situ* view of *Isoetes andicola* (I.a.) and associated species *Plantago rigida* (P.r.) at Junin, Peru. Diameter of the coin, *c.* 2.8 cm.

and embedded in paraplast. Sections were photographed and the extent of intercellular airspace was measured with Sigma Scan Measurement System software (Jardel Scientific, Sausalito, CA). Percentage of cross-sections occupied by airspace ($\bar{X} \pm$ SD, $n = 10$) was: leaves = 58.0% ± 7.9, stems = 35.1% ± 11.4, roots = 75.2% ± 4.8.

If carbon dioxide diffuses from the sediment to the photosynthetic leaf tips through intercellular airspace, we can calculate whether or not the flux of CO_2 along this pathway is sufficient to maintain observed rates of photosynthesis (Table 9.3). Here we assume that CO_2 in the gas phase of the root is in equilibrium with free CO_2 in the soil solution and that a single root supplies CO_2 to a single leaf (a close approximation in light

Table 9.3 *$^{14}CO_2$ uptake in the light and dark by chlorophyllous portions of* Isoetes andicola *leaves grown under moist or wet conditions*

		CO_2 uptake by green leaves (μmol CO_2 mg^{-1} chl h^{-1})	
		$^{14}CO_2$ fed to leaves	$^{14}CO_2$ fed to roots
		$\bar{X} \pm$ SD (n)	$\bar{X} \pm$ SD (n)
Light	(Moist)	0.74 + 0.14 (3)	39.75 + 6.14 (3)
	(Wet)	1.12 + 1.05 (2)	18.78 + 0.79 (2)
Dark	(Moist)	0.24 + 0.16 (2)	0.51 + 0.07 (2)

From Keeley *et al.* 1984.

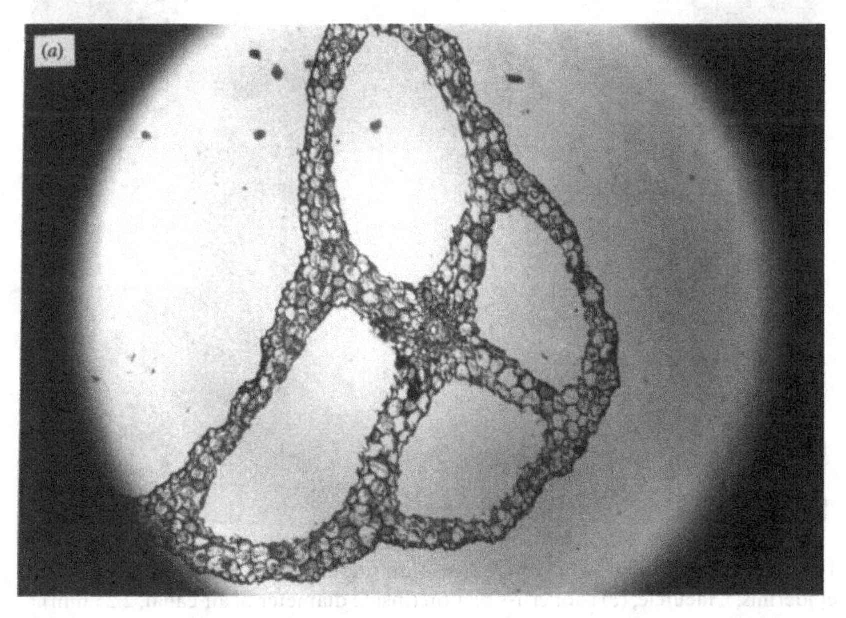

Figure 9.3. Cross-sections of *Isoetes andicola.* (*a*) Whole leaf (dimensions of largest lacunae, 0.93 × 0.40 mm. See over for Fig. 9.3(*b*) and (*c*)

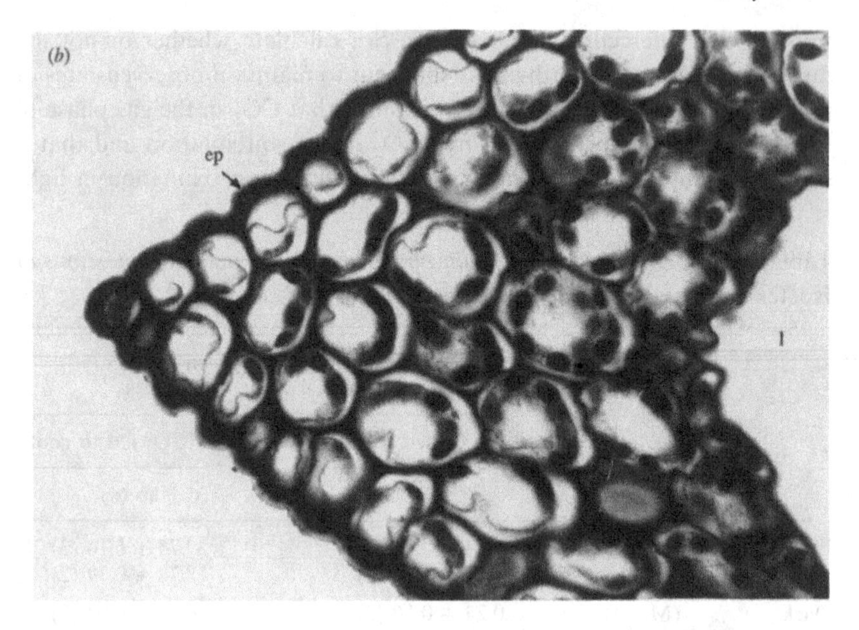

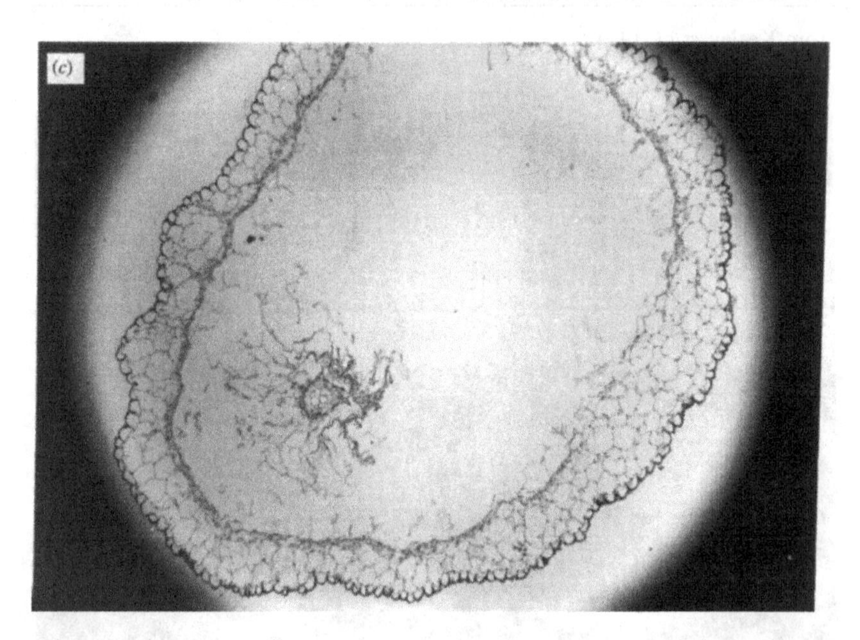

Figure 9.3. (*continued*) (*b*) Leaf cross-section (diameter of largest cell, 80 μm); ep, epidermis; l, lacunae; (*c*) root cross section (inside diameter of air canal, 2.25 mm).

Table 9.4 *Water chemistry characteristics of interstitial sediment water from sites with terrestrial* Isoetes *species*

Species	Site	Date	(*n*)	pH	Free CO_2 (mol m^{-3})	Oxygen (mol m^{-3})
Isoetes andicola	Junin, Peru, 4200 m	Nov	(3)	5.95–6.25	3.59–4.93	0.02–0
Isoetes novo-granadensis	Quito–Baeza Pass, Ecuador, 4050 m	Nov	(3)	5.60–6.00	1.46–1.77	0.16–0
		Jul	(1)	5.77	3.63	
Isoetes andina	L. Chisaca, Colombia, 3650 m	Dec	(4)	4.44–4.80	2.15–7.26	0.10–0

See Appendix 9.1 for precise localities, and Keeley & Busch 1984 for methods.
From J. E. Keeley, unpublished data.

Table 9.5 *Calculations of carbon dioxide diffusion through intercellular airspace of roots, stems and leaves of* Isoetes andicola *based on the lowest sediment carbon dioxide concentration measured in the field (Table* 9.4) *and a* CO_2 *diffusion coefficient at* 4000 *m* (0.063 *MPa*) *and* 10 °*C*

$$\text{Flux of } CO_2 = \frac{[CO_2]_{\text{sediment}} - [CO_2]_{\text{leaf}}}{r_{\text{leaf}} + r_{\text{corm}} + r_{\text{root}}}$$

$$\text{where resistance, } r = \frac{\text{Length of pathway}}{\text{Diffusion coefficient of } CO_2} \times \frac{1}{\text{Cross-sectional airspace}}$$

$$r_{\text{leaf}} = \frac{3 \times 10^{-2}\ \text{m}}{2.27 \times 10^{-5}\ \text{m}^2\ \text{s}^{-1}} \times \frac{1}{1.39 \times 10^{-6}\ \text{m}^2} = 0.95 \times 10^9\ \text{s m}^{-3}$$

$$r_{\text{corm}} = \frac{1 \times 10^{-2}\ \text{m}}{2.72 \times 10^{-5}\ \text{m}^2\ \text{s}^{-1}} \times \frac{1}{30.38 \times 10^{-6}\ \text{m}^2} = 1.44 \times 10^7\ \text{s m}^{-3}$$

$$r_{\text{root}} = \frac{7 \times 10^{-2}\ \text{m}}{2.72 \times 10^{-5}\ \text{m}^2\ \text{s}^{-1}} \times \frac{1}{2.13 \times 10^{-6}\ \text{m}^2} = 1.37 \times 10^9\ \text{s m}^{-3}$$

$$\text{Flux of } CO_2 = \frac{3.589\ \text{mol m}^{-3} - 5 \times 10^{-4}\ \text{mol m}^{-3}}{(0.95 \times 10^9 + 1.44 \times 10^7 + 1.37 \times 10^9)\ \text{s m}^{-3}}$$

$$= 1.5372 \times 10^{-9}\ \text{mol s}^{-1}\ [\text{leaf}^{-1}]$$

[1.3×10^{-2} g FW green tissue leaf^{-1} and 1.00 mg Chl g^{-1} green tissue]

Flux of CO_2 = 425 μmol CO_2 mg^{-1} Chl h^{-1}

of the plant architecture: Figure 9.1). According to Fick's Law of gas diffusion, flux of CO_2 would be a function of the concentration gradient between the root and leaf and the internal resistances along the root–stem–leaf pathway (Nobel 1983). The calculations shown in Table 9.5 utilized the lowest [CO_2] measured in the field (Table 9.4) for the source, a value of 5×10^{-4} mol m^{-3} as a reasonable estimate for the [CO_2] at the site of photosynthesis, and a diffusion coefficient for CO_2 at 4000 m and 10 °C. Resistances were calculated for a leaf, stem, and root of average length and diameter and thus, for an average leaf we calculate CO_2 diffusion from the sediment of 1.5372×10^{-9} mol s^{-1}. Using a chlorophyll concentration of 1 mg g^{-1} fresh weight of chlorophyllous tissue (Keeley *et al.* 1984) and a value of 0.049 g fresh weight per leaf ($n = 10$), of which only 27% is chlorophyllous (Table 9.2), we calculate the flux of CO_2 as 425 μmol CO_2 mg^{-1} Chl h^{-1}.

This is perhaps a conservative estimate of the rate of CO_2 diffusion from the sediment to the leaves under field conditions. Calculations used the lowest CO_2 source concentration measured in the field and a pathway from roots through 1 cm of stem was considered, even though roots are initiated acropetally so that newly initiated roots are connected almost directly to leaves. Even so, it is more than sufficient to account for the maximum photosynthetic rate of 39 $\mu mol\ mg^{-1}$ Chl h^{-1} shown in Table 9.3. However, Table 9.3 experiments were performed with a CO_2 concentration of only 0.78 mol m^{-3} in the solution bathing the roots. Using this value for the 'source', and a diffusion coefficient at 20 °C and 0.1013 MPa, we calculate 59 $\mu mol\ mg^{-1}$ Chl h^{-1}, which still could account for rates of photosynthesis measured in the laboratory (Table 9.3). Other factors not considered include the diaphragms that are spaced several millimeters apart in the leaf lacunae. Because they are perforated and very thin, though, the resistances contributed by them are insignificant when added to the resistances already considered. One oversimplification is the assumption that one root contributes carbon to one leaf, but in reality the number of leaves is typically greater than the number of roots. The calculations presented here, however, suggest that a single root is likely to be able to supply CO_2 at a rate sufficient to saturate photosynthesis in more than a single leaf.

Other mechanisms of gas movement, such as mass flow as observed by Dacey (1980) in *Nuphar*, may be operating, although it is not clear how such a system would work in this case. Fixation of CO_2 in the roots and movement through the vascular system is possible as phosphoenolpyruvate (PEP) carboxylase activity is present in the roots (J. E. Keeley, unpublished data). However, PEP carboxylase levels are many times lower in roots than in leaves (on a fresh weight basis) and the vascular cylinder represents a very minor part of the total root volume. Also, the very small vascular cylinder would make the hydration of carbon dioxide to bicarbonate, and movement in solution to the leaves, an unlikely hypothesis. In light of the discussion above, diffusion of CO_2 through internal airspace would seem to be the most parsimonious model of carbon movement from the sediment to the leaves of *Isoetes andicola* (stylites).

CAM in *Isoetes andicola*

Another aspect of the physiology of these plants is the presence of Crassulacean Acid Metabolism (CAM). Under natural conditions the chlorophyllous portions of leaves showed a diurnal change in malic acid

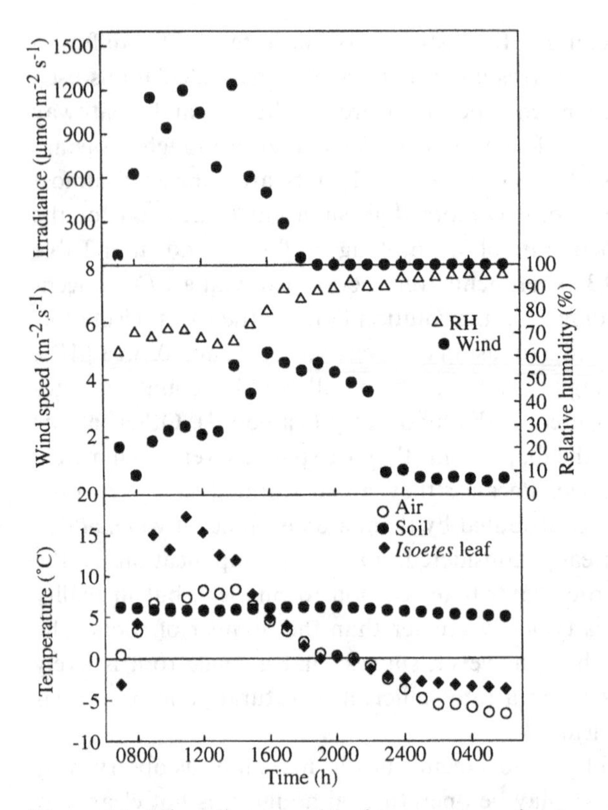

Figure 9.4. Microclimate and leaf temperatures for *Isoetes andicola* at the Junin site, Peru (November 1985). See Appendix 9.1 for exact location and Keeley & Keeley (1989) for methods.

of 40–46 μmol g^{-1} fresh weight in both summer and winter (Appendix 9.1). Due to the lack of gas exchange between leaves and the atmosphere, CAM undoubtedly was not selected for the same reason as in other terrestrial CAM plants. Potential functions of CAM would be: (i) to recycle dark-respired CO_2; (ii) to allow carbon fixation to proceed over the entire 24-hour period, thus effectively doubling the time for carbon fixation; (iii) daytime decarboxylation of malic acid in the CAM pathway may reduce photoinhibition, a particular concern for high elevation plants (Mooney *et al.* 1974).

Overnight carbon fixation is particularly noteworthy in this tropical alpine environment during the winter months when air temperatures routinely drop below freezing. Microclimatic data for one such night are shown in Figure 9.4. Despite the fact that leaf temperatures dropped to below 0 °C by 2100 h, this plant was still able to accumulate a substantial concentration of malic acid (Appendix 9.1). CAM functioning at near or below zero temperatures has also been observed in tropical alpine cacti (Keeley & Keeley 1989).

In summary, the bulk of the carbon gain in *I. andicola* is obtained from the sediment. Most of this uptake occurs during the day and it is presumably fixed via the (C_3) PCR (pentose carbon reduction) cycle reactions; ribulose bisphosphate carboxylase levels are sufficient (70–134 μmol mg^{-1} Chl h^{-1}) to account for observed light fixation rates. Overnight, CO_2 uptake continues at a low level and some of the respiratory carbon is recycled, both presumably through CAM-like carbon fixation; PEP carboxylase levels are sufficient to account for levels of acid accumulation (24–34 μmol mg^{-1} Chl h^{-1}) and dark-fixed CO_2 has been shown to be incorporated into malic acid. Since relatively little gas exchange occurs across the leaf epidermis, the photosynthetic oxygen generation is likely to result in a substantial concentration gradient between the leaf and the sediment. Outward diffusion of oxygen from the roots probably contributes to the prevention of anaerobic conditions in the sediment (Table 9.4) and to enhancing conditions for mineralization of the organic matter in the sediment (e.g. Sorrell & Dromgoole 1987).

Carbon isotopes in *Isoetes andicola*

Independent confirmation of these conclusions has recently been provided by carbon and hydrogen isotope data. Sternberg *et al.* (1985) found that the $\delta^{14}C$ of the peat *Isoetes andicola* was growing in (at the Junin site described in Keeley *et al.* 1984) was +36‰ (relative to the normal standard), a value far below the atmospheric level typical of the post-nuclear bomb testing era, i.e. post-1960 (Table 9.6). Based on this they contend that the peat was formed at some time during the mid-1950s and, since *I. andicola* cannot be older than its substrate, the formation of its biomass began at the earliest in the mid-1950s continuing up to December 1982, the date of collection. For comparison, Table 9.6 shows $\delta^{14}C$ values for several tree rings that grew between 1960 and 1970 indicating that large amounts of bomb-produced carbon-14 were incorporated into the tree biomass. Sternberg *et al.* (1985) calculated that a plant growing continuously from 1955 to 1982, fixing only atmospheric CO_2, should have a $\delta^{14}C$ value of about +370‰. Since the $\delta^{14}C$ values of the atmosphere between 1955 and 1982 were as high as or higher than +235‰ (the atmospheric level in 1982), any plant that started growth after 1955

Table 9.6 *Isotope abundances in* Isoetes andicola, *the peat it was growing in and associated species,* Plantago rigida *and* Distichia muscoides, *all collected in December 1982 from the Junin, Peru site (see Appendix 9.1). For comparison are the isotope abundances in atmospheric CO_2 and growth rings in spruce* (Picea *sp.*) *from Oregon, USA (from Sternberg* et al. *1985)*

	$\delta^{14}C$ (‰)	$\delta^{18}O$ (‰)	$\delta^{13}C$ (‰)	δD (‰)
Peat	+36		−26.6	
Isoetes andicola	+142	+18.6	−22.5	−47
Plantago rigida		+19.6	−24.5	−83
Distichia muscoides		+19.1	−22.2	−103
Atmospheric CO_2 (December 1982)	+235		−7.0	
Growth rings from spruce				
1960	+250			
1965	+900			
1970	+472			

and continued to grow until 1982, fixing only atmospheric CO_2, would have a $\delta^{14}C$ value $\geq +235$‰. The $\delta^{14}C$ value of *I. andicola* biomass was only +142‰ (Table 9.6), which is much less than what would be observed for a plant fixing atmospheric carbon dioxide since 1955, or any time after 1955, until 1982. Therefore, this plant must have obtained a large portion of its carbon by fixing CO_2 derived from the ^{14}C-depleted decomposing peat, and this must have been derived directly from the soil atmosphere, since the CO_2 evolved from the sediment would be quickly mixed with the atmosphere on these open windy sites.

Although stable carbon isotope values are often used to distinguish photosynthetic modes in terrestrial plants (e.g. ratio $^{13}C/^{12}C$), they do not readily distinguish *I. andicola* from associated non-CAM species (Keeley *et al.* 1984; Table 9.6; Appendix 9.1). One important reason for this is that the source of carbon for CAM is decomposing peat, which reflects previous fractionation events during the lifetime of the plants making up the peat. Other factors that could affect the $\delta^{13}C$ ratio of *I. andicola* are the very different diffusional resistances not encountered by normal terrestrial CAM plants. These factors could also account for the related observation that aquatic *Isoetes* species with CAM have $\delta^{13}C$ values indistinguishable from non-CAM aquatic plants with which they coexist (Keeley *et al.* 1987).

Hydrogen isotope ratios, on the other hand, readily distinguish CAM from non-CAM plants in both aquatic (Sternberg *et al.* 1984) as well as in terrestrial species (Ziegler *et al.* 1976; Sternberg & DeNiro 1983); in both environments CAM species are enriched with deuterium relative to associated non-CAM species. The δD value for *I. andicola* was 36–56‰ higher than the δD values for associated non-CAM species (Table 9.6) and this is consistent with the report of CAM for *I. andicola* (Appendix 9.1).

Characteristics of plants associated with *Isoetes andicola*

Isoetes andicola populations typically consist of hundreds of rosettes interspersed amongst other vascular plants. A strong degree of convergence is evident in the superficially similar rosette character of many of these other species e.g. *Plantago rigida* H.B.K. (Figure 9.2) and *Distichia muscoides.* Despite the marked morphological similarity with *I. andicola,* it seems likely that these species obtain the bulk of their carbon from the aerial atmosphere; their leaves have abundant stomata and a dense chlorophyllous tissue beneath the upper epidermis (e.g. Figure 9.5). Also, their leaves are almost entirely aboveground and chlorophyllous, in contrast to *Isoetes* where only the upper quarter of the leaf is above ground

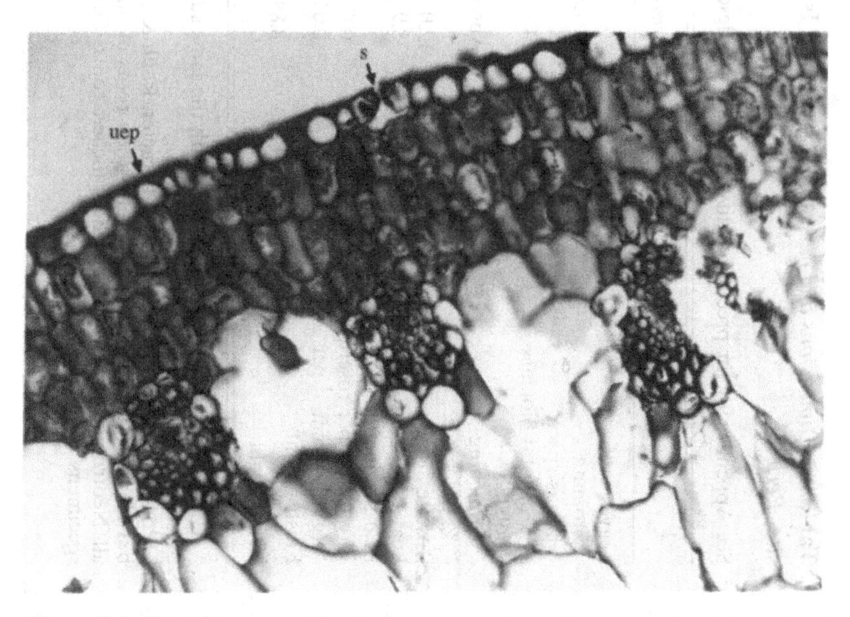

Figure 9.5. *Distichia muscoides* leaf cross-section (diameter of epidermal cell, 40 μm); uep, upper epidermis; s, stoma. Stomatal density estimated at 38 000 cm^{-2}.

Table 9.7 *Dimensions and mass of the terrestrial* Isoetes *and associated rosette-forming species from the Junin site, Peru*

See Appendix 9.1 for precise locality. Methods are described in the text.

	Isoetes andicola	*Distichia muscoides*	*Plantago rigida*
Leaves			
Stomata	Absent	Abundant	Abundant
% gas space	58	17	8
% of total biomass	24	52	33
Roots			
Branching	Absent[b]	Extensive	Extensive
% gas space	75	18	35
Rosette areal coverage (cm^2 per rosette)[a]	12.0 ± 6.7	1.3 ± 0.7	8.4 ± 3.6
Leaves per rosette	78.9 ± 28.7	23.1 ± 6.3	16.5 ± 2.1
Leaf production			
(g ODW per rosette per season)	0.473 ± 0.280	0.125[a] ± 0.043	0.100[a] ± 0.048
(g ODW cm^{-2} per season)	0.041 ± 0.021	0.096 ± 0.054	0.012 ± 0.006
$\frac{\text{Fresh weight}}{\text{Dry weight}}$	8.88	9.28	3.93

[a] $\bar{X} \pm$ SD, $n = 10$ (values within a row with the same superscript are not significantly different at $p > 0.05$, with 2-tailed t-test).

[b] Possibly there is plasticity in this character. Rauh & Falk (1959) stated that the roots of *I. andicola* were unbranched, but Hickey (1985) noted that in cultivation roots of this species were dichotomously branched and he concluded that all Neotropical *Isoetes* produced branched roots. Our designations here are based on hundreds of field-collected specimens.

and chlorophyllous (Table 9.2). Thus, aboveground chlorophyllous tissue in *Isoetes* represents only 4–6% of the total biomass, whereas a third to a half of the biomass, in *Plantago* and *Distichia*, respectively, is in the leaves (Table 9.7) and this is roughly comparable to the aboveground chlorophyllous tissue. Not only is much less of the biomass belowground, these taxa have root systems markedly unlike *Isoetes andicola*; there is a proliferation of secondary branches, in marked contrast to the complete unbranched character of *I. andicola* roots. *Plantago* and *Distichia* also have relatively little intercellular gas space in leaves and roots (Table 9.7) suggesting less potential for CO_2 diffusion from the sediment. There is no indication of CAM in *Plantago* and *Distichia* (Appendix 9.1).

Seasonal biomass production of leaves for these three species was compared by marking leaves near the beginning of the growing season (November) and then harvesting all new leaves at the end of the growing season in July. Total biomass production was significantly greater for *Isoetes* rosettes (Table 9.7). However, this is a function of the larger size of their rosettes, as there was no significant difference in leaf biomass production per cm^2 of rosette surface area. At any rate, it is somewhat surprising that an astomatous species, cut off from CO_2 exchange with the atmosphere, and with some level of CAM-like dark fixation of carbon, is capable of sustaining leaf production rates comparable to plants with a more typical mode of carbon uptake.

Other tropical Alpine terrestrial *Isoetes*

It appears that the syndrome described for stylites (*I. andicola*) is present to some extent in other tropical alpine terrestrial *Isoetes* taxa. One of the most widespread of these is *I. andina* Spruce ex Hook. of the northern Andean páramo (synonymous with *I. triquetra* A. Braun of earlier literature, e.g. Kubitzki & Borchert 1964). This species is similar to *I. andicola* in its terrestrial habitat, astomatous leaves covered by a thick cuticle, an elongate stem and large hollow unbranched roots. It is quite distinct in that it occurs only as separate rosettes that are many times larger than those of *I. andicola* (Table 9.8). In contrast to *I. andicola*, the triquetrous-shaped, spine-tipped leaves represent a much larger proportion of the biomass (Table 9.8). As with *I. andicola*, only the upper third of *I. andina* leaves is chlorophyllous, but typically the entire plant is sunk into the peat, often below ground level (J. Keeley, unpublished data). *Isoetes andina* grows in habitats similar to those of *I. andicola*, although the highly

Table 9.8 *Dimensions and mass of terrestrial* Isoetes *and associated rosette forming species from Chisaca site, Colombia*

See Appendix 9.1 for precise locality. Methods are described in the text.

	Isoetes andina	*Oreobolus obtusangulus*	*Plantago rigida*
Leaves			
Stomata	Absent	Abundant	Abundant
% gas space	27	28	8
% of total biomass	65	83	41
Roots			
Branching	Absent	Extensive	Extensive
% gas space	46	10	35
Rosette areal coverage (cm^2 per rosette)[a]	115.7 ± 46.3	1.8[a] ± 1.2	3.8[a] ± 1.9
Leaves per rosette	195.9 ± 71.3	17.4[a] ± 6.3	29.9[a] ± 14.6
Leaf production			
(g ODW per rosette per season)	2.865 ± 0.717	0.017[a] ± 0.003	0.062[a] ± 0.029
(g ODW cm^{-2} per season)	0.025 ± 0.006	0.009 ± 0.002	0.016 ± 0.007
Fresh weight / Dry weight	8.90	8.00	4.90

[a] $\bar{X} \pm$ SD, $n = 10$ (values within a row with the same superscript are not significantly different at $p > 0.05$, with 2-tailed t-test).

organic sediment is not quite as rich in CO_2 as the peat sediment of *I. andicola* (Table 9.4).

Preliminary laboratory measurements with *I. andina* indicate similarities with *I. andicola* in the form of carbon nutrition; one experiment similar to those presented in Table 9.3 showed four times greater CO_2 fixation by green leaves when CO_2 was fed to roots than when fed to leaves (10.3 vs 2.7 $\mu mol\ CO_2\ mg^{-1}\ Chl\ h^{-1}$).

In the páramo of Ecuador is another terrestrial *Isoetes*, *I. novo-granadensis* Fuchs, which probably has a similar mode of carbon nutrition. This is suggested by its astomatous leaves with thick cuticle and its habit of being buried in the sediment, similar to *I. andina*. Both of these taxa, however, have substantially less intercellular gas space in the leaves and roots (Tables 9.8, 9.9) than *I. andicola*. This may reflect less of a dependence upon diffusion of carbon dioxide from the sediment than has been demonstrated for *I. andicola*, although it may also reflect structural constraints imposed by the larger rosettes of *I. andina* and *I. novo-granadensis*. Additionally, there is evidence that both of these species have Crassulacean Acid Metabolism (Appendix 9.1).

As with *I. andicola*, there is some degree of convergence in the rosette growth habit of species associated with *I. andina* and *I. novo-granadensis*. Characteristics of the important species associated with each are presented in Tables 9.8 and 9.9. The presence of stomata in these other taxa suggests active gas exchange with the aerial atmosphere and there is no evidence of CAM (Appendix 9.1). Only one of these four species with typical CO_2 uptake from the atmosphere had greater seasonal leaf production than the associated *Isoetes* species (Tables 9.8, 9.9).

The syndrome of sediment-based carbon nutrition in terrestrial plants is apparently restricted to species of *Isoetes* in tropical alpine habitats; however, it is apparently not restricted to South America. The recently described *I. hopei* Croft from the tropical alpine parts of western New Guinea is in all likelihood an example. Based on photographs it would be difficult to distinguish it from *I. andina* and the published description (Croft 1980) suggests a similar habit; *I. hopei* is a terrestrial species with triquetrous leathery leaves lacking stomata.

Adaptive significance of sediment-based nutrition in tropical alpine habitats

Because of decreased barometric pressure, the partial pressure of carbon dioxide decliness with increasing altitude. Therefore, one might predict a

Table 9.9 *Dimensions and mass of terrestrial* Isoetes *and associated rosette forming species from Quito–Baeza site, Ecuador*

See Appendix 9.1 for precise locality. Methods are described in the text.

	Isoetes novo-granadensis	*Wernaria humilis*	*Oritrophium peruvianum*
Leaves			
Stomata	Absent	Abundant	Abundant
% gas space	34	25	38
% of total biomass	45	43	14
Roots			
Branching	Absent	Present	Very little
% gas space	39	29	35
Rosette areal coverage (cm^2 per rosette)	33.8 ± 17.8	13.4 ± 3.7	114.2 ± 55.5
Leaves per rosette	53.4 ± 40.0	116.2 ± 29.2	32.4 ± 7.4
Leaf production			
(g ODW per rosette per season)	1.100[a] ± 1.150	0.910[a] ± 0.313	0.100[a] ± 0.048
(g ODW cm^{-2} per season)	0.032 ± 0.014	0.069 ± 0.022	0.014 ± 0.006
Fresh weight / Dry weight	9.20	4.61	4.33

[a] $\bar{X} \pm$ SD, $n = 10$ (values within a row with the same superscript are not significantly different at $p > 0.05$, with 2-tailed t-test).

selective advantage to plants utilizing a decomposing peat sediment, which is enriched in carbon dioxide, as a photosynthetic source. In support of this idea, Billings & Godfrey (1967) showed that the photosynthetic rate of the temperate-alpine *Mertensia ciliata* was significantly reduced when the carbon dioxide concentration was reduced to levels approximating those at an altitude of 3100 m. They suggested this as a factor in the utilization of respiratory carbon dioxide which accumulated within the hollow stems of this plant. However, such analysis is complicated by the increased diffusion coefficient of CO_2 with a drop in barometric pressure. Calculations suggest that the increased rate of diffusion of CO_2 with altitude largely offsets the inhibitory effects of decreased partial pressure of this gas at high altitude (Gale 1972).

All tropical alpine terrestrial *Isoetes* are distributed near lakes or lagoons, although all are distributed well above the water level and, due to seasonal droughts (e.g. Table 9.1), may be subjected to soil water stress at certain times during the season. By sealing the leaves from the atmosphere this could provide an advantage under water stress conditions. In other words, if soil water stress is sufficient to cause stomatal closure in stomatous species, then a plant that obtains carbon from the sediment may be able to continue photosynthesis at times when other species are carbon-limited. At the present, evidence in support of this hypothesis is largely lacking. Most of the species studied here have leaves too small for conventional porometry. However, preliminary studies during the dry season at Junin, Peru, did show differences in midday water potentials between *Isoetes andicola* and associated species that would be consistent with this hypothesis. On 26 July 1986, midday water potentials (as measured with a Schollander-type pressure chamber, $n = 3$) were -2.63 ± 0.50 MPa for *Plantago rigida*, -2.02 ± 0.43 for *Distichia muscoides*, and -0.60 ± 0.30 for *Isoetes andicola*. These patterns were repeated on two other dates during the same month.

Evolution of sediment-based nutrition in terrestrial *Isoetes*

We hypothesize that the evolution of sediment-based carbon nutrition in terrestrial *Isoetes* taxa of tropical alpine habitats is the result of a pre-adaptation present in aquatic ancestors. Cladistic analyses (Hickey 1985) suggests an aquatic ancestry and there is reason to believe such ancestral taxa obtained carbon from the sediment.

Aquatic *Isoetes* are nearly ubiquitous in the tropical alpine region of South America and thus potential ancestral aquatic taxa are found throughout the range of Neotropical terrestrial *Isoetes*. Indeed, one to several aquatic species are present in lakes adjacent to all terrestrial *Isoetes* sites studied here. Although there is no experimental evidence that these aquatic species obtain carbon from the organic-rich sediment, there is documentation that Northern Hemisphere *Isoetes* species obtain the bulk of their carbon from the sediment (Richardson *et al.* 1984; Boston *et al.* 1987). These Northern Hemisphere *Isoetes* taxa are distributed in oligotrophic lakes that share some features with Neotropical alpine lakes. Particularly relevant is the differential between water column and sediment CO_2 concentrations; for example, a small lake near Laguna Chisaca, Colombia had a water column $[CO_2] = 0.26\ mol\ m^{-3}$, whereas the sediment $[CO_2] = 1.70\ mol\ m^{-3}$ (J. Keeley, unpublished data). Collections of aquatic *Isoetes* from Colombia, Venezuela, Ecuador, Peru and Bolivia have revealed a pattern of morphological and anatomical characteristics consistent with this hypothesized mode of carbon nutrition. All taxa lack stomata, which is to be expected in aquatic plants, but more importantly, all taxa have a very thick cuticle covering the leaf surface, which is not expected in aquatic plants, except in plants obtaining carbon from the sediment where a thick leaf cuticle would prevent the leakage of CO_2 into the water column. These aquatic *Isoetes* also have leaves that are buried up to two thirds of their length in the sediment and they have massive roots, often extending to more than 50 cm depth. Additionally, leaves, stems and roots of all of these aquatic *Isoetes* have extensive intercellular gas space (commonly the percentage of cross-sectional gas space is greater than recorded here for *I. andicola*). All of these aquatic species also possess a very well developed CAM pathway (J. Keeley, unpublished data).

If terrestrial taxa did evolve from aquatic taxa, cladistic analysis suggests that this event happened more than once. Based on a study of morphological characters, Hickey (1985) concluded that the three terrestrial taxa discussed here were derived from separate lineages. There is some chemical evidence to support this decision. In a preliminary survey of flavonoid patterns it was found that the terrestrial species, *I. andina* of Colombia and *I. novo-granadensis* of Ecuador produced chromatographic spot patterns more similar to nearby aquatic taxa than to each other or to *I. andicola* of Peru (Figure 9.6). The spot patterns produced by *I. andicola* were not obviously related to any of the dozen taxa investigated, which is consistent with Hickey's (1985) conclusion that this taxon is quite distant from other *Isoetes*.

Isoetes andina (No. 7902, collection no. of senior author) produced a chromatographic pattern very close to the aquatic *I. cleefii* Fuchs (No.

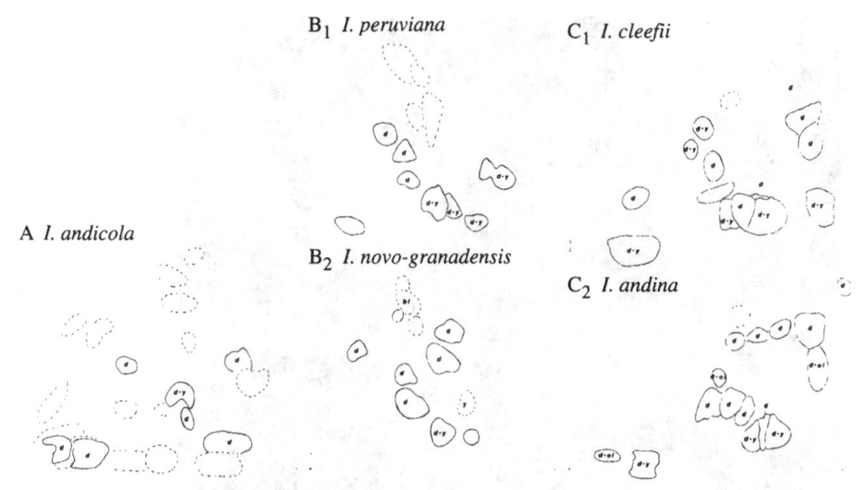

Figure 9.6. Chromatographic spot patterns for (A) terrestrial *Isoetes andicola* from Peru; (B_1) aquatic *I. cleefii* and (B_2) terrestrial *I. andina* from Colombia; (C_1) aquatic *I. peruviana* and (C_2) terrestrial *I. novo-granadensis* from Ecuador. Solid outlines indicate a flavonoid pigment appearing in UV (360 nm) and dotted outlines indicate a non-flavonoid compound that fluoresces in UV light (d, dark; y, yellow; ol, olive; bl, blue; d-y, dark changed to yellow). Origin is in the right-hand corner; vertical separation was with 15% HOAc and the horizontal axes were separated with TBA.

7876) collected from a nearby site (Figure 9.6, B) and similar to *I. karstenii* A. Braun (Nos. 7865, 7909, 10082) from a lake adjacent to the *I. andina* site. Based on morphology, these taxa are considered to be closely aligned; in Hickey's (1985) treatment, *I. cleefii* and *I. karstenii* are included in the *I. andina* alliance.

The chromatographic spot patterns generated by the terrestrial *I. novo-granadensis* (No. 10014) were quite unlike terrestrial *I. andina* but strikingly similar to the aquatic *I. peruviana* (No. 10020) Weber collected from a pond adjacent to the *I. novo-granadensis* site (Figure 9.6, C) and also very similar to another aquatic *I. killipii* (No. 10025) from a nearby lake. These taxa are also closely aligned by Hickey (1985 and personal communication).

Although the reasons are not clear, it appears that correlated with the evolution of this terrestrial syndrome is an increase in ploidy level. The common condition in the genus is $2n = 22$. *Isoetes andicola* has $2n = 44$. The aquatic *I. karstenii* has $2n = 22$ whereas the related terrestrial *I. andina* has $2n = 66$ (P. Hickey and J. Keeley, unpublished data). For *I. novo-granadensis*, $2n = 132$ (Hickey 1984) and the closest aquatic

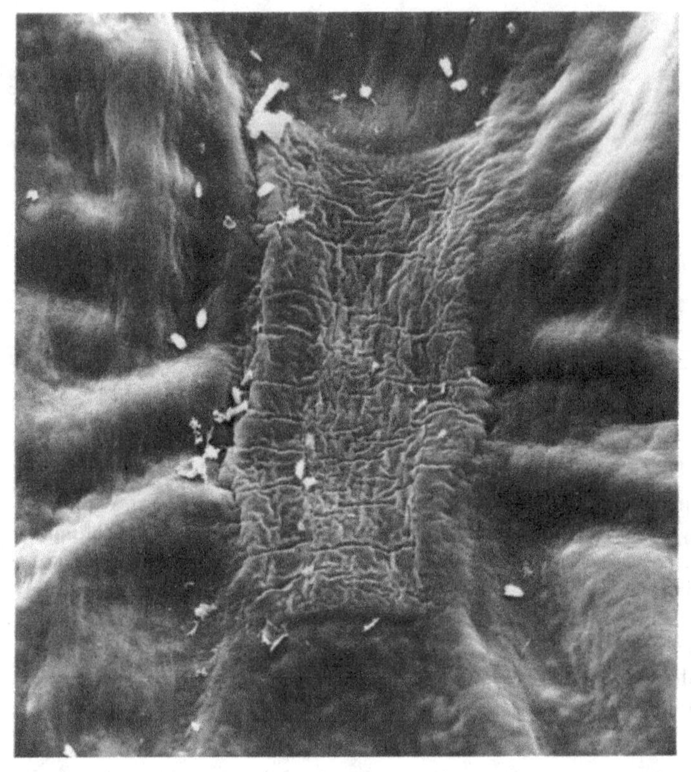

Figure 9.7. SEM (2000×) of *Isoetes andicola* leaf surface showing odd cell interpreted as an aborted guard mother cell.

relatives, *I. killipii* and *I. peruviana*, have $2n = 66$ and 44, respectively (R. Hickey and J. Keeley, unpublished data).

One observation, which could be interpreted as inconsistent with this evolutionary scenario, is the presence of a very unusual cell on the surface of *Isoetes andicola* (Figure 9.7). These cells are regularly distributed in linear arrays, similar to the arrangement of stomata in *Isoetes* taxa that possess stomata. One interpretation of these structures is that they are guard mother cells that aborted during development, which, if correct, would suggest that an ancestor of *I. andicola* had functional stomata. In Neotropical *Isoetes* stomata are present in amphibious taxa that occur in seasonally inundated habitats. If such a taxon were part of *I. andicola*'s ancestral heritage, it suggests strong selection against stomatal production in the present environment. These unusual cells (Figure 9.7), however, are not present in *I. andina* or *I. novo-granadensis* and thus evolution on to land may have followed more than one course.

Acknowledgements

The senior author is indebted to James Teeri, who provided the airline ticket that allowed for the initial visit to Peru for work with *Stylites*, and Sterling Keeley for field assistance. John Raven and Barry Osmond have made many intellectual contributions. This work has also been supported by NSF grants TFI-8100529, DEB-8004614, DEB-82-06887 and BSR-8407935, National Geographic Society grant No. 3119–15, and a fellowship from the John Simon Guggenheim Foundation to the senior author. Irwin Ting and Loretta Bates kindly provided the dual isotope porometer data on *Stylites* and Dan Walker and Judy Verveke provided the SEM.

References

Amstutz, E. (1957). *Stylites*, a new genus of Isoetaceae. *Annals of the Missouri Botanical Garden* **44**, 121–3.

Billings, W. D. & Godfrey P. J. (1967). Photosynthetic utilization of internal carbon dioxide by hollow-stemmed plants. *Science* **158**, 121–3.

Boston, H. L. & Adams, M. S. (1985). Seasonal diurnal acid rhythms in two aquatic crassulacean acid metabolism plants. *Oecologia* **65**, 573–9.

Boston, H. L., Adams, M. S. & Pienkowski, T. P. (1987). The utilization of sediment CO_2 by selected North American isoetids. *Annals of Botany* **60**, 485–94.

Cleef, A. M. (1981). *The Vegetation of the Páramos of the Colombian Cordillera Oriental*. Hirschberg: J. Cramer.

Croft, J. R. (1980). A taxonomic revision of *Isoetes* L. (Isoetaceae) in Papuasia. *Blumea* **26**, 177–90.

Dacey, J. W. H. (1980). Internal winds in waterlilies: an adaptation for life in anaerobic sediments. *Science* **210**, 1017–19.

Fuchs-Eckert, H. P. (1982). Zur heutigen Kenntnis von Vorkommen und Verbreitung der sudamerikanischen *Isoetes*-Arten. *Proceedings of the Koninklijke Nederlandse Akademie van Wetenschappen, Series C* **85**, 205–60.

Gale, J. (1972). Availability of carbon dioxide for photosynthesis at high altitudes: theoretical considerations. *Ecology* **53**, 494–7.

Gomez, L. D. (1980). Vegetative reproduction in a Central American *Isoetes* (Isoetaceae). Its morphological, systematic and taxonomical significance. *Brenesia* **18**, 1–14.

Hickey, R. J. (1984). Chromosome numbers of Neotropical *Isoetes*. *American Fern Journal* **74**, 9–13.

Hickey, R. J. (1985). Revisionary studies of Neotropical *Isoetes*. PhD dissertation, University of Connecticut, Storrs.

Karrfalt, E. E. & Hunter, D. M. (1980). Notes on the natural history of *Stylites gemmiflora*. *American Fern Journal* **70**, 69–72.

Keeley, J. E. (1981). *Isoetes howellii*: A submerged aquatic CAM plant? *American Journal of Botany* **68**, 420–4.

Keeley, J. E. (1982). Distribution of diurnal acid metabolism in the genus *Isoetes*. *American Journal of Botany* **69**, 254–7.

Keeley, J. E. (1987). The adaptive radiation of photosynthetic modes in the genus *Isoetes* (Isoetaceae). In *Plant Life in Aquatic and Amphibious Habitats*, ed. R. M. M. Crawford, pp. 113–28. Oxford: Blackwell Scientific Publications.

Keeley, J. E. & Busch, G. (1985). Carbon assimilation characteristics of the aquatic CAM plant, *Isoetes howellii*. *Plant Physiology* **76**, 525–30.

Keeley, J. E. & Keeley, S. C. (1989). Crassulacean acid metabolism (CAM) in high elevation tropical cactus. *Plant, Cell and Environment* **12**, 331–6.

Keeley, J. E., Osmond, C. B. & Raven, J. A. (1984). Stylites, a vascular land plant without stomata absorbs CO_2 via its roots. *Nature* **310**, 694–5.

Keeley, J. E., Sternberg, L. O. & DeNiro, M. J. (1987). The use of stable isotopes in the study of photosynthesis in freshwater plants. *Aquatic Botany* **26**, 213–23.

Kluge, M. & Ting, I. P. (1978). *Crassulacean Acid Metabolism*. New York: Springer-Verlag.

Kubitzki, K. & Borchert, R. (1964). Morphologische Studien an *Isoetes triquetra* A. Braun und Bemerkungen über das Verhaltnis der Gatting *Stylites* E. Amstutz zur Gattung *Isoetes* L. *Berichte der Deutschen Botanischen Gesellschaft* **77**, 227–33.

Mooney, H. A., Harrison, A. T. & Gulmon, S. L. (1974). Photobleaching in high and low elevation plants and different radiation intensities. *American Midland Naturalist* **91**, 254–6.

Nobel, P. S. (1983). *Biophysical Plant Physiology and Ecology*. San Francisco: W. H. Freeman.

Rauh, W. & Falk, H. (1959). *Stylites* E. Amstutz, eine neue Isoetaceae aus den hochandern Perus. 1. Teil: Morphologie, anatomie und entwicklungsgeschichte der vegetationsorgane. *Sitzungsberichte der Heidelberger Akademie der Wissenschaften, Mathematischen-naturwissenschaftliche Klass 1959* **2**, 85–160.

Richardson, K., Griffiths, H., Reed, M. L., Raven, J. A. & Griffiths, N. M. (1984). Inorganic carbon assimilation in the isoetids, *Isoetes lacustris* L. and *Lobelia dortmanna* L. *Oecologia* **61**, 115–21.

Sondergaard, M. & Sand-Jensen, K. (1979). Carbon uptake by leaves and roots of *Littorella uniflora* (L.) Aschers. *Aquatic Botany* **6**, 1–12.

Sorrell, B. K. & Dromgoole, F. I. (1987). Oxygen transport in the submerged freshwater macrophyte *Egeria densa* Planch. I. Oxygen production, storage and release. *Aquatic Botany* **28**, 63–80.

Sternberg, L. & DeNiro, M. J. (1983). Isotopic composition of cellulose from C_3, C_4, and CAM plants growing in the vicinity of one another. *Science* **220**, 947–8.

Sternberg, L., DeNiro, M. J. & Keeley, J. E. (1984). Hydrogen, oxygen, and carbon isotope ratios of cellulose from submerged aquatic crassulacean acid metabolism and non-crassulacean acid metabolism plants. *Plant Physiology* **76**, 69–70.

Sternberg, da S. L., DeNiro, M. J., McJunkin, D., Berger, R. & Keeley, J. E. (1985). Carbon, oxygen and hydrogen isotope abundances in *Stylites* reflect its unique physiology. *Oecologia* **67**, 598–600.
Tryon, R. M. & Tryon, A. F. (1982). *Ferns and Allied Plants*. New York: Springer-Verlag.
Wium-Andersen, S. (1971). Photosynthetic uptake of free CO_2 by the roots of *Lobelia dortmanna*. *Physiologia Plantarum* **25**, 245–8.
Ziegler, H., Osmond, C. B., Stickler, W. & Trimborn, D. (1976). Hydrogen isotope discrimination in higher plants: correlation with photosynthetic pathway and environment. *Planta* **128**, 85–92.

Appendix 9.1 *Morning and evening levels of titratable acidity (to pH 7.0 or, if indicated with an asterisk, to pH 6.4) and malic acid in photosynthetic tissues of tropical alpine terrestrial* Isoetes *and associated species (vouchers deposited in LOC)*

			Titratable acidity (μmol H^+ g^{-1} FW)		Malic acid (μmol g^{-1} FW)	
Species	Date	Country[b]	p.m. $\bar{X} \pm$ SD[a]	a.m. $\bar{X} \pm$ SD	p.m. $\bar{X} \pm$ SD	a.m. $\bar{X} \pm$ SD
BRYOPHYTA						
(Sphagnaceae)						
Sphagnum sp.						
	Dec	Colo.	50* ± 6	58* ± 2	1 ± 2	1 ± 1
	Jul	Colo.	65 ± 3	35 ± 7	15 ± 6	14 ± 5
TRACHEOPHYTA						
Lycopsida						
(Isoetaceae)						
Isoetes (*Stylites*) *andicola* (Amstutz) Gomez						
	Nov	Peru	60 ± 10	150 ± 30	15 ± 5	55 ± 15
	Jul	Peru	81 ± 7	180 ± 31	23 ± 6	69 ± 16
I. andina Hook						
	Dec	Colo.	37* ± 3	72* ± 11	40 ± 11	59 ± 3
	Jul	Colo.	22 ± 3	204 ± 42	32 ± 9	77 ± 19
I. novo-granadensis Fuchs						
	Nov	Ecua.	3* ± 4	132* ± 31	41 ± 16	82 ± 6
	Jul	Ecua.	59 ± 34	201 ± 7		
Spermopsida – Monocotyledoneae						
(*Juncaceae*)						
Distichia muscoides Nees et Meyen						
	Nov	Peru	6 ± 0	3 ± 0		
(Cyperaceae)						
Oreobolus obtusangulus Gaud.						
	Nov	Ecua.	0* ± 0	0* ± 0	15 ± 8	17 ± 5

Appendix 9.1 (*continued*)

Species	Date	Country[b]	Titratable acidity (μmol H^+ g^{-1} FW) p.m. $\bar{X} \pm$ SD[a]	Titratable acidity a.m. $\bar{X} \pm$ SD	Malic acid (μmol g^{-1} FW) p.m. $\bar{X} \pm$ SD	Malic acid a.m. $\bar{X} \pm$ SD
(Eriocaulaceae)						
Paepalanthus karsgenii Ruhl.						
	Jul	Colo.	28 ± 4	30 ± 2	3 ± 1	8 ± 7
Spermopsida – Dicotyledoneae						
(Apiaceae)						
Eryngium humile Cav.						
	Jul	Colo.	18 ± 5	26 ± 2	17 ± 4	26 ± 9
(Asteraceae)						
Espeletia grandiflora Humb. et Bonpl.						
	Jul	Colo.	95 ± 7	73 ± 11	21 ± 4	4 ± 6
Oritrophium peruvianum (Lam.) Cuatr.						
	Nov	Ecua.	9 ± 4	8 ± 4		
Wernaria humilis H.B.K.						
	Nov	Ecua.	0* ± 0	0* ± 0	2 ± 3	15 ± 9
	Jul	Ecua.	15 ± 2	14 ± 3	3 ± 5	0 ± 0
(Plantaginaceae)						
Plantago rigida H.B.K.						
	Nov	Ecua.	0* ± 0	0* ± 0	14 ± 2	25 ± 4
	Jul	Peru	38 ± 5	38 ± 4	41 ± 11	63 ± 19
	Jul	Colo.	45 ± 1	42 ± 3	39 ± 16	15 ± 1

[a] Mean ± standard deviation ($n = 3$).
[b] Sites were as follows.

Peru: Mounds of peat deposit, 60–125 cm elevation above surrounding bog, 4180 m; 100 m W of Route 3, 19.1–19.3 km NW of Junin town limit, Departamento de Junin (11° 00′ S; 76° 00′ W).

Ecuador: Boggy area, 65–390 cm elevation above shallow pool, 4040 m; below antennas at pass N of road from Quito to Baeza, 46.6 km (on old road) E of Plaza Frederico Pizarro in Quito, Provincia del Napo (00° 20′ S; 78° 10′ W).

Colombia: Heavy moss cover, 50–205 cm elevation above Laguna Chisaca, 3650 m; Páramo de Sumapaz, 30 km S of Usme, Departamento de Cundinamarca (4° 14′ S; 74° 16′ W).

10

Functional significance of inflorescence pubescence in tropical alpine species of *Puya*

GREGORY A. MILLER

Introduction

Tropical alpine environments are extreme and can generate very powerful selection pressures. Because of their simplicity they can provide valuable models for the study of evolution (Bradshaw 1971). The arborescent rosette growth form is a dominant feature across tropical alpine landscapes and has been the subject of detailed comparative ecological studies (Smith 1979, 1980, 1981; Smith & Young 1982; Young 1985 and Chapter 14). Alpine plant communities of most tropical alpine regions are distinguished from those of temperate latitudes by the presence of these giant rosette plants (Hedberg 1964; Cuatrecasas 1968; Smith 1981), with each major region possessing a unique flora. It has been assumed that the giant rosette form is an adaptive response to tropical alpine environments (Hedberg 1964; Mabberley 1973; Smith 1981), and an example of convergent evolution by different plant families to similar ecological conditions.

A notable feature of tropical alpine systems worldwide is the prevalence of pubescence, particularly among arborescent rosette plants. Most giant rosettes, including *Senecio* and *Lobelia* in Afroalpine areas (Hedberg 1964), *Lupinus, Espeletia* and *Puya* in the páramos of South America (Heilborn 1925; Smith 1981; Miller 1986), and *Argyroxiphium* in Hawaii (Carlquist 1974 and Chapter 16), produce dense pubescence on their leaves and/or inflorescences. Several possible advantages of dense pubescence in plants have been discussed in the literature, including modification of leaf energy budgets via increased reflectance and boundary layer thickness (Ehleringer *et al.* 1976, 1981; Ehleringer & Mooney 1978; Meinzer & Goldstein 1985 and Chapter 3), protection against herbivory (Levin 1973; Baruch & Smith 1979; Gates 1980) and insulation against radiative heat loss (Hedberg 1964; Smith 1974; Miller 1986).

The production of inflorescence pubescence has been described for several tropical alpine genera, but the functional significance of these hairs has not been rigorously studied (Penland 1941; Cuatrecasas 1954). In the unrelated genera *Puya* (Gilmartin 1972; Miller 1986), *Lupinus* (Penland 1941; Mani 1968) and *Lobelia* (Hedberg 1964; Mani 1968), the production of a densely pubescent, cylindrical inflorescence has evolved independently, apparently in response to the extreme tropical alpine conditions. The woolly indument on inflorescences of the tropical alpine species, *Lupinus alopecuroides* and *Lobelia telekii*, was suggested to function as insulation to protect flowers from nightly frosts, but this possibility was never tested (Hedberg 1964; Carlquist 1974). In one of the few studies where the insulative properties of floral pubescence was explicitly examined, Krog (1955) found for *Salix polaris*, an arctic vernal flowering species, that the internal temperatures of the woolly catkins was 15–25 °C higher than the surrounding air. Krog concluded that the dense hairs surrounding the flowers provided an effective insulating layer and were responsible for this large temperature difference. Flowers are vulnerable plant parts which can be damaged or killed by drastic diurnal temperature fluctuations (Jones 1992). In many tropical alpine species, the production of dense pubescence surrounding the flowers may be in response to night-time temperatures at or near freezing and may function to protect and insulate the flowers. The purpose of this chapter is to evaluate the adaptive significance of this pubescence by examining the clinal variation in the production of inflorescence pubescence observed for species of *Puya* in Ecuador (Miller 1986).

Puya species

Most ecological research on giant rosettes in tropical alpine systems has been carried out in Africa and Venezuela (Hedberg 1964; Smith 1975, 1979, 1981; Young 1982, 1985 and Chapter 14; Meinzer & Goldstein 1985; Young & Van Orden Robe 1986). In contrast, little is known of the equatorial páramo zone of the Ecuadorian Andes where species of *Puya* (Bromeliaceae) are common (Acosta-Solis 1971, 1979; Mills 1975). The genus *Puya* is principally distributed in the Andean highlands from Venezuela to Chile, with outlying species in Costa Rica, Guyana and adjacent Brazil, and northwestern Argentina (Smith 1957). Represented by a total of 175 species, *Puya* is reported to be the most primitive genus in the Bromeliaceae and the second most numerous genus in the subfamily Pitcairniodeae (Padilla 1973; Medina 1974). The name *Puya* is derived from the word for 'point' used by the Maguche Indians of Chile and is a genus comprising spiny, monocarpic acaulescent or short arborescent rosettes that produce erect, terminal inflorescences (Gilmartin 1972; Padilla 1973). While each rosette is monocarpic, several species of *Puya* show a variable frequency of vegetative propagation, producing daughter rosettes at the parent's base (Augspurger 1985; Miller 1987; Miller & Silander 1991). In Ecuador, the genus *Puya* (common name 'achupalla') shows restricted distributional limits around an elevational range of 1900–4300 m (Gilmartin 1973). These distribution patterns are fairly distinct and, while overlap does occur, the boundaries do appear to be associated with climatic differences.

Puya species are the exclusive bromeliad inhabitants of the Andean páramos, and are ecological parallels to other tropical alpine giant rosettes. Being from a family of tropical origin, these species are an exception to the numerous temperate families represented in tropical alpine situations (Vareschi 1970; Carlquist 1974; Cleef 1981). Most species of *Puya* are thought to be hummingbird pollinated (Ortiz-Crespo 1973; Augspurger 1985; G. A. Miller, personal observations) and in the páramo of Ecuador, hummingbirds consistently land on the extended bracts of *Puya clava-herculis* and *P. hamata* when visiting flowers (G. A. Miller, personal observations). This parallels findings in the Afroalpine region where sunbirds (*Nectarinia* species) are known to use the protruding bracts of *Lobelia* spp. in the same fashion (Hedberg 1964), and have been shown to be important pollinators (Young 1982).

Above 3400 m elevation in the Andean tropical alpine zones, at least six species of *Puya* are encountered. *Puya venezuelana* and *P. aristiguietae* in Venezuela (Vareschi 1970; Monasterio 1980), *P. cuatrecasii* in Colombia (Smith 1957; Cleef 1981), *P. clava-herculis* and *P. hamata* in Ecuador (Gilmartin 1972; Miller 1986), and *P. raimondii* in Peru (Padilla 1973) all grow under the extreme tropical alpine conditions, and each has responded in a similar fashion by producing densely pubescent, cylindrical inflorescences. At the lower elevational limit of *Puya* in Ecuador, the species encountered have glabrous or only slightly pubescent inflorescences. At intermediate elevations, these glabrous species are replaced by more pubescent ones, and in the high elevation páramos, the *Puya* species encountered produce densely pubescent inflorescences.

Four Ecuadorian species of *Puya* representative of this pubescence gradient with temperature were chosen to examine the functional significance of inflorescence pubescence in tropical alpine plants. *Puya aequatorialis* var. *aequatorialis* grows on steep rocky outcrops from 1900 to 2100 m and produces a glabrous inflorescence (1.0–1.5 m height). At higher elevations (2800–3200 m), *P.* aff. *vestita* is found in open montane forest patches and shows a low to intermediate level of short, reddish-brown indument on its inflorescence (1.5–2.3 m height). *Puya hamata* grows in the *Espeletia hartwegiana*-dominated páramos of northern Ecuador and produces a large (4–5 m), densely white lanate inflorescence. In the more exposed, higher elevation páramos, *P. clava-herculis* rosettes produce a globose inflorescence (1–2 m) with very dense tan to brown pubescence surrounding the flowers.

Comparative flower temperatures

The cold páramo temperatures pose several problems for the flowers of *Puya*. For example, the nightly cold may cause tissue damage and limit temperature-dependent processes such as nectar production and seed development (Baker 1975; Stiles 1978). The maintenance of suitable flower temperatures for successful nectar and seed production appears to be a requirement for the páramo species and may limit the non-páramo *Puya* species to less extreme habitats (Miller 1986). Of interest in this context is, what would be the flower temperature response of non-páramo species of *Puya* if they were relocated up into the páramo, and how would their flower temperatures compare with those of a native páramo species? Three *Puya* species, *P. hamata*, *P.* aff. *vestita* and *P. aequatorialis*, which represent different levels of inflorescence pubescence, were selected to examine the effects natural variation in pubescence had on flower temperature (Table 10.1; Figure 10.1). Individual flower temperature was defined as the internal flower tissue temperature at the ovary and was measured during a continuous 24-hour period using the procedure described by Miller (1986).

The comparative thermal responses for each species, when relocated to a common environment (the Páramo de El Angel) are presented in Figure 10.2. During this 24-hour period, *P. hamata* flowers maintained a mean elevated temperature of 2.4 °C above ambient air temperature. In contrast, the non-páramo species, *P.* aff. *vestita* and *P. aequatorialis*, maintained mean elevated flower temperatures of only 0.4 °C and 0 °C respectively above air temperature for the same period. In a similar experiment, flowers of *P. clava-herculis* in the páramos of Cotopaxi (Table 10.1) show a mean elevated flower temperature of 3.1 °C above air temperature for a 24-hour period (G. A. Miller, unpublished data). As radiation and air temperature decreased in late afternoon, flowers of both *P. aequatorialis* and *P.* var. *vestita* cooled quickly (at rates $> -3.0\,^{\circ}\mathrm{C\,h^{-1}}$) and equilibrated with air temperature within one hour (Figure 10.2*b*). In fact, flower temperatures

Table 10.1 *Species of* Puya *and research site descriptions in Ecuador*

Species	Site	Habitat	Elevation (m)	Mean annual temperature (°C)
Puya aequatorialis	Valle de la Chota 0° 30′ N 77° 55′ W	Arid montane thorn shrub	1980	16.5
Puya aff. *vestita*	Lago Cuicocha 0° 15′ N 78° 22′ W	Moist montane forest	3200	11.2
Puya hamata	Páramo de El Angel 0° 40′ N 77° 45′ W	Rain páramo	3415	10.2
Puya clava-herculis	Parque Nacional Cotopaxi 0° 38′ S 78° 30′ W	Páramo	3750	8.0
	Páramo de la Virgen 0° 20′ S 78° 10′ W	Rain páramo	3962	6.6

Mean annual temperatures were calculated from a regression analysis for 126 meteorological stations in Ecuador (after Cañadas Cruz 1983).

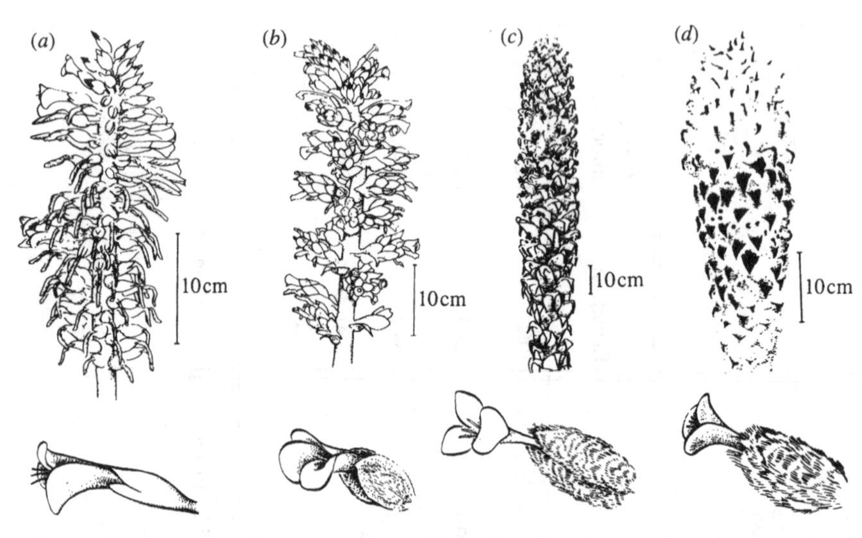

Figure 10.1. Interspecific comparison of four *Puya* species showing the variation in inflorescence morphology and pubescence. Subtending bracts were removed in lower diagram for each species to better illustrate petals and sepals. Note sepals also reflect the observed pubescence gradient (see text). (*a*) *P. aequatorialis*; (*b*) *P.* aff. *vestita*; (*c*) *P. hamata*; (*d*) *P. clava-herculis*.

of *P. aequatorialis* were closely coupled to fluctuations in air temperature and radiation throughout the period of observation. For the most part, flower temperatures of *P.* aff. *vestita* also closely tracked air temperature and radiation fluctuations, but did show a measurable temperature elevation during part of the day (Figure 10.2*b*). I attribute this temperature difference to the presence of low levels of pubescence surrounding the flowers of this species. In contrast, flower temperatures of *P. hamata* were much less sensitive to these fluctuations and showed a steady daytime increase until late afternoon, regardless of midday variation in radiation and air temperature (Figure 10.2*a*, *b*). The cold páramo temperatures are likely to present a potential barrier to the distribution of the latter two species where, under these rigorous conditions, flowers of both species would have temperatures at or near freezing and be consistently subjected to possible freezing damage and reduced developmental rates. If nectar and seed production are temperature-dependent in *Puya*, the cold páramo temperatures may limit *Puya* distribution such that only species capable of maintaining suitable flower temperatures are found in the páramo. Thus, because of the specialized modification of dense pubescence, the native páramo species appear able to persist and reproduce in the páramo.

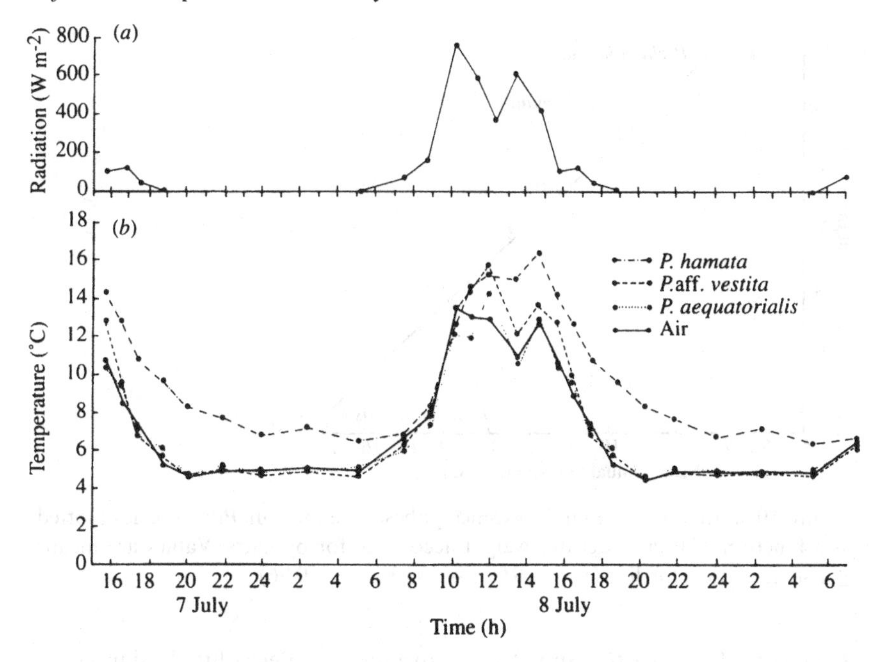

Figure 10.2. (*a*) Incoming radiation profile measured on 7–8 July 1984 in the Páramo de El Angel. (*b*) Air and mean flower temperature profiles for *P. hamata*, *P.* aff. *vestita* and *P. aequatorialis*. Flower temperatures plotted represent mean values for eight flowers for each species.

Relative investment in pubescence

Increases in the production of leaf and/or inflorescence pubescence along changing environmental gradients are widespread among many different genera and families (Ehleringer 1984). Leaf pubescence was reported to increase with aridity in species of *Espeletia* in the Venezuelan páramos (Baruch & Smith 1979) and *Encelia* in temperate deserts of North and South America (Ehleringer & Mooney 1978; Ehleringer 1984; Ehleringer *et al.* 1981). Hedberg (1964) also reported increases in leaf pubescence with elevation in Afroalpine *Senecio* species. Inflorescence pubescence has also been proposed to increase along a gradient of decreasing air temperature for species of Afroalpine *Lobelia* and Andean *Lupinus* (Hedberg 1964). The great interspecific variation in the production of inflorescence pubescence in the genus *Puya* parallels these trends. The variation in inflorescence pubescence along a gradient of decreasing air temperature was quantified using the procedure described by Miller (1986). A comparison of four species, *P. clava-herculis* (from La Virgen),

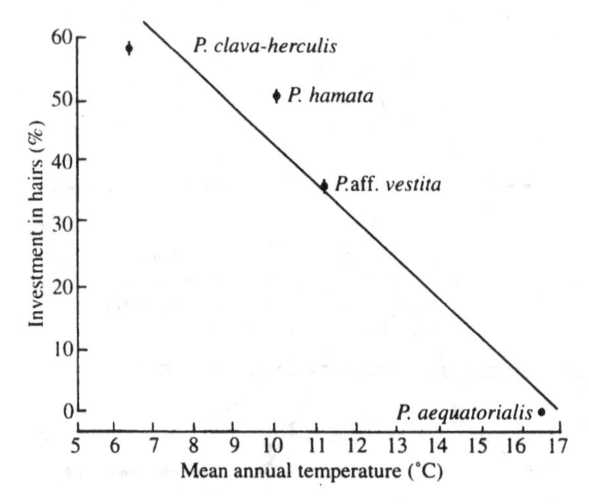

Figure 10.3. Investment in inflorescence pubescence for four *Puya* species plotted as a function of total bract dry weight accounted for by hairs. Values are means ± 1 SE. Equation for line: $y = -6.09x + 104.3$ ($r = -0.96$).

P. hamata, *P.* aff. *vestita* and *P. aequatorialis*, revealed a highly significant linear correlation ($r = -0.96$) between a species' relative investment in pubescence and mean annual temperature (Table 10.1, Figure 10.3). With a mean indument thickness of 3.8 mm, hairs accounted for 57.8% of the biomass of bracts in *P. clava-herculis*, in contrast to a 4.4 mm mean thickness and 51.4% investment in *P. hamata* and a 1.0 mm thickness and 36.6% investment of *P.* aff. *vestita*. *Puya aequatorialis* is completely glabrous, investing no energy in inflorescence pubescence.

The genus *Puya* shows a clear gradient of increased investment in pubescence presumably in response to an increasing selection pressure, in this case, decreasing mean annual air temperature. As air temperature decreases with elevation, each representative species is replaced successively by another that invests a significantly greater amount of biomass into the production of inflorescence pubescence ($p < 0.0001$, Kruskal–Wallis test).

A leaf pubescence gradient has been reported in the Venezuelan páramos for certain species of *Espeletia* (Baruch & Smith 1979). *Espeletia atropurpurea* is a nearly glabrous species restricted to mesic sites just above treeline, with leaf hairs accounting for only 4% of leaf dry weight. In contrast, *E. schultzii* is widely distributed, more drought-resistant and possesses dense whitish leaf pubescence which makes up 20.5% of leaf dry weight. Baruch & Smith (1979) considered the production of dense leaf pubescence in *E. schultzii* to be an energetically expensive character, and there is no reason to suppose otherwise for the production of inflorescence pubescence in *Puya*.

Effects of pubescence on flower temperature

To assess the functional significance of plant pubescence, experiments that manipulate pubescence density are necessary but have been performed only in a limited number of studies. Using partially shaved leaves of *Espeletia timotensis*, Meinzer & Goldstein (1985) reported that intact, pubescent leaves have higher daytime leaf temperatures than shaved leaves. Nobel (1978, 1980a, b) used a series of simulations to vary the stem apical pubescence depth in *Carnegiea gigantea*, a columnar cactus of the southwest United States. Nobel (1980a) found that maximal temperature of the apical surface decreased and the minimum temperature increased as the pubescence depth was increased from 0 to 6 mm, while further increases had little effect. He concluded that because of the moderating effect of apical pubescence and large stem diameters, the apical meristem of *C. gigantea* would not freeze as easily and thus this species ranges farther north than other columnar cactus species.

Species of *Puya* in the Ecuadorian Andes show different levels of inflorescence pubescence investment and flower temperature responses (see above). Changes in inflorescence pubescence are strongly correlated with different microclimates, and dense pubescence is suggested to be responsible for the elevated floral temperatures of páramo species. To assess the importance of dense pubescence to flower temperature, the effects of within-individual variation in inflorescence pubescence on flower temperature responses were examined in denuded and control inflorescences of *P. hamata*. All pubescence was removed around selected flowers in the manner described by Miller (1986). Flower temperature responses of denuded *P. hamata* flowers were compared with adjacent control flowers on the same inflorescence (Figure 10.4*b*). Temperature responses for control flowers of *P. hamata* in Figure 10.4*b* were identical to those observed for *P. hamata* where mean elevated temperatures of 2.4 °C were maintained above ambient air temperature for a 24-hour period. In contrast, temperatures of denuded flowers were closely coupled to fluctuating air temperature and radiation regimes and were similar to those *Puya* species with little or no pubescence. As air temperature dropped in late afternoon, denuded flowers cooled at rates $> -4.0\,^{\circ}\mathrm{C\,h^{-1}}$ versus only $-1.5\,^{\circ}\mathrm{C\,h^{-1}}$ for control *P. hamata* flowers (Figure 10.4*b*). During daylight hours, denuded *P. hamata* flower temperatures were lower than control

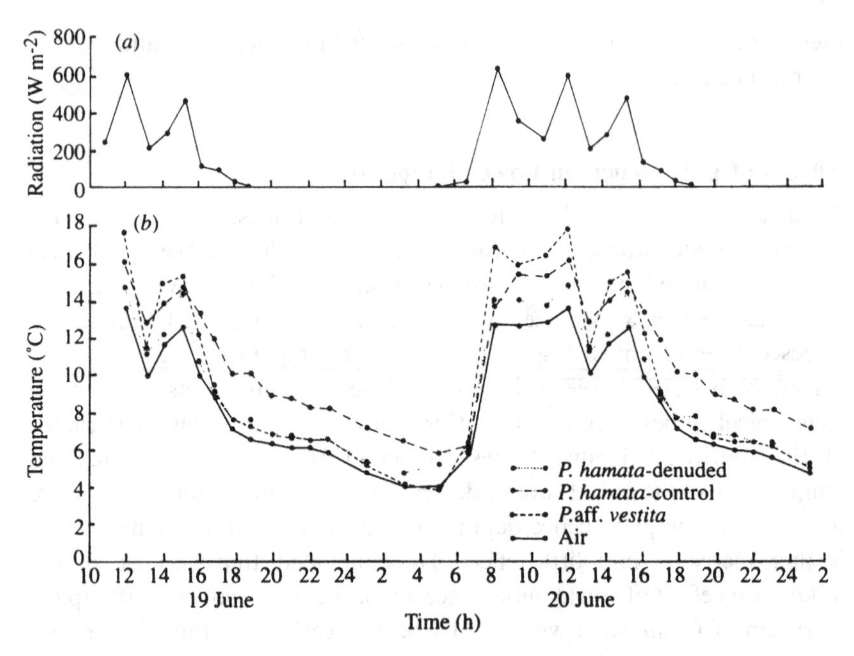

Figure 10.4. (*a*) Incoming radiation profile measured on 19–20 June 1985 in the Páramo de El Angel. (*b*) Air and mean flower temperature profiles for denuded and control *P. hamata* and control *P.* aff. *vestita*. Flower temperatures plotted represent mean values for six denuded flowers of *P. hamata* and eight flowers each of control *P. hamata* and *P.* aff. *vestita*.

flower temperatures but above ambient air temperature (Figure 10.4*b*) Denuded *P. hamata* flowers received direct solar radiation, but without pubescence and exposed to wind velocities varying from 2–5 m s^{-1} during the day, these flowers did not retain heat as well as control flowers. This difference may have been attributable to increased convective cooling rates in the absence of a protective layer of pubescence (Miller 1986). At night, denuded flowers lacking the dense insulative hairs necessary to reduce radiative heat loss rapidly equilibrated with air temperature (Figure 10.4*b*).

The dense pubescence surrounding control *P. hamata* flowers may reflect solar radiation inward, warming the flowers. In addition, daytime convective cooling rates for control flowers may have been reduced because of this protective thermal insulation. A comparison of control and denuded flower temperature responses with changing ambient air temperature in Figure 10.4*b* shows that these hairs are responsible for more than 80% of the observed temperature elevation between control *P.*

hamata flowers and air temperature at night. The remaining 20% may be attributed to the physical characteristics of the large inflorescence itself, such as high thermal heat capacity and increased boundary layer thickness (Miller 1986).

Inflorescence pubescence in *Puya* effectively reduces radiative heat loss, but does not form a 'perfect' insulator, which would maintain more uniform temperatures. This is apparent for *P. hamata* control flowers, where during the day incoming solar radiation does penetrate the pubescence layer and the flowers heat up quickly to levels well above air temperature. Hairs form a protective barrier or insulation against rapid rates of thermal radiation loss. This is shown by radiation intensity fluctuating during the day (Figure 10.4*a*), where control flowers maintained elevated internal temperatures, while denuded flowers showed greater fluctuations (Figure 10.4*b*). Flower temperature responses of *P. hamata*, presented in Figure 10.2*b*, also demonstrate this relationship, which appears analogous to the 'heat trap' effect of pubescent arctic *Salix* catkins where incoming solar radiation is reflected inward, warming the flowers, and heat loss is reduced because of the insulative effects of the hairs (Krog 1955).

Increasing the thermal mass of an inflorescence may be another possible adaptive response to decreasing ambient air temperatures. Preliminary studies suggest that, at least for four species of *Puya* (Table 10.1; Figure 10.1), inflorescence mass does increase with elevation (G. A. Miller, unpublished data). For example, *P.* aff. *vestita* produces an inflorescence that weighs 3.0–3.5 kg, in contrast to inflorescences of *P. hamata* weighing 30 kg or more. The large heat capacity and associated boundary layer of such a massive inflorescence would be expected to have a great influence on the thermal balance of the flowers (Gates 1980). Yet the removal of pubescence greatly altered flower temperatures compared with neighboring control flowers. Results such as these are significant and demonstrate that pubescence, and to a lesser degree the thermal properties of the large inflorescence, maintained elevated flower temperatures in *P. hamata*.

Time constants

I used time constants as a standardized method of comparing flower temperature responses for each species of *Puya* (Table 10.1) from different experiments. The time constant, τ, provides a measure of how closely an organism's temperature tracks air temperature in a changing environment and depends on the size and shape of the organism, its heat capacity and overall conductance or insulation (Gates 1980; Jones 1992). Sensitive tissues may be protected from daily temperature extremes by possessing large thermal time constants. This may result either from a high tissue thermal capacity or from a high degree of insulation. Larger time constants can lead to a significant reduction in the amplitude of daily tissue temperature fluctuations, and can decouple an organism from extreme oscillations in air temperature (Jones 1992). Flower time constants were calculated for each *Puya* species using the procedure described by Miller (1986) and calculations were based on the following assumptions: (i) wind velocity was <1 m s^{-1}, so convection was negligible; (ii) because calculations were made using the night-time portion of the flower temperature–response curves, radiation input was negligible or zero; (iii) there was no latent heat energy exchange.

Time constants are presented in Figure 10.5 for each of the four *Puya* species. Temperature responses for flowers of *P. hamata* and *P.* aff. *vestita* were recorded in two experiments (Figures 10.2, 10.4) and time constants were calculated in each case. For each species the respective time constant calculation yielded identical results from each experiment. This outcome strongly supports the belief that under a given set of environmental conditions, time constants are fixed for an organism. As is clearly illustrated in Figure 10.5, the densely pubescent páramo species of *Puya*

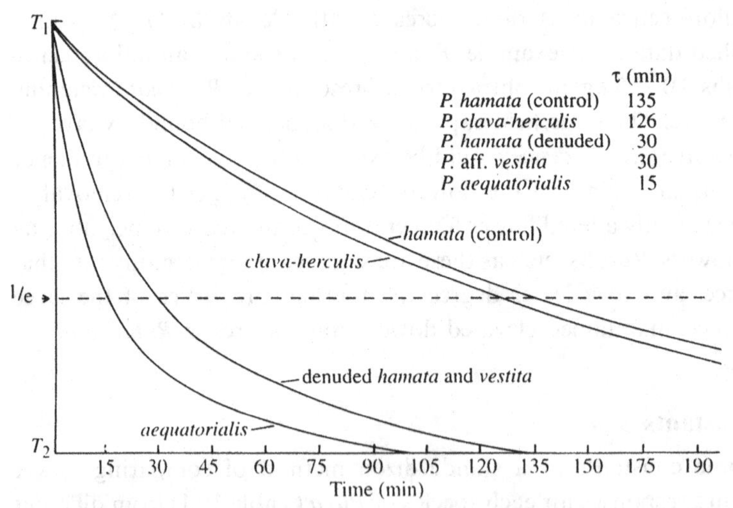

Figure 10.5. Exponential heat decay curves and time contants (τ) for flowers of four *Puya* species. Time constants were determined where the heat decay curve crossed $1/e = T_1 - 0.632\ (T_1 - T_2)$. This is the time it takes for a 63% decrease in the total difference between organism (T_1) and air temperature (T_2). Time constants are independent of the magnitude of this temperature difference (Jones 1992).

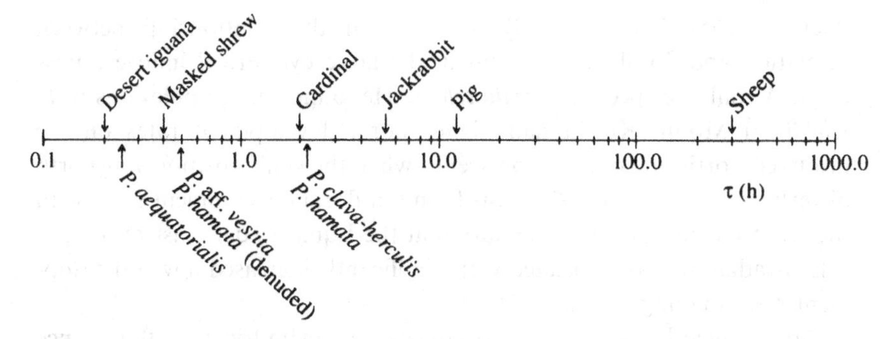

Figure 10.6. Plot of flower time constants for four *Puya* species and six animal species. Animal time constant values are taken from Gates (1980).

have time constants four to nine times larger than the less pubescent, non-páramo species. For *P. hamata,* control flowers have a time constant 4.5 times larger than denuded flowers on the same inflorescence. Thus, for a given temperature difference between flower and air temperature, it takes flowers with dense pubescence 4–9 times as long to reach thermal equilibrium with air temperature as it does flowers surrounded by little or no pubescence (i.e. denuded *P. hamata*, *P.* aff. *vestita* and *P. aequatorialis* flowers).

Time constants for six animal species are plotted for comparison with the four species of *Puya* in Figure 10.6. Although time constants generally increase with increasing mass of an organism, they also depend on their effective insulative capacity (Gates 1980). The desert iguana and cardinal are similar in size but have very different time constants, which Gates (1980) attributed to the desert iguana's lack of thermal insulation. A similar relationship can be seen along the gradient from glabrous to pubescent *Puya* species. Here, the flower time constants observed for *Puya* species with dense pubescence are nine times longer than for glabrous species, and are of the same order of magnitude (>2 h) as those for a small, well-insulated bird like the cardinal.

Flower temperature and reproductive success

Because air temperatures are low and irradiation intense in tropical alpine environments, direct insolation has an important influence on microclimates. Under a clear daytime sky, intense insolation causes high surface heating of exposed ground and vegetation, with increasing daytime temperatures expected to be favorable to plant growth and development

(Geiger 1965). Young (1984) reported on the relationship between insolation and floral development in the large cylindrical inflorescences of the Afroalpine species, *Lobelia telekii*. He found that populations of *L. telekii* on Mount Kenya had higher floral development rates on the insolated north side of an inflorescence when the sun's position was north of vertical. In addition, a shift was found in floral development rates with the seasonal shift in solar orientation at the Equator, demonstrating that solar irradiation was associated with significantly increased flower development rates (Young 1984).

Puya clava-herculis also produces a cylindrical inflorescence with densely packed flowers. In populations from Cotopaxi National Park, Ecuador (Table 10.1), inflorescences begin to bolt in February and March, flowers emerge from April to July, and seed capsules develop from July to September (G. A. Miller, unpublished data). Maturation of flowers and fruits coincides with the sun's position being north of vertical from the end of March until mid-September. The effects of natural variation in flower temperatures on fecundity were examined directly in *P. clava-herculis* using the unequal insolation on different sides of an inflorescence as a natural experiment. This differential insolation effect was due to a shadow cast on the south side of an inflorescence from April to September. The purpose was to determine the effects that unequal insolation had on north (insolated) and south (shaded) side flower temperatures, and to investigate possible correlations in fecundity.

Flower temperature responses of insolated and shaded flowers of *P. clava-herculis* were determined using the methods described by Miller (1986) and are presented in Figure 10.7. During 7.5 daylight hours (0900–1630 h), northern insolated flowers showed a mean temperature elevation of 3.4 °C $\pm$ 1.0 above the shaded southern flowers (Figure 10.7*b*). Similar results were found in *P. hamata* where north-side flowers maintained mean elevated temperatures of 2.6 °C $\pm$ 1.1 above south-side flowers for 11 daylight hours (Miller 1986). For the remainder of each 24-hour period, north and south flower temperatures were equal in both species (Figure 10.7*b*). Thus for 31% to 46% of a 24-hour day, the insolated northern exposure flowers of *P. clava-herculis* and *P. hamata* showed significantly higher temperatures than the shaded southern exposure flowers. It is apparent from these results that daytime flower temperatures were primarily influenced by irradiation levels, with increased solar exposure responsible for elevated north side flower temperatures. This variation reflects an important shift in factors influencing the thermal

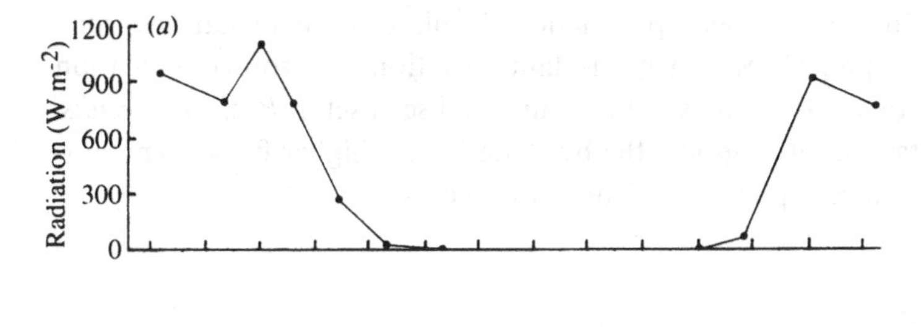

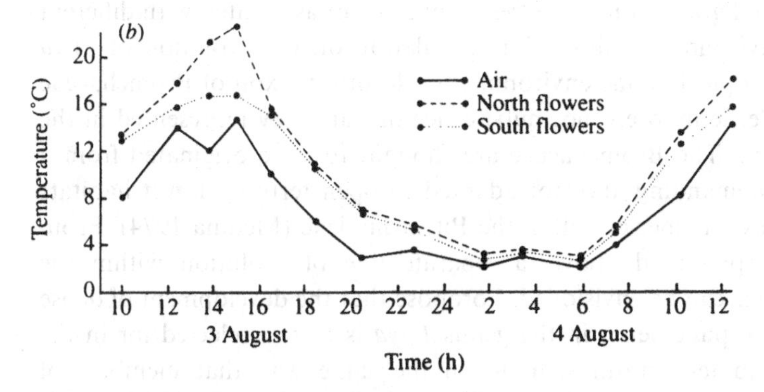

Figure 10.7. (*a*) Incoming radiation profile measured on 3–4 August 1983 in Parque Nacional Cotopaxi. (*b*) Air and mean flower temperature profiles of northern and southern exposure flowers on inflorescences of *P. clava-herculis*. Plotted temperatures are mean values from five flowers.

balance of flowers from unequal daytime irradiation, to the importance of inflorescence pubescence and mass at night.

To determine the effects of increased insolation and elevated flower temperature on reproductive success, a comparison of seed production between north and south side flowers on 11 *P. clava-herculis* inflorescences was performed. All flowering rosettes were located in Cotopaxi National Park and were sampled using the procedure described by Miller (1986). The warmer northern exposure flowers produced a significantly greater mean total seed number per inflorescence half than the south side (north = 27 378 seeds vs south = 23 218 seeds; $p = 0.04$, Wilcoxon signed-rank test). In addition, north side flowers also produced a significantly greater mean seed number per capsule than the cooler south side flowers (north = 301.9 seeds per capsule vs south = 225.1 seeds per capsule; $p = 0.03$, Wilcoxon signed-rank test).

It is apparent that there was a close association between the higher temperatures of northern exposure flowers of *P. clava-herculis* and significant increases in seed production. While other explanations, such as pollinator preference or within-plant variation in resource allocation to flowers, are possible to explain the unequal seed set in *P. clava-herculis*, these results strongly support the hypothesis that higher flower temperature can enhance a plant's reproductive success.

Conclusion

The observed gradient in pubescence investment associated with different thermal environments allows for speculation on the evolution of *Puya* species in tropical alpine environments. No other taxon of Bromeliaceae produces dense inflorescence pubescence nor are they represented in the cold páramo. The Bromeliaceae are thought to have originated from a high light demanding ancestor adapted to open terrestrial, wet habitats similar to extant species within the Pitcairniodeae (Medina 1974). From the results presented above, a separate line of evolution within the Bromeliaceae can be envisioned. I propose that the development of dense inflorescence pubescence in the genus *Puya* is a trait selected for in the cold páramo temperatures, in much the same way that members of Tillandsiodeae and Bromeliodeae developed foliar absorption and Crassulacean Acid Metabolism as specializations that allow them to tolerate dry, aerial environments. This specialization in pubescence production, perhaps associated with physiological modifications, may enable *Puya* species to live in the páramo without suffering freezing injury to flowers and reduced fecundity.

If minimum flower temperature represents a physical limit to the upward extension of *Puya* species, the production of dense pubescence by *P. hamata* and *P. clava-herculis* appears to be a functional response to the lower páramo temperatures. Increased investment in pubescence by páramo species of *Puya* maintains elevated flower temperatures and provides the flowers and developing seeds with a more favorable thermal environment. Dense pubescence has been demonstrated to have an important influence on the thermal balance of the flowers and appears to be an important factor contributing to their reproductive success. These results lend strong support to the hypothesis that dense inflorescence pubescence is an important adaptation for plants in tropical alpine environments. Further studies of the functional significance of inflorescence pubescence in other species of *Puya* are necessary and forthcoming. Additional work will provide both a better understanding of how plants have adapted to the tropical alpine environment and a foundation for an analysis of the independent evolution of inflorescence pubescence in other tropical alpine genera.

References

Acosta-Solis, M. (1971). *Los Bosques del Ecuador y sus Productos*. Quito: Editorial Ecuador.

Acosta-Solis, M. (1979). *Divisiones Fitogeográficas y Formaciones Geobotánicas del Ecuador*. Quito: Publicaciones Científicas M.S.A.

Augspurger, C. K. (1985). Demography and life history variation of *Puya dasyliriodes*, a long-lived rosette in tropical subalpine bogs. *Oikos* **45**, 341–52.

Baker, H. G. (1975). Sugar concentrations in nectars from hummingbird flowers. *Biotropica* **7**, 37–41.

Baruch, Z. & Smith, A. P. (1979). Morphological and physiological correlates of niche breadth in two species of *Espeletia* (Compositae) in the Venezuelan Andes. *Oecologia* **38**, 71–82.

Bradshaw, A. D. (1971). Plant evolution in extreme environments. In *Ecological Genetics and Evolution*, ed. R. Creed, pp. 20–50. Oxford: Blackwell Scientific Publications.

Cañadas Cruz, L. (1983). *El Mapa Bioclimático y Ecológico del Ecuador*. MAG-PRONAREG, Banco Central del Ecuador.

Carlquist, S. (1974). *Island Biology*. New York: Columbia University Press.

Cleef, A. M. (1981). The vegetation of the páramos of the Colombian Cordillera Oriental. PhD dissertation, University of Utrecht.

Cuatrecasas, J. (1954). Distribution of the genus *Espeletia*. *Proceedings of the Eighth International Congress (Paris), Section 4*, 121–32.

Cuatrecasas, J. (1968). Páramo vegetation and its life forms. *Colloquium Geographicum* **9**, 163–86.

Ehleringer, J. (1984). Ecology and ecophysiology of leaf pubescence in desert plants. In *Biology and Chemistry of Plant Trichomes*, ed. E. Rodriguez, P. L. Healey and I. Mehta, pp. 113–32. New York: Plenum Press.

Ehleringer, J., Bjorkman, O. & Mooney, H. A. (1976). Leaf pubescence: effects on absorptance and photosynthesis in a desert shrub. *Science* **192**, 376–7.

Ehleringer, J. & Mooney, H. A. (1978). Leaf hairs: Effects on physiological activity and adaptive values to a desert shrub. *Oecologia* **37**, 183–200.

Ehleringer, J., Mooney, H. A., Gulmon, S. L. & Rundel, P. W. (1981). Parallel evolution of leaf pubescence in *Encelia* in coastal deserts of North and South America. *Oecologia* **49**, 38–41.

Gates, D. M. (1980). *Biophysical Ecology*. New York: Springer-Verlag.

Geiger, R. (1965). *The Climate Near the Ground*, 4th edition. Cambridge, Mass: Harvard University Press.

Gilmartin, A. J. (1972). The Bromeliaceae of Ecuador. *Phanerogamarum Monographiae* IV. Lehre, Germany: J. Cramer.

Gilmartin, A. J. (1973). Transandean distributions of Bromeliaceae in Ecuador. *Ecology* **54**, 1389–93.

Hedberg, O. (1964). Features of Afroalpine plant ecology. *Acta Phytogeographica Suecica* **49**, 1–144.

Heilborn, O. (1925). Contributions to the ecology of the Ecuadorian páramos with special reference to cushion plants and osmotic pressure. *Svensk Botanisk Tidskrift* **19**, 153–70.

Jones, H. G. (1992). *Plants and Microclimates*, 2nd edition. Cambridge: Cambridge University Press.
Krog, J. (1955). Notes on temperature measurements indicative of special organization in Arctic and sub-Arctic plants for utilization of radiated heat from the sun. *Physiologica Plantarum* **8**, 836–9.
Levin, D. A. (1973). The role of trichomes in plant defense. *Quarterly Review of Biology* **48**, 3–15.
Mabberley, D. J. (1973). Evolution of the giant groundsels. *Kew Bulletin* **28**, 61–8.
Mani, M. S. (1968). *Ecology and Biogeography of High Altitude Insects*. The Hague: Junk.
Medina, E. (1974). Dark CO_2 fixation, habitat preference and evolution within the Bromeliaceae. *Evolution* **28**, 677–86.
Meinzer, F. & Goldstein, G. (1985). Some consequences of leaf pubescence in the Andean giant rosette plant *Espeletia timotensis*. *Ecology* **66**, 512–20.
Miller, G. A. (1986). Pubescence, floral temperature and fecundity in species of *Puya* (Bromeliaceae) in the Ecuadorian Andes. *Oecologia* **70**, 155–60.
Miller, G. A. (1987). The population biology and physiology ecology of species of *Puya* (Bromeliaceae) in the Ecuadorian Andes. PhD dissertation, the University of Connecticut, Storrs, USA.
Miller, G. A. & Silander, J. A., Jr (1991). Control of the distribution of giant rosette species of *Puya* (Bromeliaceae) in the Ecuadorian páramos. *Biotropica* **23**, 124–33.
Mills, K. (1975). Flora de la sierra. Un estudio en el Parque Nacional de Cotopaxi. *Ciencia y Naturaleza XVI*, No. 1. Quito, Ecuador.
Monasterio, M. (1980). *Estudios Ecológicos en los Páramos Andinos*. Mérida, Venezuela: Ediciones de la Universidad de los Andes.
Nobel, P. S. (1978). Surface temperatures of cacti – influences of environmental and morphological factors. *Ecology* **59**, 986–96.
Nobel, P. S. (1980a). Morphology, surface temperatures, and northern limits of columnar cacti in the Sonoran Desert. *Ecology* **61**, 1–7.
Nobel, P. S. (1980b). Influences of minimum stem temperatures on ranges of cacti in southwestern United States and Central Chile. *Oecologia* **47**, 10–15.
Ortiz-Crespo, F. I. (1973). Field studies of pollination of plants of the genus *Puya*. *Journal of Bromelian Society* **23**, 3–7.
Padilla, V. (1973). *Bromeliads*. New York: Crown Publ.
Penland, C. W. T. (1941). The alpine vegetation of the southern Rockies and the Ecuadorian Andes. *Colorado College Publications* No. 32.
Smith, A. P. (1974). Bud temperature in relation to nyctinastic leaf movement in an Andean giant rosette plant. *Biotropica* **6**, 263–6.
Smith, A. P. (1979). The function of dead leaves in *Espeletia schultzii* (Compositae), an Andean giant rosette plant. *Biotropica* **11**, 43–7.
Smith, A. P. (1980). The paradox of plant height in an Andean giant rosette species. *Journal of Ecology* **68**, 63–73.
Smith, A. P. (1981). Growth and population dynamics of *Espeletia* (Compositae) in the Venezuelan Andes. *Smithsonian Contributions to Botany* **48**.
Smith, A. P. & Young, T. P. (1982). The cost of reproduction in *Senecio keniodendron*, a giant rosette species of Mt. Kenya. *Oecologia* **55**, 243–7.
Smith, L. B. (1957). The Bromeliaceae of Colombia. *Contributions to the U.S. National Herbarium* **33**, 1–311.
Stiles, F. G. (1978). Ecological and evolutionary implications of bird pollination. *American Zoologist* **18**, 715–27.

Vareschi, V. (1970). *Flora de los Páramos de Venezuela.* Mérida, Venezuela: Universidad de los Andes.
Young, T. P. (1982). Bird visitation, seed set and germination rates in two *Lobelia* species on Mount Kenya. *Ecology* **63**, 1983–6.
Young, T. P. (1984). Solar irradiation increases floral development rates in Afroalpine *Lobelia telekii. Biotropica* **16**, 243–5.
Young, T. P. (1985). *Lobelia telekii* herbivory, mortality and size at reproduction: variation with growth rate. *Ecology* **66**, 1879–83.
Young, T. P. & Van Orden Robe, S. (1986). Microenvironmental role of the secreted aqueous solution in the Afroalpine plant *Lobelia keniensis. Biotropica* **18**, 267–9.

11

Turnover and conservation of nutrients in the pachycaul *Senecio keniodendron*

E. BECK

Pachycaul senecios from the upper Afroalpine zone produce stems up to 11 m tall which are coated with marcescent leaves. Each stem or branch is terminated by a huge leaf rosette composed of 30–120 leaves which surround the central cone-shaped leaf-bud. During the course of a year about 50–60 leaves are produced from the bud (Beck *et al.* 1980) and in principle the same number of mature leaves at the outer periphery of the rosette become senescent and die. The leaves are up to 50 cm long and 15–20 cm wide. Senescence commences at the tip which becomes yellow and subsequently necrotic. Yellowing of the lamina spreads from the tip downwards to the base. A senescing leaf, the upper part of which was already partially necrotic, was found to be still capable of photosynthesis in the lower part (E. Beck, unpublished data). This mode of gradual senescence suggests that a substantial portion of the nutrients can be mobilized and exported from the leaf before it dies. Table 11.1 shows the disappearance of nitrogen in the senescing leaf. The plant was located at 4200 m elevation in the Teleki Valley, Mount Kenya. The prolonged viability of the midribs presumably facilitates this mobilization. This type of internal recycling of nitrogen and other nutrients is common in higher plants. However, unlike the majority of the higher plants the pachycaul groundsels do not shed the dead leaves but maintain them as a dense collar around the stem. The insulating effect of this fringe appears to be required for the prevention of freezing of xylem and pith water (Hedberg & Hedberg 1979; Goldstein & Meinzer 1983). Plants of *S. keniodendron* from which dead leaves were removed had lower rates of new leaf production than intact plants (A. Smith and S. Mulkey, unpublished data).

This mantle of dead leaves may also function as a substratum for stem-borne adventitious roots and so may provide water and nutrients

Table 11.1 *Mobilization of nitrogen (N_{total} as percentage of dry weight) in relation to senescence of leaves of* Senecio keniodendron

N_{total} was analysed as described by Rosnitschek-Schimmel (1982).

	Lamina			Midrib	
	Dead tip	Upper part	Lower part	Upper part	Lower part
Green leaf	–	0.93	0.83	0.32	0.29
Senescent leaf	0.41	0.40 (yellow)	0.50 (green)	0.26	0.28

to the plant. From a specimen of *S. keniodendron*, about 2 m high, senescing leaves were removed for nutrient analysis following a vertical line down the stem. Leaf No. 1, which was at 110 cm above the ground, showed a necrotic tip while leaf No. 3 was already completely brown (Figure 11.1). Starting from the leaf base, decomposition subsequently advances towards the tip (cf. leaves Nos. 4–13 in Figure 11.1), as indicated by blackening and a slimy consistency. The leaves denoted as Nos. 14 (removed at 80 cm height, Table 11.3) and 15, respectively, were hardly discernible as individuals, consisting predominantly of vascular bundles embedded in a blackish material. The decrease of the nitrogen and phosphate content of the marcescent leaves in the zone between Nos. 1 and 5 (Table 11.2) probably reflects nutrient mobilization and export during senescence. Concomitantly, the nutrient content of the wood and pith (samples of which were collected with a stem borer) was highest in that section (cf. Table 11.3, the samples withdrawn at 100 and 110 cm, respectively). Nitrogen appeared to be exported predominantly in the form of amino acids, especially as glutamine and asparagine (Table 11.4). Only one third of the nitrogen found in the pith of that zone could not be identified and may reflect protein-bound nitrogen (Table 11.2).

In the stem section below 100 cm height, where the leaves gradually lose their original shape and size, mycelia of molds and some insect larvae were observed and the nitrogen content of the material, as related to dry weight, again rose considerably, while the concentration of phosphate increased but slightly (Tables 11.2, 11.3). This increase of nutrient concentration was due to a considerable decrease of the dry matter per leaf (about 50%) and therefore does not reflect an absolute nitrogen or phosphate accumulation. Clearly, loss of dry matter was predominantly

Figure 11.1. Senescent (Nos. 1, 2, 3) and rotting dead leaves collected from below the outermost green rosette leaf along a vertical line down the stem of a 3 m tall *Senecio keniodendron* tree. The sequence of leaf age follows the numbering, No. 12 (right) being the oldest of the leaves. Note senescence, as indicated by dried up leaf tips, proceeding from top to base whereas rotting (blackening of the material) advances in the opposite way (Nos. 6–12). Photo: E. Beck, Feb. 1986 (Mt Kenya).

caused by microbial and fungal activity. In this area of decomposition, roots were found to emerge from the stem (Figure 11.2) which in their basal regions were as thick as 1 cm in diameter. By frequent branching these roots form a network which extends through the whole layer of decaying leaves and leaf bases. The latter, in particular, by preservation of moisture produce a favorable root-bed. At a distance of about 40 cm beneath the terminal leaf rosette, only the dry and shrivelled uppermost part of the lamina and the woody leaf base can be distinguished as structures preserved from the former leaf, while the bulk of the material forms a mold-like substance which is interlaced by the persistent vascular bundles. The majority of the living adventitious roots was found in that area. As is indicated by their nitrogen and phosphate content (Table 11.2), the stem-borne roots are capable of regaining further nutrients from the soil-like material produced upon decomposition of the dead leaves.

In conclusion, it appears that in *S. keniodendron* mineralization of the

Table 11.2 *Nutrient content of decaying leaves of* Senecio keniodendron

Leaf No. 1: green with yellowish-brown tip; No. 13: black with brown tip; Nos 14, 15: black. NO_3^--N was analysed by HPLC according to Thayer & Huffacker (1980); PO_4^{3-} was analysed by the ammonium molybdate method of Murphy & Riley (1962).

Leaf No.	N_{total} as % of dry weight	NO_3^--N as % of N_{total}	PO_4^{3-} as % of dry weight
1	0.53	7.17	0.49
2	0.39	19.49	0.51
3	0.2	13.63	0.33
4	0.27	25.40	0.29
5	0.27	16.21	0.27
6	0.27	12.33	0.21
7	0.27	8.08	0.22
8	0.32	4.47	0.19
9	0.32	2.89	0.21
10	0.33	3.04	0.23
11	0.32	3.26	0.20
12	0.33	2.92	0.16
13	0.43	4.14	0.23
14	1.49	0.24	0.09
15	0.80		0.16
Stem-borne root	1.14	0.70	0.50

Table 11.4 *Composition of the fraction of free amino acids* (μmol g^{-1} *dry wt) in the pith of* Senecio keniodendron *at various stem heights*

	Stem height			
	60 cm	80 cm	100 cm	110 cm
Amino acid				
Aspartic acid	1.1	1.9	4.9	15.4
Asparagine	0.4	1.5	5.0	31.4
Glutamic acid	1.8	3.3	10.7	27.9
Glutamine	2.2	11.8	19.8	395.8
Alanine	0.8	0.6	1.1	5.6
Valine	0.1	0.2	0.7	10.3
γ-Aminobutyric acid	0.7	0.6	0.7	17.7
Lysine	0.2	0.6	4.3	17.2

Table 11.3 Senecio keniodendron: *nutrient content of the frill of the decaying leaves and the wood and the pith of the stem at the same spots*

Leaf No. 14 described by Table 11.2 is situated at 80 cm stem height. Amino acids were analysed with an automatic amino acid analyser as described by Rosnitschek-Schimmel (1982).

Position (height) of sampling (cm)	Leaf bases			Wood			Pith					
	N_{total} as % of d.w.	NO_3^--N as % of N_{total}	PO_4^{3-} as % of d.w.	N_{total} as % of d.w.	NO_3^--N as % of N_{total}	PO_4^{3-} as % of d.w.	N_{total} as % of d.w.	NO_3^--N as % of N_{total}	NH_4-N as % of N_{total}	Amino-acid-bound N as % of N_{total}	$N_{unidentif.}$ as % of d.w.	PO_4^{3-} as % of d.w.
110	1.08	0.3	0.3	2.29	0.30	n.d.	2.77	0.4	10.2	55.8	0.93	1.11
100	0.18	n.d.	0.3	0.81	0.62	n.d.	0.51	0.8	2.8	27.8	0.35	0.10
80	1.38	0.65	0.1	0.28	1.0	0.2	0.43	1.1	2.1	11.5	0.37	0.15
60	n.d.	n.d.	n.d.	0.27	1.4	0.5	0.31	1.0	2.2	6.6	0.28	0.20

d.w. dry weight, n.d., not determined.

Figure 11.2. Section of a stem of *S. keniodendron* from which the frill of dead leaves (blackish material) has been partly removed. Several adventitious roots, protruding from the stem are visible, which have been dug out from the rotten leaf material providing the root bed. Photo: as Fig. 11.1.

dead plant material, which usually takes place in the litter and humus layers of the soil occurs while the leaves are still adhering to the living stem. The lack of persistent moisture and of chemical reactions with the soil's minerals might be responsible for the incomplete decomposition of the dead leaves. However, a considerable portion of the mineralized nutrients appear to be reabsorbed by the plant via the adventitious roots produced from the stem. Thus a nutrient cycle without involvement of the soil has been developed by the pachycaul groundsels, allowing for substantial growth of the plant while retaining part of the dead plant material as an insulating mantle layer for protection against frost.

References

Beck, E., Scheibe, R., Senser, M. & Mueller, W. (1980). Estimation of leaf and stem growth of unbranched *Senecio keniodendron* trees. *Flora* **170**, 68–76.

Goldstein, G. & Meinzer, F. (1983). Influence of insulating dead leaves and low temperature on water balance in an Andean giant rosette plant. *Plant, Cell and Environment* **6**, 649–56.

Hedberg, I. & Hedberg, O. (1979). Tropical alpine life-forms of vascular plants. *Oikos* **33**, 297–307.

Rosnitscheck-Schimmel, I. (1982). Effect of ammonium and nitrate supply on dry matter production and nitrogen distribution in *Urtica dioica. Zeitschrift für Pflanzenphysiologie* **108**, 329–41.

Thayer, J. R. & Huffaker, R. C. (1980). Determination of nitrate and nitrite by high pressure liquid chromatography: comparison with other methods for nitrate determination. *Analytical Biochemistry* **102**, 110–19.

Murphy, J. & Riley, J. P. (1962). A modified single solution method for the determination of phosphate in natural waters. *Analytica Chimica Acta* **27**, 31–36.

12

Soil nutrient dynamics in East African alpine ecosystems

H. REHDER

Introduction

The nutrient relations of tropical alpine areas are poorly known despite the importance of these processes for understanding spatial and temporal variation in ecosystem productivity (Speck 1983). Nitrogen mineralization – the release of inorganic nitrogen from the substrate – is of particular importance in this respect, and has proved useful in assessing ecosystem dynamics in the European Alps (Rehder 1970, 1976a, b, 1982; Rehder & Schäfer 1978). In this chapter I address the problem of nitrogen mineralization and its impact on primary productivity in the alpine zone of Mount Kenya.

Materials and methods

Samples areas were established in the two main alpine plant communities of the Teleki Valley, on the western flank of Mount Kenya: '*Senecio keniodendron–Lobelia telekii* community' (Area 1) on steeper slopes and '*Lobelia keniensis–Senecio brassica* community' (Area 2) on moderate slopes and valley bottoms (see Table 12.1; also Hedberg 1964; Beck *et al.* 1981; Rehder 1975, 1983; Rehder *et al.* 1981, 1988). The soil at area 1 was a loamy 'mountain brown gley soil' while the soil at area 2 was a blackish peaty 'mountain wet gley soil' (Beck *et al.* 1981). Mean maximum temperatures were lower at the soil surface in area 2 (Table 12.2).

Nitrogen mineralization was measured with a field incubation test (Ellenberg 1964, 1977; Gerlach 1973). Volumetric soil samples were taken from three depths between 0 and 15 cm, weighed, sieved (4 mm screen), mixed and homogenized by hand. One part of each sample was transported as soon as possible in a cooler to the laboratory in Nairobi for determination of inorganic nitrogen (Nm). Samples were extracted with

Table 12.1 *Characteristics of the two study areas (see Table 12.6 for estimated phytomass of dominant species)*

	Area 1	Area 2
Elevation (m)	4160	4130
Slope (degrees)	25	7
Exposure	North	North
Area (m^2)	100	50
Cover by *Senecio keniodendron*	10%	0
Cover by ground vegetation	90%	90%

Table 12.2 *Daily maximum, minimum and mean temperatures (°C) averaged for 3 March–7 April 1979. Data were taken at 1 cm and 24 cm*

	1 cm	24 cm
Area 1		
Maximum	12.5	4.4
Minimum	−1.7	1.9
Mean	5.4	3.2
Area 2		
Maximum	8.0	3.7
Minimum	−1.8	1.9
Mean	3.1	2.9

1% $KAl(SO_4)_2$ solution, using a distillation apparatus (Bremner and Keeney 1965), in which the NH_4^+-N and the NO_3^--N are successively titrated. Water content of the fresh samples was determined after drying part of it at 105 °C. The inorganic nitrogen concentrations were converted from mg 100 g dry weight to g m^{-2} applying the mean values from all soil sample volume dry weights.

Another portion of each homogenized sample was put into a polyethylene bag, put inside a second bag, and buried on the site. It was removed after a period of 5–7 weeks of incubation for a second Nm determination as described above. This provided an estimate of accumulation rate of nitrogen from mineralization.

Table 12.3 *Dates for incubation experiment*

Period	Replicate	Date of initiation	Duration (days)
I	1	2 Mar 1979	36
	2	14 Mar 1979	36
	3	26 Mar 1979	36
II	1	10 Jun 1979	44
	2	28 Jun 1979	44
	3	12 Jul 1979	44
III	1	2 Sep 1979	42
	2	16 Sep 1979	42
	3	1 Oct 1979	51
IV	1	18 Jan 1980	44
	2	3 Feb 1980	44
	3	21 Feb 1980	44

Table 12.4 *Average content of carbon and nitrogen (% of dry weight), nitrogen as percentage of carbon content; dry weight per volume (d.w.) and pH at* 0–15 *cm depth*

	Area 1	Area 2
%C	6.19	13.52
%N	0.42	0.98
N as % of C	6.85	7.44
d.w. per volume	73.4	44.3
pH	5.2	6.1

These measurements were repeated three times during each of four periods spanning rainy and dry seasons of 1979–80 (Table 12.3).

Results and discussion

Soils at area 2 had higher C, N and pH than did soils at area 1 (Table 12.4). Volumetric dry weight was lower on area 2 than area 1. The results of the incubation study (Table 12.5) appear to be correlated with seasonal variation in soil moisture. Period I fell at the beginning of the rainy season

Table 12.5 *Results of the incubation study*

A. Mean initial values of NO_3^--N. B. Mean initial values of Nm. C. Mean accumulation of NO_3^--N after field incubation. D. Mean accumulation of Nm after field incubation. E. Mean water content for the incubated samples.

	I	II	III	IV	Mean
A. Initial NO_3^--N ($g\ m^{-2}$)					
Area 1	0.03	0.34	0.12	0.21	0.17
Area 2	0.02	0.09	0.00	0.24	0.09
B. Initial Nm ($g\ m^{-2}$)					
Area 1	0.37	0.90	0.40	0.70	0.59
Area 2	0.60	0.95	0.19	0.82	0.65
C. Accumulation of NO_3^--N ($g\ m^{-2}$)					
Area 1	0.47	0.42	0.11	0.16	0.29
Area 2	0.08	0.17	0.08	0.16	0.12
D. Accumulation of Nm					
Area 1	0.91	0.65	0.22	0.29	0.52
Area 2	0.76	1.67	0.26	0.32	0.75
E. Soil water content (% of fresh wt)					
Area 1	40	38	28	29	34
Area 2	60	62	53	50	56

and period II at the end of the rainy season; period III was at the end of the dry season and period IV was at the beginning. The most conspicuous decrease in Nm accumulation occurred during the dry season. The initial Nm values are greatest in periods II and IV, probably because of higher accumulation in the preceding wet season. The importance of nitrification in the mineralization process generally is greater in the better-drained soils of area 1 than in the wet soils of area 2.

The mean values for Nm accumulation (Table 12.5) were used to estimate total annual yield of Nm. For area 1 this was 4.5 g Nm m^{-2}; for area 2 it was 6.5 g Nm m^{-2}. These annual estimates were then used to make rough estimates of net primary productivity at the two sites. Preliminary estimates were made of the relative contribution of dominant species to above- and belowground biomass at each site (Table 12.6). Preliminary data on total N content for each species were then used to estimate the weight of nitrogen incorporated into above- and belowground biomass per year, assuming that the nitrogen cycle is in equilibrium. These data in turn were used to estimate total biomass production. Total annual estimated primary productivity for area 1 was 682 g m^{-2}, with 289 g m^{-2}

Table 12.6 *Estimation of annual phytomass production (g m^{-2}) for the two study areas, based on estimates of annual nitrogen mineralization, standing biomass and plant tissue nitrogen content. A, above ground; B, below ground*

	g d.w. as % of total d.w.		% N content for tissues		Total g N		Annual production	
Dominant species	A	B	A	B	A	B	A	B
Area 1								
Senecio keniodendron	8.5	12.7	0.8	0.4	0.068	0.057	57.9	86.7
Festuca pilgeri	21.2	42.4	0.6	0.6	0.127	0.254	144.5	289.3
Alchemilla argyrophylla	10.6	2.1	1.0	0.7	0.106	0.015	72.3	14.3
Alchemilla johnstonii	2.1	0.4	1.3	0.9	0.027	0.004	14.3	2.7
Total	42.4	57.6			0.658		289	393
Area 2								
Senecio brassica	9.6	13.4	0.8	0.5	0.072	0.067	93.2	129.9
Lobelia keniensis	9.6	13.4	0.8	0.5	0.077	0.067	93.2	129.9
Festuca pilgeri	9.6	19.2	0.6	0.6	0.058	0.115	93.2	186.1
Alchemilla johnstonii	1.9	0.4	1.3	0.9	0.025	0.004	18.5	3.9
Ranunculus oreophytus	7.6	15.3	1.4	0.5	0.106	0.077	73.8	148.3
Total	38.3	61.7			0.668		372	598

in aboveground productivity. This aboveground figure is comparable to that of 300 g m^{-2} obtained for an alpine tundra in Europe (Rehder 1970, 1976a, b, 1982; Rehder & Schäfer 1978). Area 2 had an estimated annual primary productivity of 970 g m^{-2}, with aboveground productivity of 372 g m^{-2}.

These results, although preliminary, do suggest the great importance of topography and seasonality in controlling fundamental ecosystems processes such as mineralization in tropical alpine habitats.

Acknowledgements

The investigations of 1979–80 were sponsored by the 'A, W, Schimper-Stipendium' of the 'H. u. E. Walter-Stiftung', those of 1980/81 and 1983 by the Deutsche Forschungsgemeinschaft. I am very obliged also to the Kenyan Ministry of Tourism and Wildlife (Ministry of Environment and Natural Resources), the Botany and Soil Science Departments of the University of Nairobi, especially to Prof. Dr J. O. Kokwaro and Prof. Dr

S. O. Keya, and to our field assistants, especially Mr C. Cheseney Chebei, for their reliable and untiring cooperation in the field and in the laboratory. I also thank my colleagues Prof. De E. Beck, Dr H. Gilck and G. Gebauer for their assistance.

References

Beck, E., Rehder, H., Pongratz, P., Scheibe, R. & Senser, M. (1981). Ecological analysis of the boundary between the afroalpine vegetation types 'Dendrosenecio woodlands' and '*Senecio brassica–Lobelia keniensis* community' on Mt. Kenya. *Journal of the East Africa Natural History Society* **172**, 1–11.

Bremner, J. M. & Keeney, D. R. (1965). Steam distillation methods for determination of ammonium, nitrate and nitrite. *Analytica Chimica Acta* **32**, 485–95.

Ellenberg, H. (1964). Stickstoff als Standortsfaktor. *Berichte der Deutschen Botanische Gesellschaft* **77**, 82–92.

Ellenberg, H. (1977). Stickstoff als Standortsfaktor, insbesondere für mitteleuropäische Pflanzengesellschaften. *Oecologia Plantanum* **13**, 1–22.

Gerlach, A. (1973). Methodische Untersuchungen zur Bestimmung der Stickstoffnettomineralisation. *Scripta Geobotanica* (Göttingen), **5**.

Hedberg, O. (1964). Features of afroalpine plant ecology. *Acta Phytogeographica Suecica* **49**, 1–144.

Rehder, H. (1970). Zur Ökologie, insbesondere Stickstoffversorgung subalpiner und alpiner Pflanzengesellschaften im Naturschutzgebiet Schachen (Wettersteingebirge). *Diss. Bot.* 6. Lehre: Cramer.

Rehder, H. (1975). Tropische Wuchsformen und ihre Lebensräume im südlichen Kenia. *Naturwissenschaftliche Rundschau* **28**, 250–8.

Rehder, H. (1976a). Nutrient turnover studies in alpine ecosystems. I. Phytomass and nutrient relations in four mat communities of the Northern Calcareous Alps. *Oecologia* **22**, 411–23.

Rehder, H. (1976b). Nutrient turnover studies in alpine ecosystems. II. Phytomass and nutrient relations in the Caricetum firmae. *Oecologia* **23**, 49–62.

Rehder, H. (1982). Nitrogen relations of ruderal communities (Rumicion alpini) in the Northern Calcareous Alps. *Oecologia* **55**, 120–9.

Rehder, H. (1983). Untersuchungen zur Stickstoffversorgung der afroalpinen Vegetation am Mount Kenya. *Verhandlungen der Gesellschaft für Ökologie* (Festschrift für Heinz Ellenberg) **11**, 311–27.

Rehder, H. & Schäfer, A. (1978). Nutrient turnover studies in alpine ecosystems. IV. Communities of the Central Alps and comparative survey. *Oecologia* **34**, 309–27.

Rehder, H., Beck, E., Kokwaro, J. O. & Scheibe, R. (1981). Vegetation analysis of the upper Telekei Valley (Mount Kenya) and adjacent areas. *Journal of the East Africa Natural History Society* **171**, 1–8.

Rehder, H., Beck, E. & Kokwaro, J. O. (1988). The afroalpine plant communities of Mt. Kenya (Kenya). *Phytocoenologia* **16**, 433–63.

Speck, H. (1983). *Mount Kenya Area.* African Studies Series. Berne, Switzerland: Universität Bern.

13

An overview of the reproductive biology of *Espeletia* (Asteraceae) in the Venezuelan Andes

PAUL E. BERRY and RICARDO N. CALVO

Introduction

Communities of *Espeletia*, known locally as '*frailejones*', constitute the dominant and most striking physiognomic elements of the high montane páramo vegetation of the northern Andes (Figure 13.1). Different species occupy a variety of habitats, ranging from cloud forest, as low as 1500 m in elevation, to close to the upper limit of plant growth, at 4700 m. With over 130 currently recognized species and a wide diversity of life forms and other morphological features, *Espeletia* has been noted as one of the foremost examples of adaptive radiation in plants (Carlquist 1974).

One of the keys to understanding the ways that groups such as *Espeletia* have successfully colonized different habitats in high, tropical mountains lies in a study of their reproductive systems. Vegetative reproduction in *Espeletia* is rare and occurs to a limited extent in just a few species that produce axillary rosettes close to the ground (Cuatrecasas 1979). Seed dispersal is also very restricted, due to the absence of a pappus or other specialized dispersal structure (Cuatrecasas 1976; Smith 1981; Guariguata 1985). Thus, the breeding and pollination systems are the major determinants of gene flow in *Espeletia* and constitute an important factor affecting the genetic structure of populations.

The little previous information on the reproductive biology of *Espeletia* was obtained from studies undertaken in the páramos of Edo. Mérida, Venezuela. Values of natural seed set ranging from 33 to 68% were reported by Smith (1981) for three different populations of *E. schultzii*; other reports of natural seed set were 66% in *E. spicata* (Estrada 1983) and 25% in *E. timotensis* (Guariguata 1985). Monasterio (1983) demonstrated the existence of irregular, superannual flowering episodes in *E. spicata* and *E. timotensis* and suggested that four co-occurring species in the high páramos maintained separate flowering periods so as to avoid

Figures 13.1–13.4. Habitat and growth forms of *Espeletia*. Fig. 13.1. Superpáramo community dominated by stem rosettes of *E. moritziana*, 4300 m, Páramo de Piedras Blancas, Mérida, Venezuela. Fig. 13.2. Stem rosette of *E. timotensis*. Fig. 13.3. Sessile rosette of *E. angustifolia*. Fig. 13.4. Tree of *E. humbertii*.

pollinator competition. Four species of *Espeletia* studied by Smith (1981) flowered during the wet season, and synchronous, superannual flowering cycles were documented in *E. timotensis* and in the monocarpic *E. floccosa*.

By analysing a broad spectrum of species over the entire altitudinal range of the genus, we attempted to address a series of fundamental questions regarding the reproductive biology of *Espeletia*. Are species self-compatible, and what is their degree of self- and cross-pollination? What are the pollinating agents of the different species, and how effective are they? Do pollination systems change with altitude? Are some species pollinator-limited? Do sympatric species overlap in flowering time, and to what extend does interspecific hybridization occur? To answer the questions posed above, data were collected on the breeding systems, pollination mechanisms and reproductive phenology of 13 *Espeletia* species.

In the Himalayas and in the Andes of central Chile, insect diversity and abundance decrease strongly with altitude (Mani 1962; Arroyo *et al.* 1982). Assuming that a similar pattern occurs in the tropical Andes, we predicted that high altitude species of *Espeletia* would show higher levels of self-compatibility than lower elevation ones, a greater degree of pollinator limitation manifested by reduced seed set, or else a change to abiotic pollination.

Species studied

Thirteen species of *Espeletia* were chosen to represent the ecological and morphological diversity present in the genus (Table 13.1). The lowest altitude species, *E. neriifolia* (at 2200 m), and the highest altitude one, *E. moritziana* (at 4280 m), were studied along with a series of species from intermediate elevations. Growth forms included branched trees (*E. badilloi, E. neriifolia*: Figure 13.4), stem rosettes (*E. marcescens, E. moritziana, E. schultzii, E. spicata, E. timotensis*: Figure 13.2), stemless rosettes (*E. angustifolia, E. lindenii*: Figure 13.3), low, branched rosettes (*E. atropurpurea, E. semiglobulata*) and tuberous rosettes (*E. batata, E. floccosa*). Three species are monocarpic (*E. floccosa, E. lindenii, E. marcescens*) and the rest are polycarpic. Capitulum morphology varied widely, including flower number (*c.* 50 in *E. badilloi* to over 1500 in *E. moritziana*), ligule development (conspicuous in most, but reduced or vestigial in *E. atropurpurea* and the four highest elevation species), color (yellow, white, dull green), and in the orientation of the head (upright in most species to nodding in the four high elevation species).

The species were analysed in the field in three different localities in Venezuela: Hacienda Yeremba, Edo. Aragua (lat. 10° 25′ N, long. 67° 13′ W), for *E. neriifolia*; Páramo de San José, Edo. Mérida (8° 20′ N, 71° 18′ W), for *E. angustifolia, E. atropurpurea, E. badilloi, E. lindenii* and *E. marcescens*; and Páramo de Piedras Blancas, Edo. Mérida (8° 54′ N, 70° 51′ W), for the remaining species. The elevations where individual species were studied are included in Table 13.1.

Breeding systems

All *Espeletia* are monoecious, with functionally male disc flowers and female ray flowers. Due to the close proximity of these two kinds of

Table 13.1 *Characteristics of the breeding and pollination systems of 13 species of* Espeletia *from the Venezuelan Andes*

Species	Habitat (Altitude)	Self-compatibility Index[a]	Pollinators (Level of activity)[b]	Open/cross seed set
E. neriifolia (H.B.K.) Wedd.	Cloud forest (2200 m)	0.057	Small to large bees, wasps, flies and small diverse insects (High)	1.24
E. badilloi Cuatr.	Cloud forest (2560 m)	0	Small bees and small diverse insects (Low)	NA[d]
E. marcescens Blake	Cloud forest (3000 m)	0.005[c]	Medium to large bees and small flies (High)	NA
E. lindenii Sch. Bip.	Páramo (2765 m)	0.004	Small to large bees, wasps and beetles (High)	0.85
E. atropurpurea A. C. Smith	Páramo (2880 m)	0.012[c]	Small to large bees (High)	NA
E. angustifolia Cuatr.	Páramo (2940 m)	0.012[c]	Small to medium-sized bees and flies (Low)	NA
E. schultzii Wedd.	Páramo (3915 m)	0	Hummingbirds, small to large bees and flies (High)	0.81
E. floccosa Standl.	Páramo (4065 m)	0.096	Hummingbirds, small to medium-sized bees, flies and beetles (Medium)	0.93

E. semiglobulata Cuatr.	Páramo and superpáramo (4070 m)	0.084	Small flies (Low); wind	0.46
E. batata Cuatr.	Páramo and superpáramo (4120 m)	0.032	Small and large bees, wasps and flies (High)	0.56
E. spicata Sch. Bip.	Superpáramo (4200 m)	0.147	Wind; small flies (Low)	0.56
E. timotensis Cuatr.	Superpáramo (4220 m)	0.115	Wind; small insects (Low)	0.32
E. schultzii Wedd.	Superpáramo (4250 m)	0.027	Hummingbirds, large bees and small flies (Medium)	0.10
E. moritziana Sch. Bip.	Superpáramo (4280 m)	0.089	Wind; small flies (Low)	0.44

[a] Ratio of mean seed set in self-pollinations to cross-pollinations.
[b] See text for levels of visitor activity.
[c] Results of open-pollinations were used in lieu of cross-pollinations, which were not obtained for these species.
[d] Cross-pollination results lacking, but we recorded open-pollinated seed set of 0.11 in *E. badilloi*, 0.41 in *E. marcescens*, 0.82 in *E. atropurpurea* and 0.49 in *E. angustifolia*.

flowers, however, the possibility of within-capitulum selfing (geitonogamy) is high, especially since pollen dispersal and stigma receptivity overlap considerably within capitula.

To determine levels of self-compatibility, controlled pollinations were performed in the 13 species. Five or more individuals were used for each test, except for certain crosses for which sufficient flowering individuals were not available. Treatments included hand self-pollination in bagged capitula, hand cross-pollination, and open-pollination (i.e. natural seed set).

All achenes from test capitula were examined individually for the presence or absence of embryos, since well-formed achenes are formed in *Espeletia* whether or not the ovule has been fertilized. The mean proportion of filled achenes per plant was determined for each test. A Self-Compatibility Index was calculated by dividing the values of seed formation in self-pollination by the corresponding value of cross-pollination, in those species where both values were obtained. Following Bawa (1974) and Zapata & Arroyo (1978), a species with an index of 0.2 or less was considered self-incompatible.

Our results indicate that the Venezuelan species of *Espeletia* are self-incompatible, with a slight tendency towards self-compatibility in the highest elevation species (Table 13.1). Thus, outcrossing is required for successful pollination of *Espeletia* species. In most species studied, the Self-Compatibility Index is zero or close to zero, as in *E. badilloi*, *E. schultzii* and others. *Espeletia floccosa* and the four highest elevation taxa, on the other hand, have notably higher values of self-compatibility (0.09–0.15). These higher overall levels of self-compatibility were largely due to the presence of a few individuals that set nearly as many seed in self-pollination as in crosses, whereas the majority of the sample was strongly self-incompatible. A similar phenomenon was reported by Bawa (1979) for some lowland tropical rain forest tree species, but the cause of this pattern is not clear.

Pollination

Observations of floral visitors were made over fixed time periods (10 min each) and under different weather conditions until major pollinators could be determined for each species. In each period, the total number of capitula under observation and the number of visits by each type of potential pollinator were recorded. This allowed us to calculate a visitation rate, consisting of the number of visits per capitulum per minute (Arroyo *et al.* 1985). Based on these rates, we assigned three relative levels of visitation: High (>0.01 visits per capitulum per minute), equivalent to one or more visits every 100 minutes; Medium (between 0.01 and 0.001), or one visit every 100 to 1000 minutes; and Low (<0.001), less than one visit every 1000 minutes. Visitors were classified into major groups, as hummingbirds, large, medium or small bees, wasps, flies, small diverse insects, or wind. Table 13.1 lists the overall levels of visitation rates of all potential pollinators for each species.

In *Espeletia*, pollen is openly exposed in the disc flowers by the non-functional stigmas, and all animal visitors were seen to remove pollen. However, the importance of different visitors as pollinators varied greatly, depending on their foraging habits, interplant mobility and the efficiency of pollen removal.

Bees were the most common visitors, found on nine of the 13 species studied. Large bees (*Bombus* spp.) were the major pollinators of most *Espeletia*; these insects visited large numbers of capitula, transported large pollen loads and moved rapidly between plants. *Apis* was an important pollinator only in cloud forest and low elevation species of *Espeletia* (Figure 13.5). Small halictid bees were very common visitors (Figure 13.6), but they remained up to several minutes on single capitula and flew much less frequently between plants than the larger bees.

Hummingbirds were important pollinators of two species of *Espeletia* occurring between 3900 and 4250 m. *Oxypogon guerinii* visited both *E. floccosa* and *E. schultzii* near the upper limits of the páramo belt. Individuals of *Oxypogon* extracted small amounts of nectar (0.43 ± 0.15 μl in the male flowers of *E. schultzii*, $n = 11$) by perching on the edge of the capitulum, rather than hovering (Figure 13.8); in the process, the bird's breast, neck or bill often brushed against the stigmas and anthers, effectively transporting pollen from one capitulum to another. *Oxypogon* visited *E. schultzii* up to the latter's altitudinal limit at 4300 m and remained active even in light rain and snow flurries, when no insect activity was recorded. *Espeletia batata* occurs together with *E. schultzii* and has nearly identical capitula. However, it was never seen to receive visits from *Oxypogon*, apparently because the capitula are solitary and are borne on weak scapes that do not support the weight of the hummingbird.

Bibionid flies (*Dilophus* sp.) were common visitors of *Espeletia* flowers (Figure 13.7), especially at high elevations where bees or other insects were scarce. These small flies insert their conspicuous probosci into open male flowers to extract nectar, coating themselves with pollen in the process. However, they are very poor fliers and often remain up to an hour on a single capitulum. As geitonogamous pollinators, these Bibionidae could

Figures 13.5–13.9. Pollinating agents in *Espeletia*. Fig. 13.5. *Apis mellifera* on capitulum of *E. neriifolia*, 2200 m, Hacienda Yeremba, Aragua, Venezuela. Fig. 13.6. Halictid bee on a capitulum of *E. floccosa*, 4060 m, Páramo de Piedras Blancas. Fig. 13.7. Bibionid flies (*Dilophus* sp.) on a capitulum of *E. batata*, 4150 m, Páramo de Piedras Blancas. Fig. 13.8. Female hummingbird of *Oxypogon guerinii* visiting a capitulum of *E. schultzii*, 4150 m, Páramo de Piedras Blancas. Note the grasping rather than hovering posture of the bird while probing for nectar. Fig. 13.9. Large, drooping capitula of *E. moritziana*, a wind-pollinated species, 4300 m, Páramo de Piedras Blancas. Note the reduced ligules and the exposed pollen on the disc flowers.

be effective pollinating agents, but in the self-incompatible *Espeletia*, their very restricted interplant movements severely limit their effectiveness as cross-pollinating agents.

The effectiveness of pollinators in different species of *Espeletia* was estimated by comparing the level of seed set in open-pollinated plants versus that of hand cross-pollinated plants (Bierzychudek 1981; see Table 13.1). The ratio of these two values has been used as a measure of pollinator limitation in self-incompatible species, indicating strong pollinator limitation when close to zero and efficient pollination when close to one (Zapata & Arroyo 1978). This type of comparison has been criticized when different numbers of flowers are used in the two tests and because enhanced seed set by hand cross-pollination in one year may lead to reduced seed set in subsequent years (Bawa & Webb 1984); our use of this index was to compare degrees of pollinator limitation in related species over a single reproductive episode.

In the four lowest elevation species studied (Table 13.1), as well as in *E. floccosa*, *E. lindenii* and the low elevation population of *E. schultzii*, seed set was not pollinator-limited, a result consistent with the high levels of visitation rates and the variety of insect visitors recorded in these populations. In contrast, seed set of *E. batata* was moderately pollinator-limited, despite an overall high visitation rate. This is due to the high proportion of visits by bibionid flies, which are ineffective pollinating agents. The low open pollination value of *E. badilloi* (0.11) is consistent with the low visitation rate observed. The high elevation population of *E. schultzii* had an intermediate visitation rate with most visits by bibionid flies, which limited the amount of outcrossed pollen, resulting in a much lower proportion of open versus crossed seed (0.10) than the lower elevation population (0.81). Smith (1981) also reported a marked reduction in the fecundity of high altitude populations of *E. schultzii* in the same study area.

Espeletia moritziana, *E. semiglobulata*, *E. spicata* and *E. timotensis* are the only Venezuelan species that occur exclusively above 4000 m in severe 'superpáramo' habitats. The possibility of wind pollination in these species was suggested by their distinctive capitula, which are large and pendulous, with reduced, inconspicuous ligules (Figure 13.9). Visitation rates to all four species were extremely low, and the only visitors were bibionid flies or small curculionid beetles, whose capacity for interplant pollen transfer is extremely limited. Bumblebees and hummingbirds were often present in mixed populations of high altitude *Espeletia*, but they were observed visiting only *E. schultzii* and never nearby individuals of the above species.

Wind pollination in the superpáramo *Espeletia* species in Venezuela was tested by enclosing capitula of the different species in nylon mesh bags with 1 × 1 mm openings, excluding insects but not all airborne pollen (Bawa & Crisp 1980). The resulting seed set was then compared with that of adjacent open-pollinated capitula (Table 13.2). *Espeletia batata* and *E. schultzii*, two entomophilous species, were also tested as controls. The ratio between seed set of the mesh-bagged treatment and that of the open-pollinated heads indicate the effectiveness of wind pollination in the species studied.

The results of the wind pollination tests show that there is virtually no seed formation in mesh-bagged heads for the entomophilous control species (Table 13.2). By contrast, in the four test species, the seed set in

Table 13.2 *Results of open and wind pollination tests in six species of* Espeletia

Species[a]	Open pollination		Wind pollination		Wind/open pollination index
	No. of achenes (No. of plants)	Mean % filled achenes/plant	No. of achenes (No. of plants)	Mean % filled achenes/plant	
E. batata	2380 (17)	0.37	369 (3)	0	0
E. schultzii	2168 (22)	0.50	804 (6)	0.01	0.02
E. moritziana	12180 (33)	0.38	14481 (34)	0.23	0.61
E. semiglobulata	2142 (23)	0.19	603 (6)	0.09	0.47
E. spicata	11014 (45)	0.37	7592 (30)	0.35	0.95
E. timotensis	10981 (27)	0.27	7022 (21)	0.16	0.59

[a] Altitudes of populations studied as in Table 13.1; *E. schultzii* was studied at the lower elevation site.

mesh-bagged heads ranged from 47 to 95% of that obtained in the corresponding open-pollinated heads, indicating an effective pollination via wind-carried pollen. Lower fecundity levels in the wind pollination tests than in the corresponding open pollination tests was expected due to the obstruction of pollen by the mesh net enclosing the capitula.

The pollen grains of the superpáramo *Espeletia* from Venezuela are the only ones in the genus with markedly reduced spines (Salgado-Labouriau 1982), a modification consistent with wind pollination in other genera of the Asteraceae (Wodehouse 1959). In addition, the pollen–ovule ratios of the superpáramo species of *Espeletia* were the highest of the 13 species (P. E. Berry and R. Calvo, unpublished data), which is expected for inefficient pollination systems like anemophily (Cruden 1977; Faegri & van der Pijl 1979). In *E. moritziana*, over 14 million pollen grains are produced per capitulum. Nonetheless, the presence of short ligules and the persistence of nectaries which still produce small quantities of nectar in the disc flowers, as well as occasional visits by small insects, suggest that anemophily is of recent origin in the species of superpáramo *Espeletia*. Cuatrecasas (1986) considers these high elevation species to be the most advanced in the genus, and the areas they presently occupy have come into existence only in the last 10 000 years, when glaciers receded from present-day superpáramos (Salgado-Labouriau 1981; Schubert 1981).

Phenology

Seasonal phenology of flowering and fruiting was studied in seven species of *Espeletia* that occur above 4000 m in the Páramo de Piedras Blancas, Mérida, Venezuela. This is a relatively dry páramo, with an average yearly precipitation of 800 mm, and a bimodal pattern of rainfall; the dry season extends from December until April or May, with over 75% of the yearly precipitation falling between May and November. Although each species in this high páramo tends to occupy a specific habitat type (Monasterio 1981), contact areas exist where up to five species were found growing together. Since generalist pollination systems occur in all seven species, separation of flowering periods could be a mechanism of reproductive isolation to prevent possible deleterious effects of interspecific pollination.

To investigate the duration and overlap of flowering and fruiting times, we made monthly observations of 25–30 marked individuals per species over a 3-year period. Individuals were located at random along transects crossing dense stands of each species, along a 5 km stretch of road from El Aguila to Piñango.

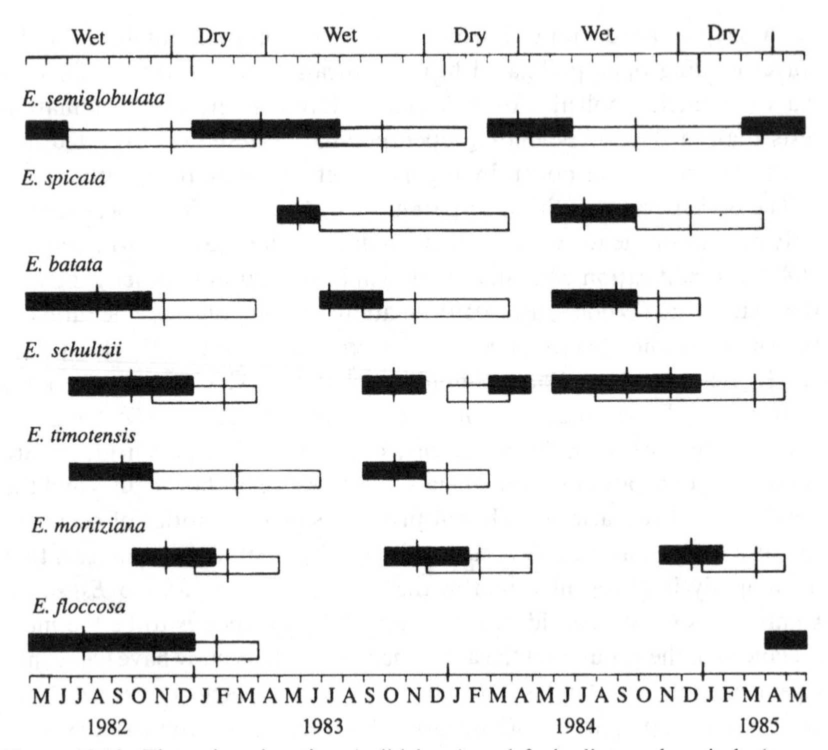

Figure 13.10. Flowering duration (solid bars) and fruit dispersal periods (open bars) of seven species of *Espeletia* in the Páramo de Piedras Blancas over a 3-year period. Vertical lines indicate peak levels of activity of flowering and fruiting.

Figure 13.10 presents the peak and the total flowering and fruiting period for each of the seven species. Flowering occurs mostly in the wet season. Only *E. semiglobulata* and *E. moritziana* flowered partly during the dry season. Flowering periods extended from 3 to 6 months, leading to overlapping periods of anthesis among most species. Flowering peaks also coincided in some species, such as *E. schultzii* and *E. timotensis*, and *E. batata* and *E. floccosa*. In 1982, individuals of five different species were flowering simultaneously in October, although the peak flowering period was displaced among four of the five.

Three of the species studied showed superannual flowering periods. *Espeletia floccosa* is monocarpic (semelparous); seven individuals flowered in 1982, and no further flowering occurred until 1985, when eight individuals began to bloom. Among the polycarpic (or iteroparous) species, *E. spicata* did not flower in 1982; only a few scattered plants bloomed in 1983, and then the population flowered massively in 1984.

The inverse situation occurred in *E. timotensis*, which flowered heavily in 1982, with only a few individuals in flower in 1983 and none in 1984.

Fruit dispersal was more extended than flowering, and peak dispersal periods occurred at the end of the wet season and during the dry season (Figure 13.10).

The extended flowering periods recorded (2–6 months) were in part due to variability between individuals in the populations, but also to the extended duration of anthesis in single capitula of the high elevation *Espeletia*. In *E. semiglobulata* (142 ± 25 flowers per head, $n = 13$), receptive stigmas were present in individual capitula for 21 ± 3.3 days. Individual female flowers remained receptive for 13.4 ± 3.2 days. Single male flowers lasted 2–4 days, but the overall pollen dispersal period was 19.9 ± 3.4 days. In *E. spicata* (380 ± 192 flowers per head, $n = 18$), total stigma receptivity lasted 33.8 ± 4.2 days, and male flowers dispersed pollen over a period of 15–23 days. Individual capitula of *E. moritziana* were observed to last up to 45 days in anthesis.

Hybridization

In the course of our field studies, numerous cases were found of individuals with intermediate character combinations between different pairs of *Espeletia* species. Familiarity with the few species present in our study area, the occasional occurrence of most intermediate individuals, and their almost invariable presence along contact zones between larger populations of their presumed parental species, suggested that these individuals were interspecific hybrids. Smith (1981) previously reported the occurrence of numerous putative hybrids between *E. schultzii* and *E. weddellii* (the latter now recognized as *E. batata*; J. Cuatrecasas, personal communication), whenever the two species occurred together in the Páramo de Piedras Blancas. He also noted the existence of a probable hybrid between *E. schultzii*, a páramo stem rosette species, and *E. humbertii*, a branched forest tree, at a site where both species grew in close proximity.

A series of interspecific crosses was made to test if viable hybrid seed could be produced in different species combinations. Table 13.3 lists the results of seed formation in controlled crosses between four species of *Espeletia*. In five of the eight crosses made, over 50% filled achenes were obtained, values similar to those obtained in intraspecific cross-pollinations. Only *E. schultzii* failed to set seed when used as the female parent in crosses with *E. moritziana* and *E. timotensis*, but it produced 52–55% seed set when used as the male parent with the same species. Thus,

Table 13.3 *Levels of seed formation in interspecific crosses of four species of* Espeletia

Cross ♀ × ♂	Number of achenes (number of plants)	Proportion of filled achenes
E. moritziana × *E. timotensis*	1655 (4)	0.86
E. timotensis × *E. moritziana*	1014 (3)	0.85
E. spicata × *E. timotensis*	706 (2)	0.68
E. timotensis × *E. spicata*	639 (1)	0.32
E. schultzii × *E. timotensis*	325 (2)	0.01
E. timotensis × *E. schultzii*	1608 (4)	0.52
E. moritziana × *E. schultzii*	972 (3)	0.55
E. schultzii × *E. moritziana*	381 (3)	0

post-mating isolating mechanisms are largely absent between the species tested.

In a more extensive crossing program, Berry *et al.* (1988) obtained viable seed from crosses involving five species of *Espeletia* and presumed natural hybrids between *E. batata* and *E. schultzii*. The only cross that failed to set seed was *E. schultzii* × *E. timotensis*, as in this study. In several instances, higher percentages of hybrid seed were produced than in the controls of intraspecific cross-pollination, and in all combinations obtained, hybrid seed germinated under controlled conditions at almost the same rate as seeds from intraspecific crosses, and normal seedlings were produced.

The extent of natural interspecific hybridization in *Espeletia* can be inferred by field studies in areas where the species are well known, and intermediate individuals can be easily detected. Marked differences in capitulum size, capitulescence structure, ligule color and dimensions, growth form and leaf morphology among many of the sympatric species makes detection of putative hybrids relatively easy.

Espeletia schultzii is the most widespread species of *Espeletia* in the Venezuelan Andes, occurring from 2600 to 4300 m, and as such comes into contact with several other species of the genus. It is common in the area of Mucubají and the Páramo de Piedras Blancas, where all major ecological studies of *Espeletia* have been carried out in Venezuela, and the species are well known (e.g. Baruch 1979; Monasterio 1981; Smith 1981; Goldstein & Meinzer 1983; Meinzer *et al.* 1985). Based on the criteria of obvious character intermediacy and the sympatric occurrence

Table 13.4 *Putative natural interspecific hybrids detected in* Espeletia[a]

Hybrids involving *Espeletia schultzii* (*E. schultzii* ×):	
E. angustifolia	*E. moritziana*
E. batata	*E. spicata*
E. humbertii	*E. timotensis*
E. lindenii	
Other combinations:	
E. badilloi × *E. neriifolia*	
E. moritziana × *E. spicata*	
E. pittieri × *E.* cf. *marcescens*	
E. semiglobulata × *E. spicata*	
E. spicata × *E. timotensis*	
E. timotensis × *E. moritziana*	

[a] Voucher specimens deposited in the U.S. National Herbarium (US) and at the Universidad Simón Bolívar, Caracas.

of the supposed parental species, putative hybrids involving *E. schultzii* were detected in the combinations listed in Table 13.4. *Espeletia schultzii* is a polycarpic stem rosette species; the different hybrid combinations found include a monocarpic species (*E. lindenii*), as well as a branched tree species (*E. humbertii*). Table 13.4 also lists the putative interspecific hybrids involving nine species other than *E. schultzii.* Altogether 13 different hybrid combinations are reported.

Discussion

Recent studies on the reproductive biology of tropical plant communities have been limited mainly to lowland or lower montane forests (Bawa 1974, 1979; Zapata & Arroyo 1978; Sobrevila & Arroyo 1982; Bawa *et al.* 1985a, b). This is also true for detailed studies of particular taxonomic groups such as *Inga* (Koptur 1983, 1984), *Heliconia* (Kress 1983) and *Cordia* (Opler *et al.* 1975). It is now clear that in woody forest species, obligate outcrossing occurs in a large majority of the species studied. Tropical high mountain plant communities, however, are subject to a series of severe environmental conditions which generally results in a reduced species diversity both in plants and their potential pollinators (Mani 1962). This had led to hypotheses that predict higher amounts of autogamy or apomixis at high altitudes, especially near the upper vegetational limits (Billings 1974; Arroyo *et al.* 1981).

The results summarized in this chapter establish that 13 species of *Espeletia* occurring across a broad altitudinal gradient in Venezuela are obligate outcrossers, and just a slight tendency exists towards self-compatibility in the high páramo species. Pollination systems, on the other hand, shift dramatically from entomophily and ornithophily in cloud forest and páramo species, to anemophily in the species that are restricted to the high superpáramo. The severe periglacial conditions of the superpáramo, which experiences almost daily freeze–thaw cycles and frequent snowfalls during the wet season (Monasterio 1981), limits the abundance and diversity of pollinating insects, to the extent that zoophylous species like *E. schultzii* that enter into the superpáramo zone show low seed set due to reduced pollinator activity. Although anemophily is rare in the Asteraceae (Leppik 1977), Smith & Young (1982) indicated that *Senecio keniodendron*, another tropical giant rosette species from Mount Kenya, may also be wind-pollinated, while the lower elevation relative *S. brassica* is insect-pollinated (A. Smith, personal communication). Recently, Carr *et al.* (1986) have also reported the existence of strong self-incompatibility systems in *Argyroxiphium* from Hawaii.

A series of morphological adaptations of the superpáramo species of *Espeletia* have occurred that appear to favor wind pollination. The capitula are drooping and well elevated above the leaf crown, and they have reduced ligules, short-spined pollen and high pollen–ovule ratios. More important, individuals typically occur in dense monospecific stands in open habitats, as in most wind-pollinated plants (Whitehead 1983). On the other hand, the anemophilous species of *Espeletia* lack plumose stigmas and dry season flowering that Faegri & van der Pijl (1979) consider as part of the 'anemophilous syndrome'. However, the unusually long periods of stigma receptivity and pollen dispersal in the capitula of high elevation *Espeletia* greatly increase the chances of the occurrence of sunny or rainless periods while the plant is in flower.

The occurrence of irregular, superannual flowering cycles in two wind-pollinated species, *E. spicata* and *E. timotensis*, has important implications for their reproductive success. Due to the typically leptokurtic pattern of pollen dispersal in wind-pollinated species (Levin & Kerster 1974; Whitehead 1983), a higher density of flowering individuals should lead to higher pollination success and subsequent seed set. It is therefore predictable that the average fecundity of individuals in a population will be markedly higher in mast-flowering years than in years where just a few individuals are in flower. Berry & Calvo (1989) have recently demonstrated that individuals flowering during non-mast years have severely reduced seed set compared to plants in mast-flowering populations. To what extent this density-dependent seed set should lead to increased synchronization of flowering years is unknown, due to our ignorance of the genetic and environmental components of flowering behavior in these species.

Crossing experiments indicate that there are few post-mating barriers to interspecific hybridization in *Espeletia*, at least as far as seed formation and germination are concerned. The widespread occurrence of probable natural hybrids between different species also indicates the lack of reproductive barriers between species. Over 30 species of *Espeletia* that have been studied cytologically share the same chromosome number of $n = 19$ (Powell & King 1969; Powell & Cuatrecasas 1970). This chromosomal uniformity is unusual for such a large group in the Heliantheae and is probably an indication of the close affinity and recent origin of the entire genus (Robinson *et al.* 1982). Thus, the crossing and chromosomal results do not offer support for the proposal of Cuatrecasas (1976) to separate *Espeletia* into seven different genera.

What, then, prevents widespread hybrid swarms from forming? Most species of *Espeletia* in the high páramos of Venezuela are restricted to specific substrates or habitat types, where they occur in large, monospecific populations (Monasterio 1981); this greatly restricts interspecific gene flow, since it is unlikely for pollen to travel long distances to reach individuals of other species. A second restriction on interspecific gene flow is the separation of flowering times, as in the wind-pollinated species *E. semiglobulata* and *E. moritziana*, which maintain completely separate flowering periods. Finally, the establishment and survival of hybrid progeny may be limited by a strong habitat selection. Nonetheless, there are numerous contact zones between species, and it is in these areas that hybrids are normally found; wind pollination and generalistic insect pollination in *Espeletia* facilitate interspecific pollination in these cases. In the case of *E. batata*, this species does form extensive hybrid swarms whenever it occurs with *E. schultzii* (Berry *et al.* 1988).

Espeletia has undergone extensive speciation in the short time that páramo habitats have come into existence, and widespread natural interspecific hybridization may have been a significant mechanism for increasing the degree of genetic recombination in the group. During the Pleistocene glacial periods, the páramo vegetation belt descended as much as 1500 m compared with its present level (Hammen 1974), greatly increasing its area and causing previously isolated páramos to become united. As the páramo environments retreated to higher elevations during the interglacials, certain recombinants from interspecific crosses could have survived and become isolated, leading to the formation of new morphological species.

Espeletia belongs to a small number of plant groups that have developed particular adaptations to high tropical mountain environments, most notably in the stem rosette habit (Hedberg & Hedberg 1979). Similar groups, such as *Argyroxiphium* in Hawaii and *Senecio* and *Lobelia* in East Africa, have also come to dominate the landscape of the highest areas of the world where vascular plants are able to survive. Our sampling of a broad diversity of *Espeletia* species has shown them to be a predominantly sexually reproducing, outcrossing group with two main pollination systems; insects and wind. Berry (1987) studied 15 species of herbaceous cushion plants and stemless rosettes in the superpáramo of Mérida, Venezuela, and found that all were self-compatible; 12 of them were autogamous. Thus, *Espeletia* is unusual in maintaining high outcrossing levels in superpáramo habitats. Its unique ability to establish dense stands of essentially arboreous individuals in the harsh, periglacial zone of the Venezuelan Andes between 4000 and 4700 m is an important feature that has favored the adoption of wind pollination in an otherwise typically entomophilous genus. Further studies of the reproductive biology of these and other high altitude, tropical composites should yield information that will improve our understanding of the processes that have led to some of the most spectacular instances of adaptive radiation among higher plants.

Acknowledgements

Our páramo studies were financed by the following organizations in Venezuela: Consejo Nacional de Investigaciones Científicas y Tecnológicos (CONICIT, Project S1-1408), Universidad Simón Bolívar (Decanato de Investigaciones) and Universidad de Los Andes (Graduate Program in Ecology), to which we wish to express our gratitude. We are also thankful to Juan Silva, Maximina Monasterio, Eduardo Cartaya, and José Gómez for their support at different stages of our research. Carol Horvitz and Claudia Sobrevila kindly offered numerous helpful comments on an earlier draft of this paper.

References

Arroyo, M. T. K., Armesto, J. J. & Villagrán, C. (1981). Plant phenological patterns in the high Andean cordillera in Central Chile. *Journal of Ecology* **69**, 205–23.

Arroyo, M. T. K., Primack, R. & Armesto, J. J. (1982). Community studies in pollination ecology in the high temperate Andes of central Chile. I. Pollination mechanisms and altitudinal variation. *American Journal of Botany* **69**, 82–97.

Arroyo, M. T. K., Primack, R. & Armesto, J. J. (1985). Community studies in pollination ecology in the high temperate Andes of central Chile. II. Effect of temperature on visitation rates and pollination possibilities. *Plant Systematics and Evolution* **149**, 187–203.

Baruch, Z. (1979). Elevational differentiation in *Espeletia schultzii* (Compositae), a giant rosette plant of the Venezuelan paramos. *Ecology* **60**, 85–98.

Bawa, K. S. (1974). Breeding systems of tree species of a lowland tropical community. *Evolution* **28**, 85–92.

Bawa, K. S. (1979). Breeding systems of tree species of a lowland tropical community. *New Zealand Journal of Botany* **17**, 521–4.

Bawa, K. S. & Crisp, J. E. (1980). Wind-pollination in the understorey of a rain forest in Costa Rica. *Journal of Ecology* **68**, 871–6.

Bawa, K. S. & Webb, C. J. (1984). Flower, fruit and seed abortion in tropical forest trees: implications for the evolution of paternal and maternal reproductive patterns. *American Journal of Botany* **71**, 736–51.

Bawa, K. S., Bullock, S. H., Perry, D. R., Colville, R. E. & Grayum, M. E. (1985a). Reproductive biology of tropical lowland rain forest trees. II. Pollination systems. *American Journal of Botany* **72**, 346–56.

Bawa, K. S., Perry, D. R. & Beach, J. H. (1985b). Reproductive biology of tropical lowland rain forest trees. I. Sexual systems and incompatibility mechanisms. *American Journal of Botany* **72**, 331–45.

Berry, P. E. (1987). Los sistemas reproductivos y mecanismos de polinización del género *Espeletia* en los Andes venezolanos. *Anales IV Congreso Latinoamericano de Botánica II*, pp. 25–33.

Berry, P. E., Beaujon, S. & Calvo, R. (1988). La hibridización en la evolución de los frailejones (*Espeletia*, Asteraceae). *Ecotrópicos* **1**, 11–24.

Berry, P. E. & Calvo, R. N. (1989). Wind pollination, self-incompatibility, and altitudinal shifts in pollination systems in the high Andean genus *Espeletia* (Asteraceae). *American Journal of Botany* **76**, 1602–14.

Bierzychudek, P. (1981). Pollinator limitation of plant reproductive effort. *American Naturalist* **117**, 837–40.

Billings, D. W. (1974). Arctic and alpine vegetation: plant adaptations to cold summer climates. In *Arctic and Alpine Environments*, ed. J. D. Ives & R. G. Barry, pp. 403–43. London: Methuen.

Carlquist, S. (1974). *Island Biology*. New York: Columbia University Press.

Carr, G., Powell, E. A. & Kyhos, D. W. (1986). Self-incompatibility in the Hawaiian Madiinae (Compositae): an exception to Baker's rule. *Evolution* **40**, 430–4.

Cruden, R. W. (1977). Pollen–ovule ratios: a conservative indicator of breeding systems in flowering plants. *Evolution* **31**, 32–46.

Cuatrecasas, J. (1976). A new subtribe in the Heliantheae (Compositae): Espeletiinae. *Phytologia* **35**, 43–61.

Cuatrecasas, J. (1979). Growth forms of the Espeletiinae and their correlation to vegetation types. In *Tropical Botany*, ed. K. Larsen & L. Holm-Nielsen, pp. 397–410. London: Academic Press.

Cuatrecasas, J. (1986). Speciation and radiation of the Espeletiinae in the Andes. In *High Altitude Tropical Biogeography*, ed. F. Vuilleumier and M. Monasterio, pp. 267–303. Oxford: Oxford University Press.

Estrada, C. (1983). Dinámica del crecimiento y reproducción de *Espeletia* en el páramo desértico. Tésis de Maestría, Universidad de Los Andes, Mérida, Venezuela.

Faegri, K. & van der Pijl, L. (1979). *The Principles of Pollination Biology*. New York: Pergamon Press.

Goldstein, G. & Meinzer, F. C. (1983). Influence on insulating dead leaves and low temperatures on water balance of Andean giant rosette plants. *Plant, Cell and Environment* **6**, 649–56.

Guariguata, M. (1985). Dispersión, dinámica del banco de semillas y germinación en *Espeletia timotensis* Cuatr. (Compositae). Trabajo de grado, Universidad Simón Bolívar, Caracas, Venezuela.

Hammen, R. van der (1974). The Pleistocene changes of vegetation and climate in tropical South America. *Journal of Biogeography* **1**, 3–26.

Hedberg, I. & Hedberg, O. (1979). Tropical-alpine life-forms of vascular plants. *Oikos* **33**, 297–307.

Koptur, S. (1983). Flowering phenology and floral biology of *Inga* (Fabaceae: Mimosoideae). *Systematic Botany* **8**, 354–68.

Koptur, S. (1984). Outcrossing and pollinator limitation of fruit set: breeding systems of neotropical trees (Fabaceae: Mimosoideae). *Evolution* **38**, 1130–43.

Kress, W. J. (1983). Self-incompatibility systems in Central American *Heliconia* (Heliconiaceae). *Evolution* **37**, 735–44.

Leppik, E. E. (1977). The evolution of capitulum types of the Compositae in the light of insect–flower interaction. In *The Biology and Chemistry of the Compositae*, ed. V. H. Heywood, J. B. Harborne and B. L. Turner, pp. 61–89. London, Academic Press.

Levin, D. A. & Kerster, H. W. (1974). Gene flow in seed plants. *Evolutionary Biology* **7**, 139–220.

Mani, M. S. (1962). *Introduction to High Altitude Entomology: Insect Life above Timberline in the Northwestern Himalayas*. London: Methuen.

Meinzer, F. C., Goldstein, G. H. & Rundel, P. W. (1985). Morphological changes along an altitude gradient and their consequences for an Andean giant rosette plant. *Oecologia* **65**, 278–83.

Monasterio, M. (1981). Las formaciones vegetales de los páramos venezolanos. In *Estudios Ecológicos en los Páramos Andinos*, ed. M. Monasterio, pp. 93–158. Mérida, Venezuela: Ediciones de la Universidad de los Andes.

Monasterio, M. (1983). Adaptaciones de especies al trópico frío: el caso de *Espeletia* en el páramo desértico. Trabajo de ascenso, Universidad de Los Andes, Mérida, Venezuela.

Opler, P. A., Baker, H. G. & Frankie, G. W. (1975). Reproductive biology of some Costa Rican *Cordia* species (Boraginaceae). *Biotropica* **7**, 234–47.

Powell, A. M. & Cuatrecasas, J. (1970). Chromosome numbers in Compositae: Colombian and Venezuelan species. *Annals of the Missouri Botanical Garden* **57**, 374–9.

Powell, A. M. & King, R. M. (1969). Chromosome numbers in the Compositae: Colombian species. *American Journal of Botany* **56**, 116–21.

Robinson, H., Powell, A. M., King, R. M. & Weeden, J. F. (1982). Chromosome numbers in Compositae XII: Heliantheae. *Smithsonian Contributions to Botany* **52**.

Salgado-Labouriau, M. L. (1981). Paleoecología de los páramos venezolanos. In *Estudios Ecológicos en los Páramos Andinos*, ed. M. Monasterio, pp. 159–69. Mérida, Venezuela: Ediciones de la Universidad de Los Andes.

Salgado-Labouriau, M. L. (1982). Pollen morphology of the Compositae of the northern Andes. *Pollen and Spores* **24**, 397–452.

Schubert, C. (1981). Aspectos geológicos en los Andes venezolanos: historia, breve sintesis, el Cuaternario y bibliografía. In *Estudios Ecológicos en los Páramos Andinos*, ed. M. Monasterio, pp. 29–46. Mérida, Venezuela: Ediciones de la Universidad de Los Andes.

Smith, A. P. (1981). Growth and population dynamics of *Espeletia* (Compositae) in the Venezuelan Andes. *Smithsonian Contributions to Botany* **48**, 1–45.

Smith, A. P. & Young, T. P. (1982). The cost of reproduction in *Senecio keniodendron*, a giant rosette species of Mt. Kenya. *Oecologia* **55**, 243–7.

Sobrevila, C. & Arroyo, M. T. K. (1982). Breeding systems in a montane tropical cloud forest in Venezuela. *Plant Systematics and Evolution* **140**, 19–37.

Whitehead, D. R. (1983). Wind pollination: some ecological and evolutionary perspectives. In *Pollination Biology*, ed. L. Real, pp. 97–108. Orlando, Florida: Academic Press.

Wodehouse, R. P. (1959). *Pollen Grains*. New York: Hafner Publishing Co.

Zapata, T. R. & Arroyo, M. T. K. (1978). Plant reproductive ecology of a secondary deciduous tropical forest in Venezuela. *Biotropica* **10**, 221–30.

14
Population biology of Mount Kenya lobelias

TRUMAN P. YOUNG

Introduction

Tropical alpine environments impose unique selection pressures on plants, producing a number of special adaptations. Among these are caulescent and acaulescent rosettes, nyctinasty, semelparity, and resistance to nightly frost. For each of these adaptations there are a number of possible evolutionary explanations. To discriminate between among explanations, and to understand better tropical alpine environments in general, integrated long-term studies of particular species are necessary. Such studies include morphology, ecological physiology, demography and reproductive biology. Long-term population biology studies of tropical alpine plants are rare; most are presented in this volume. These studies are helping us to understand the nature of adaptation in tropical alpine environments.

Giant rosette plants make ideal research subjects for a number of reasons. They are conspicuous and characteristic members of virtually all tropical alpine communities. Their morphologically discrete form makes their growth and individual dynamics easy to quantify. Lastly, they are relatively long-lived, enabling us to examine the effects of long-term changes in local environment.

Since 1977, I have been studying two closely related giant rosette species on Mount Kenya, *Lobelia telekii* and *L. keniensis*. I present here a summary of the first 7 years of that study, concentrating on the comparative population biology of these two species.

A number of factors make Mount Kenya lobelias particularly attractive as research subjects. (i) The alpine environment of Mount Kenya is one of the most intensively studied ecosystems in the tropics (Hedberg 1957, 1964; Coe 1967; Coe & Foster 1972; Young & Peacock 1985; Young 1990a). (ii) Giant rosette lobelias are found over much broader ecological and geographic ranges than other giant rosette genera. Giant rosette lobelias are found not only throughout East Africa's alpine areas, but also in the forests of East Africa, South America, the East Indies, India, and China, and the montane and alpine environments of Hawaii (Mabberley 1975a). (iii) Mount Kenya's alpine lobelias exhibit many of the special adaptations characteristic of tropical alpine plants, including giant rosette growth habit (Smith & Young 1987), nyctinasty (Hedberg 1964), and resistance to nightly frost (Beck *et al.* 1982; Young & Robe 1986). My particular interest has been the study of the evolution of semelparity, and for this Mount Kenya lobelias have proven ideal (Young 1990b).

Mount Kenya lobelias

Lobelia telekii and *L. keniensis* are closely related. Both are placed in the section Rhyncopetalum of the genus *Lobelia* (Hedberg 1957; Mabberley 1975a). Rare hybrids do occur; reproductive individuals are intermediate in all respects and are easily identified (Hedberg 1957; T. P. Young, personal observation). In addition to the one described by Hedberg, I have examined two in recent years (one shown to me by Alan P. Smith). Although Hedberg describes the seeds of the hybrid as 'completely aborted', I was able to germinate 11 out of 300 (3.7%) from one such hybrid. Despite occasional hybridization, *L. telekii* and *L. keniensis* remain easy to distinguish in the field (Figure 14.1).

Lobelia keniensis belongs to the *L. deckenii* group. Hedberg (1957) provisionally considered *L. keniensis* a full species, but Thulin (1984) preferred to make it a subspecies of *L. deckenii.* In any case, the distinct form *keniensis* is endemic to alpine Mount Kenya, and the *L. deckenii* group is endemic to alpine East Africa, with a different form on each mountain. *Lobelia telekii* is endemic to alpine East Africa.

In 1984, I collected a third species of giant lobelia on Mount Kenya. Previously known only from Mount Elgon and the Aberdare Mountains, *Lobelia aberdarica* on Mount Kenya is restricted to the northern slopes at elevations of 3000–3500 m. I limit my discussion here to the more common *L. telekii* and *L. keniensis.*

Distributional patterns

Lobelia telekii occurs mainly on dry rocky slopes at elevations of 3500–5000 m. *Lobelia keniensis* occurs mainly on moist valley bottoms at elevations of 3300–4600 m. These two species separate along a gradient

Figure 14.1. Photographs of reproductive *Lobelia telekii* (*a*) and *L. keniensis* (*b*) in the Teleki Valley, Mount Kenya. The scale in both is 1 m. The *L. telekii* inflorescence in the background is *c.* 2 m tall.

related to soil moisture. Data from 18 sites throughout the upper Teleki Valley on Mount Kenya show that dry season soil moisture is strongly correlated with estimates of percentage vegetation cover (Figure 14.2; $r = 0.86$, $p < 0.01$). Repeat surveys showed that percentage vegetation cover remained fairly constant seasonally ($< 10\%$ change). Soil analyses indicate that the availability of most mineral nutrients increases along this gradient (Young 1984a). The gradient from low to high vegetation cover is therefore one of generally increasing resource availability. Data from wet and dry weather periods indicate that water is probably the key resource for both species (Young 1984a). Large-scale transect data show that *L. telekii* is most frequent in dry, less vegetated habitats, and that *L. keniensis* is restricted to moist, more vegetated habitats (Figure 14.3*a*). Drier habitats are more common higher on the mountain. *L. telekii* is more abundant above 4000 m and *L. keniensis* is more abundant below 4000 m.

The upper distributional limit of *L. keniensis* on Mount Kenya is curious. Common along the bottom of the Teleki Valley up to 4100 m, it then disappears abruptly. Further up the valley *L. keniensis* is represented only by dense populations of small (*c.* 5 cm diameter) rosettes and (until 1983) isolated patches of adult plants in moist drainages off the main valley bottom. A similar pattern occurs in other major valleys on Mount Kenya. The Teleki Valley rises only very gradually (10–20 m over 1 k) above the upper limit of continuous *L. keniensis*, although rain gauge data indicate that rainfall decreases dramatically along this gradient (Young 1984a).

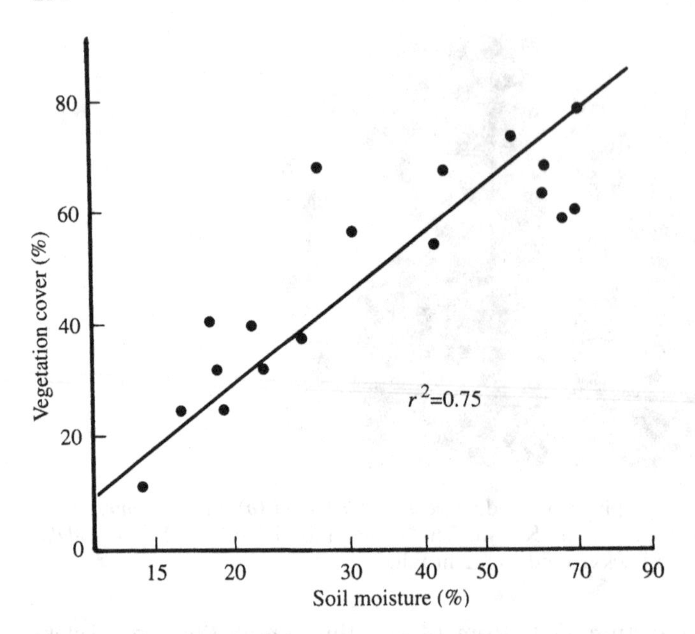

Figure 14.2. The relationship between dry season soil moisture (log scale) and percentage vegetation cover in 18 sites in the upper Teleki Valley.

A photograph taken in 1948 shows abundant reproductive *L. keniensis* plants at the head of the Teleki Valley, 1 km above their current continuous distribution (Hedberg 1964). This photograph indicates that the local distribution of *L. keniensis* was once much greater than it is today. A clue to its disappearance at the upper end of its distribution was provided by a recent drought on Mount Kenya. Near the end of this drought (July 1984), I found that most of the isolated surviving patches of adults high in the valley had recently died (Figure 14.4). By mid-1985 all remaining adults in these patches had been eaten to the ground by hyrax (A. P. Smith, personal communication). I suggest that a similar or more severe drought 15–25 years ago eliminated most *L. keniensis* adults from the drier upper parts of valleys on Mount Kenya (see Young & Smith, Chapter 18). As mentioned above, several sites now have numerous seedlings that may eventually replace the lost populations.

Morphology

Lobelia telekii and *L. keniensis* are morphologically similar in a number of ways. Both produce rosettes that grow up to 50 cm in diameter. Mature rosettes consist of 150–300 expanded leaves each 18–28 cm long. All the

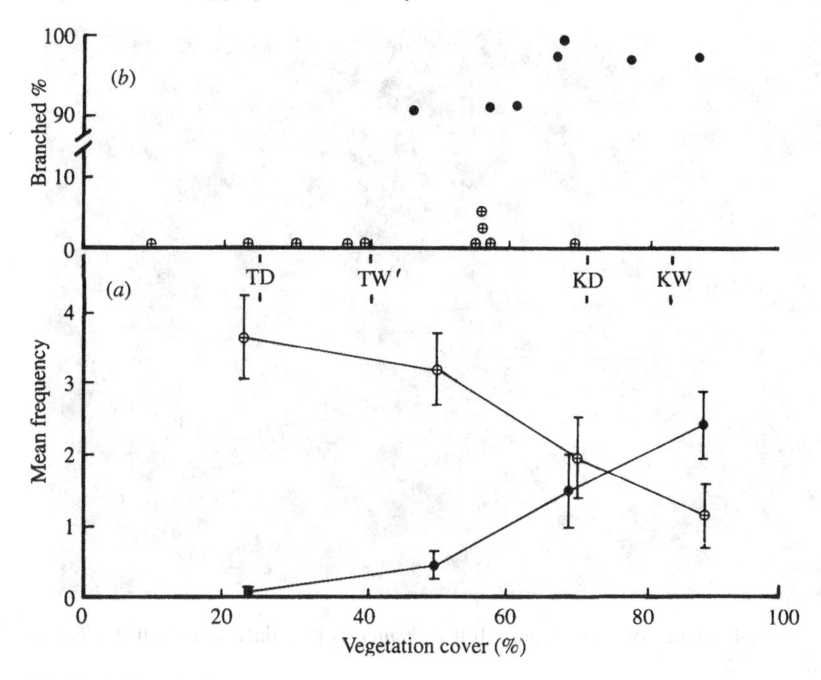

Figure 14.3. The frequency ±SE (*a*) and branching (*b*) of *L. telekii* and *L. keniensis* along a soil moisture gradient. Frequency was measured as the number of 1 m^2 quadrats out of ten along a 30 m transect containing each species. Data are from 42 transects. Branching data are from populations of 20–100 reproductive plants. Arrows indicate the locations along the environmental gradient of demographic plots for *L. telekii* drier (TD) and wetter (TW) sites, and *L. keniensis* drier (KD) and wetter (KW) sites.

mature leaves of a particular rosette are similar in size. I have shown leaf length (LL) to be a reliable measure of total rosette size (Young 1984a) and will use LL throughout to indicate rosette size. Lifespan of mature leaves is 6–9 months in either species. Cotyledon-stage seedlings may not produce their first true leaf until more than a year after germination. Taylor & Inouye (1985) report a similar phenomenon for seedlings of the subalpine rosette plant *Frasera speciosa* in Colorado, USA. At the other extreme, large rosettes with many leaves produce more than one leaf per day.

In the center of each rosette is a dense leaf bud and a single apical meristem. Rosette dissections showed that for every expanded leaf there were one to two leaves in the leaf bud. Total leaf lifespan from initiation to death is therefore up to 24 months. Leaf growth curves are sigmoid, and leaves attain full size soon after separating from the hard, conical bud.

Figure 14.4. Death of an isolated adult *L. keniensis* population high in the Teleki Valley, 1985.

The production of floral structures almost certainly proceeds under an accelerated schedule (inflorescences may contain up to 5000 flowers). Nonetheless, many months elapse between the initiation of reproductive structures in the bud and the appearance of the inflorescence.

The rosette grows from the end of a cylindrical subterranean stem that may reach 5 cm in diameter. Both this stem and the leaves are protected by a strong bitter latex (Mabberley 1975b). There are usually a relatively few (4–6) major roots associated with each rosette. These roots, and in *L. keniensis*, the stems, have numerous slender branch roots throughout their length. The main roots appear to serve as storage organs, as do the subterranean stems. The main roots are up to 1.5 cm in diameter and up to 5 m long. The roots are fairly superficial, growing mostly in the upper 20 cm of soil.

In both *Lobelia* species, reproductive rosettes produce a single terminal inflorescence and then die. However, if a plant has branched to produce a multiple-rosette clone prior to reproduction, it may survive to reproduce repeatedly. *Lobelia telekii* is almost always (> 99%) unbranched and the resources of the entire plant, including the roots, go into producing a large inflorescence. The entire plant dies, and *L. telekii* is therefore semelparous.

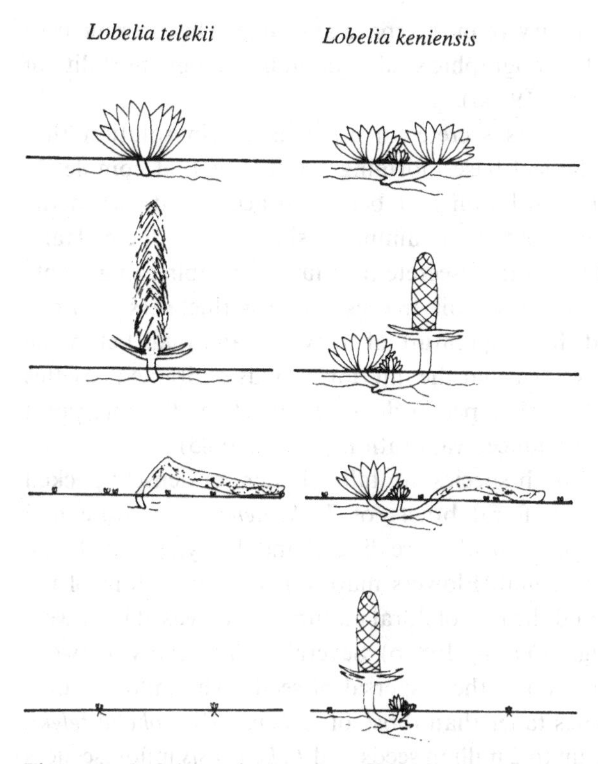

Figure 14.5. Schematic drawing of the life histories of semelparous *L. telekii* and iteroparous *L. keniensis*. Successive stages follow down the figure (from Young 1984a).

Lobelia keniensis is usually (90–100%) branched, with occasional unbranched plants being more common in drier habitats (Figure 14.3*b*). *Lobelia keniensis* divides its resources between a moderate-sized inflorescence and the maintenance of the root system and clone that represent potential future reproduction. *Lobelia keniensis* is iteroparous. Therefore growth form is intimately related to life history in Mount Kenya lobelias (Figure 14.5).

Branching of the subterranean stem in *L. keniensis* occurs at the site of old leaf-bases, and is apparently independent of the reproduction of rosettes both in time and in space. New leaf buds were never found less than 10 cm from the base of a living rosette, but were as likely to occur near reproductive as near vegetative rosettes. Many more rosette buds are initiated than ever survive to reproduce. On one excavated *L. keniensis* stem, every old leaf base along one side had initiated a new bud (T. P. Young, personal observation). At most, only one or two of these approximately

100 buds could possibly survive in the space available. This extreme case may be anomalous, but demographic studies do indicate high mortality of small rosette buds (Young 1984a).

Older *L. keniensis* rosettes secrete an aqueous pectin solution that accumulates between the leaf bases, forming a reservoir that completely covers the leaf bud. This solution first begins to accumulate when the rosettes are several years old, at a minimum size of LL 7.0 cm. Hairs present at the bases of leaves may secrete this fluid. On subfreezing nights the top centimeter of this reservoir freezes up to a thickness of 1 cm. Experiments show that the temperature of the water and buds below the ice never falls below 0 °C, but that the interior of buds of drained rosettes can cool as low as −6 °C. The pectin does not affect the freezing point of the solution, but may reduce evaporation (Young 1985).

The inflorescences of both species are tall cylinders of densely packed protandrous flowers. The floral bracts of *L. keniensis* are ovate and glabrous, while those of *L. telekii* are linear and hairy, probably an adaptation to its drier habitat. Flowers mature from the bottom of the inflorescence upward, and the rate of floral maturity is increased in flowers exposed to direct sunlight (Young 1984b). Several months elapse between the appearance of flowers and the dispersal of seeds. The inflorescences of *L. telekii* are 2–4 times taller than those of *L. keniensis*. *Lobelia telekii* inflorescences produce up to 2 million seeds and *L. keniensis* inflorescences produce up to 600 000 seeds. The flowers and pods of *L. keniensis* are larger than those of *L. telekii*, and produce more seeds per pod. However, *L. telekii* produces up to 5000 flowers per inflorescence, compared with less than 1000 flowers per inflorescence in *L. keniensis*. Both species are primarily bird-pollinated. Both *Lobelia* species are capable of self-pollination. Exclusion of pollinators had no effect on seed set, but did significantly reduce seed germination rates (Young 1982). The amount and frequency of reproduction increase with increasing resource availability in Mount Kenya lobelias (Young 1990b).

Seed and seedling biology

The seeds of *L. telekii* are oval and *c.* 1 mm in diameter, weighing an average of 0.078 mg each. The seeds of *L. keniensis* are slightly larger, and each bears a narrow (*c.* 1 mm) wing along one side. The size of this wing is highly variable. The seeds of *L. keniensis* weigh an average of 0.196 mg each.

For both species, lots of 100 seeds from each of 10 inflorescences were weighed and germinated. Heavier lots of *L. telekii* seeds had higher

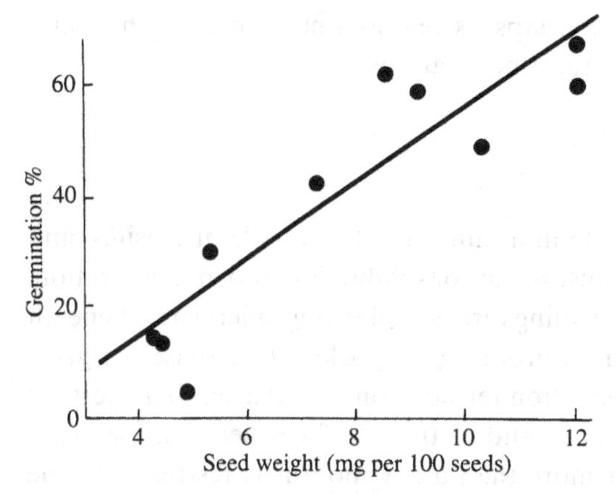

Figure 14.6. The relationship between mean seed weight and percentage germination in groups of 100 seeds from each of 10 *L. telekii* inflorescences.

germination rates than lighter lots (Figure 14.6; percentage germination = 0.695 (mg per 100 seeds) – 13.5; $r = 0.90$; $p < 0.005$). In *L. telekii*, both seed weights and germination rates were weakly correlated with the size of the inflorescence from which they were taken ($r = 0.59$ and 0.53, $p = 0.08$ and 0.12, respectively). In *L. keniensis*, seed weight was only very weakly correlated with germination rate ($r = 0.25$, $p = 0.48$), perhaps because much of the variation in seed weight was due to variation in wing size, rather than differences in the embryos themselves. Neither seed weight nor germination rate was significantly correlated with inflorescence height or the number of rosettes on the plant from which the seeds were taken (r values ranged from -0.35 to $+0.17$).

Initial studies of both species with seed traps and seed bags, and other observations, lead me to believe that most seeds fall within a meter or so of inflorescences, and that seed dormancy in nature is minimal, with seeds rarely remaining dormant through their first rainy season. In germination experiments, most seeds germinated 1–3 weeks after being placed on moist filter paper (Young 1982), requiring moist conditions for more than a few days to break dormancy. However, stored seeds can remain viable for several years. Seeds collected in 1983 and placed in cool dry storage were placed on moist filter paper for 5 weeks in 1990. Both *L. telekii* (45% germination; 113 out of 250 seeds) and *L. keniensis* (29% germination; 43 out of 150 seeds) showed considerable viability after 7 years storage.

Data from seedling plots in the field make it clear that only a small fraction of the seeds successfully germinate in nature, as opposed to

60–90% in the laboratory. Perhaps as few as one seed in a thousand successfully germinates in natural populations.

Seedling 'safe sites'

Lobelia telekii seeds germinate in a minority of available microsites, and survive in even fewer. Sites must be free of solifluction or dense vegetation. I have found and marked seedlings in the following microsites: beneath *Alchemilla* shrubs, in stabilized ground near rocks, at the bases of grass tussocks, at the edges of solifluction terraces, on the adjacent bare vertical walls of the solifluction terraces, and at the interfaces between the latter two microsites. Survival for more than a few months is restricted to the vertical and interface microsites, and occasionally at the bases of grass tussocks. The locations of older, but still small, rosettes confirm these observations. Larger rosettes are often found on the tops of solifluction terraces. Whether these rosettes initially became established in such sites is not known.

Lobelia keniensis seeds germinate only in the following microsites: in grass tussocks, under *Alchemilla* shrubs, beneath the prostrate leaves of the flat rosette plant *Haplocarpha ruepellii*, on steep unvegetated stream banks, and on flat open solifluction soils. On open solifluction soils there are occasionally large numbers of germinating seeds; I have marked over 100 in an area of less than 200 cm^2. *Lobelia keniensis* seedlings survive for more than a year only at sites near the edges of solifluction soils.

In both species, mortality of young seedlings was exponential (Young 1984a), and due to desiccation, frost heaving, and herbivory by insects, snails and rodents (compare Table 18.3 in Chapter 18). In addition, *L. keniensis* seedlings sometimes died after being submerged for several weeks during the rainy season.

Demography of Mount Kenya lobelias

Summary of the first 3.5 years of demography

I monitored over 2400 individuals (including >600 seedlings) of *L. telekii* and *L. keniensis* continuously from 1978 to 1986. Demographic analyses of the first 40 months of this study, along with details of the study sites and methods, are presented elsewhere (Young 1984a, 1985). Below is a summary of those earlier results.

1 The study period was divided into three: a wet period from February 1978 to May 1979, a transition period from June 1979 to January 1980, and a dry period from February 1980 to June 1981.
2 Two populations each of *L. telekii* and *L. keniensis* were regularly monitored, one in a relatively wet site and one in a relatively dry site. Because *L. keniensis* occurs in much wetter habitats than *L. telekii*, the 'dry' *L. keniensis* sites were considerably wetter than the 'wet' *L. telekii* sites (see Figure 14.3*a*).
3 In both species, survivorship and growth rates were lower in the dry period than in the wet period and lower in the drier sites than in the wetter sites. However, growth rates in the *L. telekii* 'wet' sites were comparable to those in the much wetter *L. keniensis* 'dry' sites. Growth rates were generally low, and Mount Kenya alpine lobelias first reproduce only after several decades of growth from seed.
4 There was a minimum size at reproduction for rosettes of both species. Rosette size at reproduction was significantly correlated with inflorescence size, more dramatically in *L. telekii* than in *L. keniensis*. Inflorescence size tripled over the observed range of flowering rosette sizes in *L. telekii*, but increased only 25% over a similar range of rosette sizes in *L. keniensis*. Single *L. keniensis* rosettes began to branch at LL 16.0, the same as the minimum size for flowering.
5 The largest rosettes of *L. telekii* were heavily preyed upon by hyrax (*Procavia johnstoni mackinderi*), and the smallest rosettes suffered mortality from other factors, resulting in a bimodal size-specific pattern of survivorship.
6 *Lobelia telekii* inflorescence size was greater in the wetter sites than in the drier sites. Inflorescence size was relatively constant in *L. keniensis*, but plants in the wetter sites had more rosettes and flowered more frequently than plants in the drier sites.
7 Seedling survivorship declined exponentially and at a similar rate (0.74–0.79 per month) for both species. For non-seedling rosettes, *L. telekii* mortality was higher than *L. keniensis* mortality. Mortality of clonal *L. keniensis* plants was very low, occurring only in the drier sites and during the dry period.
8 In clonal *L. keniensis* plants, small rosettes grew much faster than single rosettes of the same size. However, among medium-sized rosettes, clonal rosettes had much higher mortality and much greater variance in growth rate than single rosettes. I attributed this to an exploratory growth strategy in *L. keniensis*. When conditions were good, smaller rosettes grow and flourish, supported by the larger

rosettes in the clone. However, when conditions worsen, or when growing rosettes fail to secure a suitable location for future growth, the clonal plant reduced or partially recovered its investment in these adventitious rosettes, allowing them to shrink and die.

9 The number of constituent rosettes in individual clonal *L. keniensis* plants varied considerably. Large clones tended to become smaller and small clones tended to become larger. However, the average number of rosettes per plant remained unchanged throughout the study period.

10 Slower growing *L. telekii* rosettes were more likely to die, more likely to be eaten by hyrax, and tended to flower at a smaller size than faster growing rosettes. Similar patterns may hold for *L. keniensis.*

11 Calculations of age from growth rates were significantly biased by differential mortality. This and the non-random variation in size at reproduction make it very difficult to accurately estimate the age at reproduction (generation time), or the age of any rosette of a particular size.

12 It appears that hyrax preferentially attack *Lobelia* plants under stress, both seasonally and locally. This applies to both vegetative and reproductive plants.

Five and a half years of demography

Here I report a summary of $5\frac{1}{2}$ years' demography of Mount Kenya lobelias, nearly doubling the length of the previous study period. It is clear from this new analysis that even a relatively long study describes only a thin slice of the dynamics of a long-lived plant living in an environment with long-term environmental fluctuations.

A protracted drought has dominated *Lobelia* biology on Mount Kenya. After a period of wet weather lasting until mid-1979, Mount Kenya experienced below-average rainfall through the end of the surveys reported here (July 1983). Even at the early stages of this drought there were the significant reductions in *Lobelia* growth and survivorship outlined above. By comparing the demographic data from the previous report to those for the entire $5\frac{1}{2}$ years reported here, I show several striking developments that could not have been predicted from the previous study.

Lobelia telekii demography

Lobelia telekii populations underwent dramatic reductions between May 1981 and July 1983 (Figure 14.7). Over the previous 3 years, size structure remained relatively unchanged, with recruitment from smaller size classes

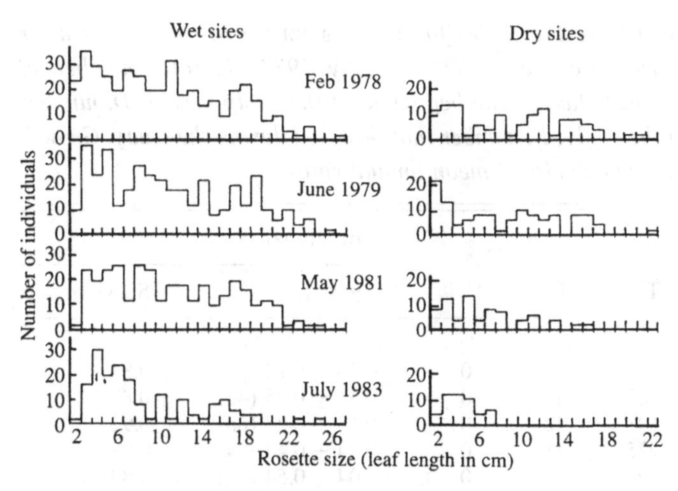

Figure 14.7. Changes in the size distribution of *L. telekii* rosettes from February 1978 to July 1983 in drier and wetter plots. The drier plots totalled 100 m^2, and the wetter plots totalled 200 m^2.

replacing rosettes lost to semelparous reproduction and other mortality. The disappearance of larger rosettes began to be evident before May 1981, especially in the drier sites, but over the following 26 months the elimination of larger rosettes was much more pronounced. In the drier sites, not a single rosette larger than LL 8.0 cm was alive in July 1983. In striking contrast, the size distribution of rosettes smaller than LL 8.0 cm remained virtually unchanged throughout the study in both sites.

The disappearance of larger rosettes is also shown by demographic data (Table 14.1). Larger rosettes experienced high mortality and reduced growth rates. The death of larger rosettes was probably due to a combination of drought and hyrax attack (discussed more fully in Chapter 18). Among smaller rosettes herbivory was relatively rare, and the most common factors associated with death were wilting, bud death, and burial under small mudslides (Table 18.3 in Chapter 18). More than 10% of the rosettes in the LL 2.5–5.5 size classes died after being buried. This represents 65% of all cases for which an associated factor was recorded in those size classes. *Lobelia telekii* occurs on slopes with little vegetation cover, where such mudslides are not uncommon after rain.

Clearly, the *L. telekii* populations reported here experienced tremendous changes during this study, changes that have resulted in a highly unstable size distribution. I estimate that it will be at least 15 years, and probably 30 or more, before there is flowering again in the drier *L. telekii* sites.

Table 14.1 *Size-specific life tables for* Lobelia telekii *in wetter and drier sites for the period February 1978 to July 1983. T, total number of individuals in the size class at the beginning of the study period; D, number that died during the study; F, number that flowered during the study. Growth and survivorship are calculated mean annual rates*

Size LL (cm)	T	D	F	Mean annual rates: LL ± SE (*n*)	Mean annual rates: Survivorship
Wetter sites					
⩽2.0	62	39	0	0.33 ± 0.13 (23)	0.84
2.5–3.5	82	39	0	0.31 ± 0.25 (43)	0.89
4.0–5.5	108	40	0	0.13 ± 0.27 (68)	0.92
6.0–8.5	125	41	0	0.04 ± 0.44 (84)	0.93
9.0–11.5	109	53	0	−0.04 ± 0.54 (56)	0.89
12.0–14.5	72	44	1	0.35 ± 0.64 (27)	0.84
15.0–17.5	74	44	13	0.34 ± 0.46 (17)	0.85
18.0–20.5	61	35	20	0.58 ± 0.42 (6)	0.86
⩾21.0	41	10	30	0.45 (1)	0.95
Drier sites					
⩽2.0	35	26	0	0.30 ± 0.22 (9)	0.78
2.5–3.5	23	15	0	0.16 ± 0.25 (8)	0.82
4.0–5.5	23	12	0	0.06 ± 0.21 (11)	0.82
6.0–8.5	43	18	0	−0.24 ± 0.23 (24)[a]	0.91
9.0–11.5	31	27	0	−0.71 ± 0.45 (4)	0.69
12.9–14.5	29	28	0	−0.91 (1)	0.54
15.0–17.5	29	27	2	(0)	
18.0–20.5	18	11	7	(0)	
⩾21.0	6	5	1	(0)	

[a] Excluded one individual that grew +6.0 cm (1.1 cm/year), 5.8 standard deviations above the mean (included, mean becomes −0.22 ± 0.35).

Local extinction in these plots is still a possibility, depending on whether the smaller rosettes continue to thrive. I expect greatly reduced flowering for many years to come even in the wetter sites.

The phenology of *L. telekii* has already begun to show the effects of large rosette mortality, with little flowering since July 1980 (Figure 14.8). Fifty-three per cent of all flowering between February 1978 and July 1983 occurred in the first 7 months of 1980 (Figure 14.8*b*). I estimate that it takes 12–18 months for initiated inflorescences to appear, so this reproductive burst was initiated sometime in early 1979, at the end of a sustained rainy period. This illustrates the difficulty, in these slow-growing plants with many months between initiation of flowering and seed

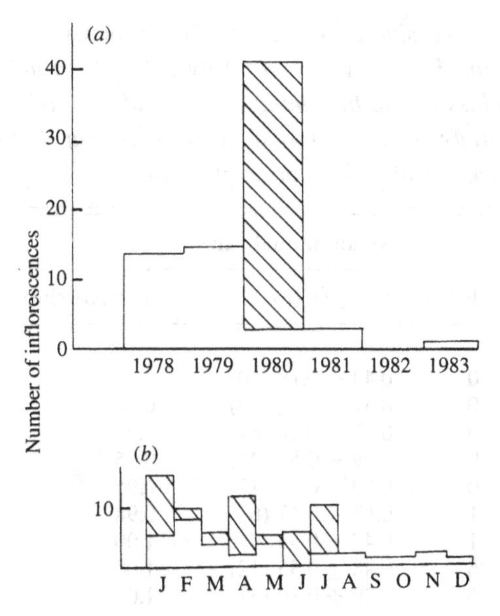

Figure 14.8. Flowering phenology of *L. telekii* by year (*a*) and by month (*b*). The hatched portion represents the inflorescences produced in the first 7 months of 1980.

dispersal, of evolving reproductive strategies related to the weather conditions seedlings are likely to encounter (see Taylor & Inouye 1985).

In semelparous species the age of first reproduction (generation time) is equal to lifespan. Several factors make it difficult to estimate lifespan in *L. telekii*. First, seedling growth is very slow, and I have no reliable estimates of transition times from germination to the smallest rosette sizes. Second, yearly variation in growth rates is high (Young 1984a), reducing confidence in estimates in mean growth rates. Third, the greater mortalities of slower growing rosettes strongly bias age estimates (Young 1985). For these reasons, any estimate of lifespan must be tentative. Nonetheless, I estimate that *L. telekii* plants flower and die at 40–70 years of age, with faster growing rosettes (in moister sites and microsites) flowering both younger and at a larger size than slower growing rosettes.

Lobelia keniensis demography

Pre-reproductive, single-rosette *L. keniensis* plants are morphologically similar to *L. telekii* plants, and their demographies are similar (Table 14.2), with the drought reducing survivorship and growth rates. However,

Table 14.2 *Size-specific life tables for single-rosette* Lobelia keniensis *in wetter and drier sites for the period February 1978 to July 1983. T, total number of rosettes in each size class at the beginning of the study period; D, number that died during the study; F, number that flowered during the study. Growth and survivorship are calculated mean annual rates.*

Size LL (cm)	T	D	F	Mean annual rates LL ± SE (*n*)	Survivorship
Wetter sites					
⩽2.0	36	26	0	0.43 ± 0.18 (10)	0.79
2.5–3.5	36	7	0	0.31 ± 0.17 (29)	0.96
4.0–5.5	79	12	0	0.11 ± 0.24 (67)	0.97
6.0–8.5	60	15	0	0.09 ± 0.50 (45)	0.95
9.0–11.5	23	5	0	0.20 ± 0.53 (17)	0.95
12.0–14.5	8	1	1	0.67 ± 0.47 (6)	0.98
15.0–17.5	10	0	1	0.47 ± 0.33 (9)	1.00
18.0–20.5	18	0	6	0.48 ± 0.35 (12)	1.00
⩾21.0	8	0	5	0.39 ± 0.19 (3)	1.00
Drier sites					
⩽2.0	34	26	0	0.31 ± 0.12 (8)	0.77
2.5–3.5	72	26	0	0.27 ± 0.20 (46)	0.92
4.0–5.5	94	11	0	0.25 ± 0.20 (82)	0.98
6.0–8.5	50	5	0	0.15 ± 0.36 (45)	0.98
9.0–11.5	13	3	0	0.45 ± 0.48 (10)	0.95
12.0–14.5	10	3	0	0.26 ± 0.49 (7)	0.94
15.0–17.5	22	4	0	0.38 ± 0.43 (18)	0.96
18.0–20.5	22	0	3	0.34 ± 0.52 (20)	1.00
⩾21.0	6	0	1	0.13 ± 0.50 (5)	1.00

unlike *L. telekii* populations, *L. keniensis* populations have remained fairly stable. I believe that *L. keniensis* survived the drought on Mount Kenya better than *L. telekii*, for two reasons. First, *L. keniensis* occurs in more mesic sites, where soil moisture is higher and where dense vegetation slows the rates of soil drying. However, it must be remembered that not too long ago, *L. keniensis* populations in the upper Teleki Valley suffered devastating loss, quite possibly related to drought (see above). A more important factor in their success is that *L. keniensis* populations are usually far from hyrax colonies, and so suffer little hyrax herbivory.

Reduced growth rates in the LL 4.0–8.5 cm size classes coincide with conspicuous cohorts of smaller rosettes (Figure 14.9). This may indicate that competition within size classes is stronger than among size classes, a situation that may be more likely when competition occurs mainly below

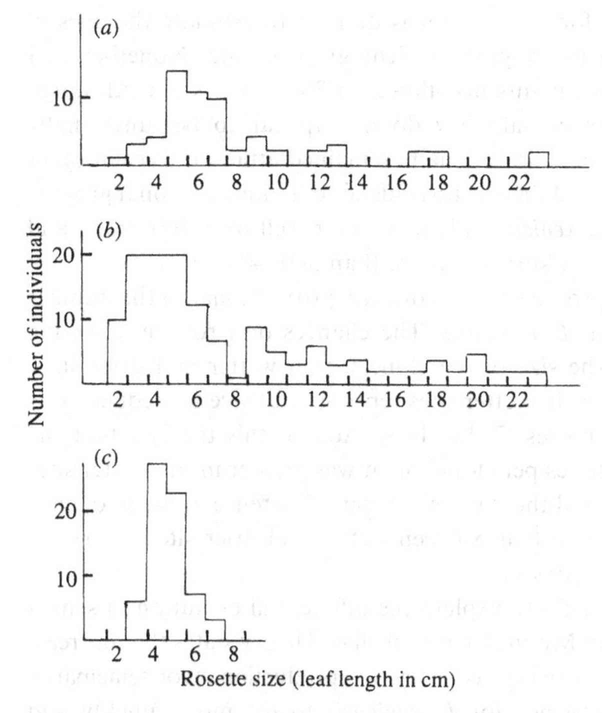

Figure 14.9. Size distributions of *L. keniensis*, showing cohorts of small rosettes in wetter sites (*a*), in drier sites (*b*), and in the site at the head of the Teleki Valley where only seedlings occur (*c*).

ground (Wilson 1988). So far mortality rates have remained relatively low (Table 14.1).

Clonal *L. keniensis* plants, most of which have reproduced at least once, continued to exhibit highly fluctuating numbers of constituent rosettes (see Young 1984a). I have not yet been able to discern long-term patterns of clone size increase or clonal senescence from these fluctuating patterns. Surprisingly, the average number of rosettes per plant continued to remain relatively constant through a drought that had strong negative effects on single-rosette plants of both species, demonstrating the clonal plants' superior ability to deal with environmental stress.

Mortality of clonal plants increased between 1981 and 1983, but remained low. Three out of 96 clonal plants in the wetter sites died between 1978 and 1983, for a mean annual survivorship of 0.994. Eight out of 141 clonal plants in the drier sites died between 1978 and 1983, for a mean annual survivorship of 0.989. Although survivorship was lower in the drier sites, this difference was not statistically significant.

For *L. keniensis*, as for *L. telekii*, it is difficult to estimate the ages of individuals, even with many years of demographic data. Nonetheless, I suggest that *L. keniensis* plants first flower at 30–60 years old. Although *L. keniensis* plants are clonal, they do not appear to be functionally immortal. All rosettes of a clone usually remained attached to each other throughout their lives, and there is no evidence of unlimited clonal growth. I suggest that many *L. keniensis* plants live for well over 100 years, and would be surprised if any survived more than 300–400 years.

The frequency of reproduction was directly proportional to the number of rosettes per plant in *L. keniensis*. The chances of a rosette flowering were independent of the size of the clone within wetter and drier sites. However, clonal rosettes in wetter sites reproduced twice as frequently as clonal rosettes in drier sites (Table 14.3). Add to this the fact that the average number of rosettes per clonal plant was greater in the wetter sites than in the drier sites, and there is a significant difference in the frequency of reproduction per individual between wetter and drier sites (every 5.8 vs every 11.7 years, $p < 0.001$).

These patterns may help to explain the differential evolution of semelparity and iteroparity in Mount Kenya lobelias. Theory states that decreasing survivorship and decreasing frequency of reproduction favor semelparity (Young 1981). The tendency for *L. keniensis* to die more quickly and flower more rarely in drier sites may have driven the evolution of semelparity in *L. telekii*, which occurs in the driest habitats (Young 1990).

Comparison of 3.3 and 5.5 year demographies

A comparison of the life table presented in Young (1984a, Tables 3 and 4) and Tables 14.1 and 14.2 here, reveals an intriguing pattern (Figures 14.10, 14.11). Virtually all medium and larger rosette size classes of both species had lower mean annual growth rates and lower mean annual survivorships over the total 5.5 year period than over the initial 3.3 year period. Larger plants grew more slowly and died more frequently over the last 2+ years of the study than over the first 3.3 years. Of 53 such comparisons (Figures 14.10 and 14.11) for rosettes larger than LL 3.5 cm, 39 (74%) represent lower survivorship or growth rates, and only four (8%) represent greater survivorship or growth rates, when comparing the 5.5 and 3.3 year data. This was expected, given the ongoing drought and continued hyrax herbivory.

However, the pattern for the smallest rosettes is completely opposite. Rosettes smaller than LL 4.0 cm actually grew faster and had higher

Table 14.3 *The reproductive frequency of* Lobelia keniensis *clonal plants of different sizes in wetter and drier sites. All values were calculated over the full* 5.5 *year period*

Number of rosettes per plant	Total number of plants	Total number of inflorescences	Number of inflorescences per plant	Number of inflorescences per rosette
Wetter sites				
2	13	3	0.24	0.12
3	9	6	0.66	0.22
4	11	10	0.89	0.23
5	9	8	0.91	0.18
6	7	5	0.72	0.12
7	4	5	1.20	0.17
8	7	10	1.43	0.19
9	4	5	1.25	0.13
10	1	1	1.00	0.10
11	3	3	1.33	0.12
12	3	6	2.00	0.16
13	1	2	2.00	0.15
Total	72	65		0.16
Drier sites				
2	16	3	0.19	0.094
3	19	4	0.21	0.070
4	16	3	0.19	0.047
5	7	3	0.43	0.086
6	7	1	0.14	0.024
7	15	13	0.87	0.124
8	7	5	0.72	0.089
9	3	2	0.66	0.074
10	1	1	1.00	0.010
12	3	6	2.00	0.167
15	2	4	2.00	0.133
Total	96	45		0.091

survivorship during the last 2+ years than they had previously. Of 16 such comparisons, 11 (69%) represented greater growth or survival, and only three (19%) represented lower growth or survival. This difference between the larger and the smallest rosettes is significant ($\chi^2 = 26.1$, $p < 0.001$). This difference was not due to biases introduced by the demonstrated differential mortality of slower growing rosettes (Young 1985). My records show that many of the smallest individual rosettes grew very slowly, if at all, throughout the 1978–81 surveys, and quickly thereafter. In summary, as the situation got worse for medium and large

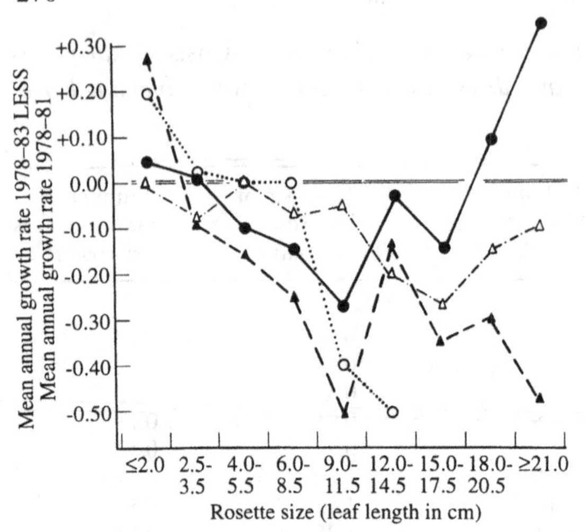

Figure 14.10. Size-specific difference in growth rate estimates between the first 3.3 years of the study and the full 5.5 years, for rosettes of both *Lobelia* species. ●, *L. telekii* wetter sites; ○, *L. telekii* drier sites; ▲, *L. keniensis* wetter sites; △, *L. keniensis* drier sites.

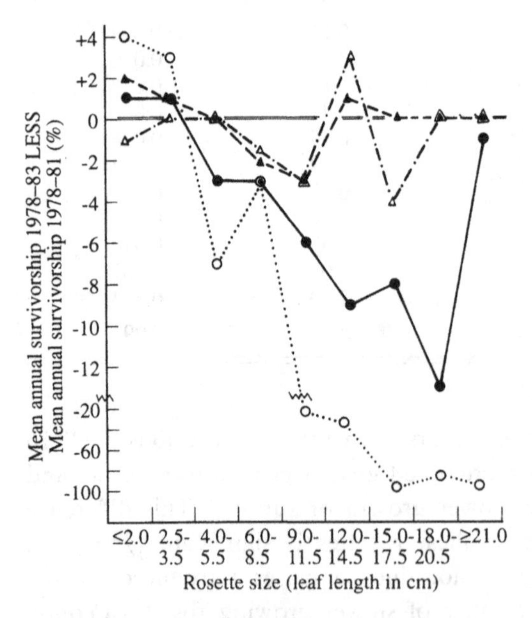

Figure 14.11. Size-specific differences in mortality rates between the first 3.3 years of the study and the full 5.5 years for single rosettes of both species. ●, *L. telekii* wetter sites; ○, *L. telekii* drier sites; ▲, *L. keniensis* wetter sites; △, *L. keniensis* drier sites.

rosettes, it got better for the smallest rosettes. This was not the case early in the drought, when all size classes suffered (Young 1984a).

I hypothesize that the relative success of very small rosettes in both Mount Kenya *Lobelia* species was due to the competitive release provided by the death of larger rosettes during a time of drought. This is supported by the fact that the surprising success of the smallest rosettes was more apparent in *L. telekii* (which had higher mortality in general), and was especially apparent in the drier sites, where large rosette mortality was greatest. The fact that the surviving larger rosettes did not similarly benefit implies that the nature of competition during drought among larger rosettes was qualitatively different from the competition between larger and smaller rosettes. In experimental studies in the Andes and on Mount Kenya, the removal of adults greatly increased growth and survival of seedlings in the giant rosette genera *Espeletia* (Smith 1983) and *Senecio* (Smith & Young, Chapter 15). I believe a similar phenomenon has occurred naturally among Mount Kenya lobelias in recent years.

Such competitive release would explain the distinct cohorts of small *L. keniensis* rosettes that appear to have become established years ago around the time of a major die-off of adults in the upper Teleki Valley (Figure 14.9) and the upper parts of other valleys. The fact that these cohorts occur not only in the upper valleys but also among apparently healthy adult populations implies that adult mortality in those populations may have been significant at the same time as the *L. keniensis* die-off higher on the mountain.

Lobelia keniensis plants in the wetter site cohort are larger than those in the drier site cohort (Figure 14.9). Since growth rates are higher in wetter sites individuals of the same age would be expected to be larger in wetter sites than in drier sites. These cohorts are the result of events that years ago altered *L. keniensis* distribution and population structure.

The population biologies of both Mount Kenya giant *Lobelia* species have been greatly influenced by rare and extreme demographic perturbations. That these perturbations have long-term effects lasting decades demonstrates the importance of considering historic factors in understanding contemporary ecological patterns.

Acknowledgements

This work was supported by grants from the Smithsonian Institution (to Alan P. Smith), the Arthur K. Gilkey Memorial Fund of the American Alpine Club, and the Public Health Service (grant No. 5T32GM-07517-02). Research was carried out under the auspices of the Office of the President of the Republic of Kenya, and the Kenya National Parks. Many people assisted in this project, chief among them John Omirah Miluwi and Martin Otiento. I especially wish to thank the late Alan P. Smith, who first brought me to Africa, and whose spirit guides me still.

References

Beck, E., Senser, M., Scheibe, R., Steiger, H.-M. & Pongrantz, P. (1982). Frost avoidance and freezing tolerance in afroalpine 'giant rosette' plants. *Plant, Cell and Environment* **5**, 215–22.

Coe, M. J. (1967). *Ecology of the alpine zone of Mount Kenya*. The Hague: Junk.

Coe, M. J. & Foster, J. B. (1972). Notes on the mammals of the northern slopes of Mount Kenya. *Journal of the East Africa Natural History Society* **131**, 1–18.

Hedberg, O. (1957). Afroalpine vascular plants: a taxonomic revision. *Symbolae Botanicae Uppsaliensis* **15**.

Hedberg, O. (1964). Features of Afroalpine plant ecology. *Acta Phytogeographica Suecica* **49**, 1–149.

Mabberley, D. J. (1975a). The giant lobelias: pachycauly, biogeography, ornithophily, and continental drift. *New Phytologist* **74**, 365–74.

Mabberley, D. J. (1975b). The giant lobelias, toxicity, inflorescence and tree building in the Campanulaceae. *New Phytologist* **75**, 289–95.

Smith, A. P. (1983). Population dynamics of Venezuelan *Espeletia*. *Smithsonian Contributions to Botany* **48**, 1–45.

Smith, A. P. & Young, T. P. (1987). Tropical alpine plant ecology. *Annual Review of Ecology and Systematics* **18**, 137–58.

Taylor, O. R. & Inouye, D. W. (1985). Synchrony and periodicity of flowering in *Frasera speciosa* (Gentianaceae). *Ecology* **66**, 521–7.

Thulin, M. (1984). Lobeliaceae. In *Flora of Tropical East Africa*, ed. R. Polhill, 65p. Rotterdam: A. A. Balkema.

Wilson, J. B. (1988). The effect of initial advantage on the course of plant competition. *Oikos* **51**, 19–24.

Young, T. P. (1981). A general model of comparative fecundity for semelparous and iteroparous life histories. *American Naturalist* **118**, 27–36.

Young, T. P. (1982). Bird visitation, seed set, and germination rates in two *Lobelia* species on Mount Kenya. *Ecology* **63**, 1983–6.

Young, T. P. (1984a). Comparative demography of semelparous *Lobelia telekii* and iteroparous *Lobelia keniensis* on Mount Kenya. *Journal of Ecology* **72**, 637–50.

Young, T. P. (1984b). Solar irradiation increases floral development rates in afro-alpine *Lobelia keniensis*. *Biotropica* **16**, 243–5.

Young, T. P. (1985). *Lobelia telekii* herbivory, mortality, and size at reproduction: variation with growth rate. *Ecology* **66**, 1879–83.

Young, T. P. (1990a). Mount Kenya forests: an ecological frontier. In *Mount Kenya Area: Differentiation and Dynamics of a Tropical Mountain Ecosystem*, ed. M. Winiger, pp. 197–201. Geographica Bernensia, African Studies Series A8.

Young, T. P. (1990b). Evolution of semelparity in Mount Kenya lobelias. *Evolutionary Ecology* **4**, 157–71.

Young, T. P. & Peacock, M. M. (1985). Vegetative key to the alpine vascular plants of Mount Kenya. *Journal of the East Africa Natural History Society* **185**, 1–9.

Young, T. P. & Robe, S. V. O. (1986). Microenvironmental role of a secreted aqueous solution in afro-alpine *Lobelia keniensis*. *Biotropica* **18**, 267–9.

15

Population biology of *Senecio keniodendron* (Asteraceae) – an Afroalpine giant rosette plant

ALAN P. SMITH and TRUMAN P. YOUNG

Introduction

Senecio keniodendron (Asteraceae) is an abundant and widespread giant rosette species endemic to the alpine zone of Mount Kenya (lat 0°), Kenya. Along with *Senecio brassica*, it forms a dominant component of the alpine plant community, and shows a high degree of morphological convergence with giant rosette genera in other tropical alpine areas (Hedberg & Hedberg 1979; Smith & Young 1987; Smith, Chapter 1). Here we summarize studies of *S. keniodendron* population biology carried out between 1977 and 1985, and briefly compare these results with those both for *S. brassica*, and for the convergent Andean genus *Espeletia* (Asteraceae).

Senecio keniodendron occurs most commonly on upper slopes and ridges from *c.* 3700–4600 m elevation. Adult densities generally range from 1 to 10 plants per 100 m^2, with greatest densities on talus slopes between 4000 and 4200 m. The mean height of *S. keniodendron* plants is significantly positively correlated with slope angle (Figure 15.1; $r = +0.60$; $p < 0.001$). *Senecio keniodendron* is replaced by *S. brassica* on lower slopes and valley floors; *S. brassica* has a lower elevational range than *S. keniodendron* – *c.* 3400–4400 m. The upper limit of growth for *S. keniodendron* (4600 m) may be due to a combination of drought and freezing stress (see also Perez 1987, on Andean *Espeletia*). Figure 15.2 shows the surface response curves for the densities of *S. keniodendron* ($p < 0.04$) and *S. brassica* ($p < 0.01$) im the Teleki Valley, relative to elevation and slope. The two species typically show very little spatial overlap along topographic gradients (Figure 1.9, in Smith, Chapter 1).

Senecio keniodendron can grow to 6 m tall, and produces extensive secondary xylem resulting in a woody stem. The inflorescences are terminal, and the terminal rosette dies after flowering. Several lateral rosettes develop around the base of the dying terminal rosette. Two to

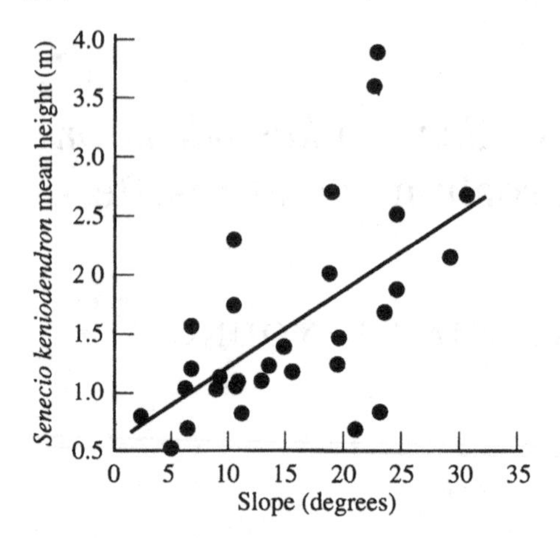

Figure 15.1. Relationship between slope angle and mean *Senecio keniodendron* height of the Teleki Valley, Mount Kenya.

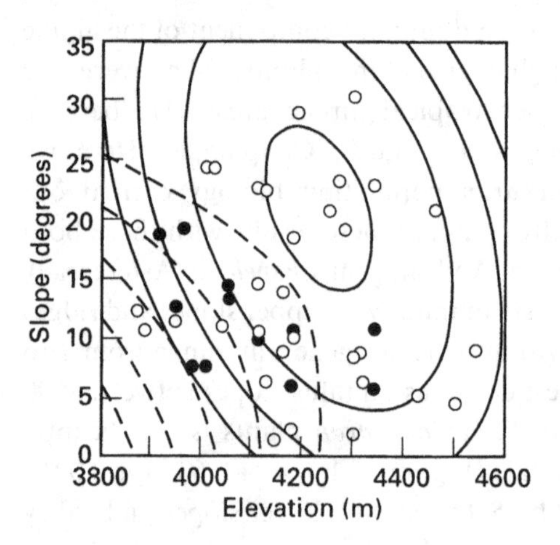

Figure 15.2. Response curves for *S. keniodendron* (——) and *S. brassica* (– – –) with respect to elevation and slope angle. The points represent 42 transects throughout the Teleki Valley. ●, sites in which there were hybrids between *S. keniodendron* and *S. brassica*.

four of these typically survive, and form lateral branches; reproduction is therefore followed by branch production, providing a clear record of past flowering. Dead rosette leaves are retained on the stem for many years. At elevations above 4200 m they may be retained for the life of the plant, but at lower elevations the lowermost leaves typically decay or burn off. Hyrax sometimes eat off portions of dead leaves. Initially, experimental removal of dead leaves resulted in reduced night-time stem temperatures and reduced rates of leaf production. However, it was also followed by production of dense corky bark (S. Mulkey, unpublished data) apparently providing compensatory insulation against freezing stress and/or fire.

Reproduction

Flowering tends to occur synchronously throughout the entire range of *S. keniodendron*; these 'mast years' appear to occur at intervals of 5 years and longer (Smith & Young 1982). Well-documented flowering episodes occurred in 1929, 1957, 1974, 1979 and 1985. The cues for this synchrony are unknown. Capitula are rayless, nodding, and yellow-brown in color; pollen is powdery, and readily blown away by wind. No obvious insect or bird pollination has been observed, although flies are occasionally seen resting on the undersides of the inflorescences. The breeding system is not known for this species.

Estimated achene production per plant during a reproductive year varies from 550 (minimum for plants reproducing for the first time) to over 3 million (maximum for plants reproducing for the third or fourth time with multiple reproductive rosettes: see Smith & Young 1982). Most plants reproduce four times or less during their lifetime. Reproductive plants are more likely to die during the following year than are non-reproductive plants of the same size (Smith & Young 1982). However, reproductive plants that survive produce more achenes in the next reproductive episode, because reproduction is followed by branching, and several rosettes on a given plant can flower simultaneously. Pre-dispersal seed predation by hyrax and weevils has been observed (Young & Smith, Chapter 18). Achenes mature approximately 5–6 months after flowering. The achenes have a pappus, and appear to be broadly dispersed by strong upslope and downslope winds characteristic of Mount Kenya. In 1980 large flocks of yellow-fronted canaries (*Serinus canicollis*) fed on *S. keniodendron* achenes after dispersal (Young & Evans 1993).

Germination

Repeated surveys of 1 m × 1 m permanent quadrats between 1977 and 1985 (A. P. Smith, unpublished data) suggest that germination begins 1–3 months after dispersal if the soil is moist. During the dry season (typically January–March, with a shorter dry season in July–August) germination ceases except on seepage slopes and stream banks. Germination is very patchy among microsites: it appears to be lowest under dense, nearly closed-canopy stands of conspecific adults on talus slopes, and highest on wet slopes with abundant cover of prostrate shrubs (especially *Alchemilla johnstonii*, Rosaceae) and tussock grasses. Densities of cotyledon-stage seedlings can reach 500–600 per m^2 on wet slopes at 4200 m.

Microhabitat correlations of this sort could result either from differential dispersal or from differential germination. In order to test for sensitivity of germination to microhabitat, an experimental study was initiated in the field at 4200 m in October 1985.

Soil was collected from the upper 5 cm at two sites which differed in density of cotyledon-stage seedlings (see above): a north-facing talus slope with dense, closed-canopy stands of adult *S. keniodendron* and very few seedlings ('talus-slope'), and a moist north-facing slope with little rock, abundant cover by *Alchemilla johnstonii* and *Festuca pilgeri*, and abundant seedlings ('grassy-slope'). The soil was heat-sterilized for 30 minutes in a pressure cooker to eliminate viable seeds. Twenty-four plastic pots (6.5 cm, 9 cm deep) were filled with soil from the talus slope, and 24 were filled with soil from the grassy slope. Mature *S. keniodendron* achenes were collected from 20 adult plants at 4200 m, and were thoroughly mixed. One hundred achenes were then planted in each of the pots, excluding hollow and shrivelled achenes. For each soil type the following treatments were initiated, with six pots per soil type per treatment.

1 Achenes planted 5 mm beneath the soil surface
2 Achenes planted on the soil surface with no additional cover
3 Achenes planted on the soil surface, and then covered with a layer of coarse gravel approximately 1 cm deep
4 Achenes planted on the soil surface, and then covered with a layer of green leaves and branches of *Alchemilla johnstonii* approximately 1 cm deep

The experiment was initiated on 11 October 1985 and terminated on 30 May 1986. The pots were placed in an open, level site at 4200 m under aluminum insect screening to eliminate seed rain on the treatments. The pots remained moist throughout the rainy season (October–late December), were allowed to dry out during the dry season (late December–mid-April) and remained moist after the rains began again in late April.

Some germination began in mid-December, but ceased during the dry season. Most germination occurred during the following April and May. Highest germination occurred under rock and litter cover (Table 15.1) and both full exposure on the soil surface and burial reduced germination. The two soil types differed slightly in germination responses: germination

Table 15.1 *Germination of* Senecio keniodendron *achenes under experimental conditions in the field*

Treatment	Talus slope soil	Grassy slope soil
Achenes buried	1.8 ± 1.6	2.8 ± 4.0
Fully exposed	2.5 ± 2.1	11.3 ± 7.2
Under rock	23.7 ± 7.5	12.8 ± 5.3
Under *Alchemilla*	19.2 ± 4.4	23.7 ± 7.8

Data are mean number of achenes per pot, out of 100 planted per pot, which germinated during the $7\frac{1}{2}$ months of the experiment, $n = 6$ for each treatment; mean ±SD in this and following tables.

of fully exposed achenes was greater on the grassy-slope soil than on the talus-slope soil.

Seedling establishment

Repeated surveys of marked juveniles in permanent quadrats between 1977 and 1985 suggest that most seedlings died within a few months of germination, with most mortality concentrated in periods of low rainfall. For the few individuals that survive this early stage, there can be great variation in growth rate among plants within the same site: most juveniles grow very little or shrink in size, and eventually die after many years; only a few that germinate in favorable microsites grow rapidly.

To clarify these microenvironmental effects on seedling establishment and growth, a transplant experiment was initiated in late March 1978. The experiment was in Hohnel Valley on a south-facing slope at 4150 m, a site dominated by *S. keniodendron* adults, and on the adjacent valley floor about 100 m away, a site dominated by adults of *S. brassica*. Juvenile plants of *S. keniodendron* were excavated from the slope site, and were transplanted to one of nine treatments:

A South-facing slope (*S. keniodendron* habitat)

1. Control (normal vegetation with *S. keniodendron* adults, *Alchemilla johnstonii*, *Festuca pilgeri* and *Lobelia telekii*)
2. Adult *S. keniodendron* retained, other species removed from two plots approximately 10 m × 10 m; juveniles transplanted to bare soil
3. Adult *S. keniodendron* cut down and removed from two plots approximately 10 m × 10 m; all other species retained

4 All vegetation removed from two plots approximately 10 m × 10 m; juveniles transplanted to bare soil
5 Same treatment as in 4, but bare soil covered with rocks to simulate talus slope conditions ('talus treatment'); juveniles were surrounded by, but not covered by rocks.

B Valley floor (*S. brassica* habitat)
6 Control (analogous to treatment 1 above)
7 Analogous to treatment 3 above, but *S. brassica* adults were removed
8 Same as treatment 4 above
9 Same as treatment 5 above

Treatment 2 was not repeated on the *S. brassica* site owing to time constraints.

There were 72 juveniles per treatment (total 648 plants). Within a given treatment juveniles were transplanted into eight replicate 1 m^2 quadrats, with nine plants per quadrat in order to facilitate relocating the plants for surveys. Four quadrats were located in each of the replicate 10 m × 10 m treatment plots. Maximum leaf length and total number of leaves were measured for each plant initially and after one year. Leaf length and leaf number were multiplied to give an approximate index of plant size ('total leaf length': cf Young 1984).

There was high mortality on the 'bare soil' and 'adults only' treatments (Table 15.2); in both cases there was extensive frost heaving of transplants due to diurnal freeze–thaw cycles. This source of mortality was eliminated by retaining all or most vegetation ('control' and 'other species only' treatments) and by covering bare soil with rocks ('talus' treatment). Elimination of adult *S. keniodendron* plants by itself resulted in a significant increase in mortality: mean number of mortalities per 1 m^2 quadrat was 1.2 ± 0.9 for control plants, versus 3.0 ± 1.3 for 'other species only' treatment (adult *S. keniodendron* removed, other species retained), $F = 9.80$, d.f. = 14, $p < 0.01$. Removal of *S. brassica* did not have this effect: 4.0 ± 1.4 mortalities per m^2 for controls versus 5.9 ± 2.8 per m^2 for 'other species only' treatment, $F = 2.86$, d.f. = 14, $p > 0.20$. This suggests that conspecific adults may provide a nurse tree effect on vegetation sites (cf. Smith 1984). However, growth was much lower on vegetated sites, with or without conspecific adults, than on the 'talus' treatment, where competitive effects and frost heaving were both eliminated. Perez (1989, 1991) has shown that both large rocks and tree rosette *Espeletia* increase soil moisture in the microenvironment.

Growth and survivorship were both generally lower in the *S. brassica*

Table 15.2 *Juvenile* Senecio *transplant experiment; mortality and change in plant size over a one year period*

	Control	Bare soil	Talus	Adults only	Other species only
S. keniodendron habitat					
Initial leaf length (cm)	6.7 ± 1.7	7.2 ± 2.0	6.7 ± 1.7	6.8 ± 1.8	6.8 ± 1.8
Initial leaf number	7.0 ± 2.6	6.8 ± 2.5	6.5 ± 2.5	7.1 ± 2.7	6.5 ± 2.7
Δ leaf length (cm)	−2.7 ± 1.9	−2.3 ± 3.2	0.5 ± 3.0	1.6 ± 2.5	−2.6 ± 2.3
Δ leaf number	−0.3 ± 2.1	6.2 ± 7.2	8.0 ± 5.8	3.4 ± 4.8	0.02 ± 2.4
Δ total leaf length (cm)	−20	9	60	40	−17
% mortality	15	72	31	71	36
S. brassica habitat					
Initial leaf length	6.7 ± 1.7	6.8 ± 1.9	6.5 ± 1.7	Not done	6.6 ± 1.4
Initial leaf number	6.9 ± 2.6	6.8 ± 3.3	6.4 ± 2.7	–	6.6 ± 2.9
Δ leaf length (cm)	−3.2 ± 1.9	−2.5 ± 1.6	−1.8 ± 2.7	–	−2.6 ± 1.9
Δ leaf number	−1.3 ± 2.8	−1.2 ± 5.8	1.1 ± 3.8	–	1.4 ± 2.5
Δ total leaf length (cm)	−27	−22	−7	–	−12
% mortality	46	86	57	–	65

Figure 15.3. *S. keniodendron* juvenile transplant experiment in October 1985, $6\frac{1}{2}$ years after initiation. (*a*) A representative plant in the control treatment (*S. keniodendron* habitat); leaves are approximately 4 cm long, slightly shorter than in 1978. (*b*) Two surviving transplants in the 'talus' treatment (*S. keniodendron* habitat); ruler to right of rosettes is 30 cm long; plants of this size can flower.

site than for analogous treatments in the *S. keniodendron* site, suggesting that the clear habitat segregation between the two species probably is enforced, at least in part, in the juvenile stage. Although competition could play a role in poor performance of juveniles on *S. brassica* sites, comparison of 'talus' treatments in the two sites suggests that competition is not necessary to enforce the difference in performance.

Figure 15.3 compares transplants in control and 'talus' treatments in October 1985. The surviving plants in the control treatments had not changed size greatly in $6\frac{1}{2}$ years, while several plants in the 'talus' treatment were approaching reproductive size.

Adult growth and mortality

Sixty adults were marked in each of eight sites, differing in elevation and topography, in Teleki Valley. Height, number of rosettes and leaf size were recorded in May 1977 and were remeasured periodically. Production of new leaves was recorded every 4–8 weeks from May 1977 to February 1981. Leaf dry weight was determined for all study plants, permitting conversion from leaf production per week to dry weight per week. Rainfall was measured at the Teleki Valley Ranger Station (4180 m) every day. Monthly rainfall varied from 21 mm during January 1980 to 284 mm in May 1978. Mean number of new leaves per rosette per week varied from 1.9 ± 0.4 for a talus slope population at 4200 m during May 1978 to 0.9 ± 0.2 for a ridgetop population at 4200 m during a dry period (July–August 1980). Regressions of leaf expansion rate against rainfall (Table 15.3) suggest that growth rate may be in part limited by rainfall.

Analysis of nearest neighbor relationships (Table 15.4; see Pielou 1962) suggests that growth can be influenced by competition with conspecific adults in some sites: the sum of the heights of neighboring adults in centimeters (y) increases as the distance in cm between them (x) increases:

$$y = a + b \ln x$$

Mean height growth for study populations over 2 years (May 1977–May 1979) was 11.6 ± 14.9 cm at 3900 m ($n = 5$), 2.4 ± 8.9 cm on a north-facing talus slope at 4200 m ($n = 47$), -2.4 ± 7.0 cm on a grassy north-facing slope at 4200 m ($n = 52$), 3.9 ± 4.1 cm on a north-facing talus slope at 4350 m ($n = 52$) and 5.3 ± 8.0 cm on a south-facing talus slope at 4500 m ($n = 45$) (cf. Hedberg 1969). Within each of these populations some individuals grew as much as 20–40 cm in 2 years, and some individuals decreased in height. These rates of height growth suggest that

Table 15.3 *Regressions of rate of new leaf production per week* (y) *against rainfall* (x) *in mm for* Senecio keniodendron

Population	r^2	r	a	b	$p <$
3900 m					
North-facing grassy	0.12	0.344	1.098	0.001	0.20
4200 m					
South-facing talus	0.30	0.549	1.220	1.24×10^{-3}	0.05
North-facing talus	0.58	0.764	1.094	0.002	0.01
North-facing grassy	0.52	0.719	1.110	1.51×10^{-3}	0.01
Valley floor	0.25	0.504	1.247	0.001	0.05
Ridge	0.35	0.591	1.043	0.001	0.05
4340 m					
North-facing talus	0.15	0.382	1.298	0.001	0.15
4500 m					
South-facing talus	0.25	0.496	1.283	0.001	0.05

$y = ae^{bx}$; d.f. = 17 in all cases.

Table 15.4 *Nearest neighbor relationships for* Senecio keniodendron *in May 1977*

The sum of the heights of pairs of neighboring adults in cm (y) was regressed against the distance between the pairs in cm (x): $y = a + b \ln x$.

Population	r^2	r	a	b	d.f.	p
3900 m						
North-facing grassy	0.001	0.03	398.4	7.1	41	>0.50
4200 m						
South-facing talus	0.35	0.59	−483.5	191.9	45	<0.01
North-facing talus	0.12	0.35	187.7	65.2	45	<0.02
North-facing grassy	0.03	0.16	62.6	60.9	40	>0.20
Valley floor	0.06	0.25	−7.0	63.8	46	<0.15
Ridge	0.44	0.66	−465.1	135.1	47	<0.01
4340 m						
North-facing talus	0.46	0.68	−526.4	162.8	47	<0.01
4500 m						
South-facing talus	0.08	−0.28	385.0	−23.9	42	<0.10

plants that are 4–5 m tall may be as old as several hundred years. Use of adult height growth rates may greatly underestimate age for most plants because most individuals spend many years (perhaps decades in some cases) as suppressed juveniles, without increasing in height; only rare individuals in particularly favorable microsites grow rapidly as juveniles (e.g. Figure 15.3). Conversely, if most individuals that survive to adulthood occur in such favorable microsites, adult age may be overestimated when based on mean growth rates (cf. Young 1985).

Growth is loosely but consistently correlated with reproduction (Table 15.5; see Smith & Young 1982). In general, rosettes which grew faster between May 1977 and August 1978 were more likely to reproduce during the subsequent flowering episode (beginning in September 1978 and continuing into 1979) than were slower growing plants of the same size. Plants which were branched, and had therefore reproduced in the past, had slower growing rosettes than unbranched plants, and were less likely to flower in the future. Note that the highest growth rates (Table 15.5) occurred in the highest elevation population, near the upper elevational limit of the species at 4500 m. At this site there were almost no potential interspecific competitors; conspecific neighbors were generally so far apart that competition was probably unlikely (see results for the 4500 m population in Table 15.4). Thus both interspecific and intraspecific competition may have been reduced, perhaps contributing to higher growth rates.

The primary sources of adult mortality during the study were elephant feeding (Mulkey *et al.* 1984), reproduction (Smith & Young 1982) and fire. No fires occurred at study population sites; however, observations of recently burned sites elsewhere suggest that fire can cause as much as 50% mortality among adults. The frequency and intensity of fires have not been well documented. It is clear, however, that fires are most common on the dry northern and eastern slopes of Mount Kenya (Bill Woodley, personal communication). In the absence of elephant damage, adult mortality ranged from 0 to 3% per year for study populations, with no consistent trends along elevational or topographic gradients (A. P. Smith, unpublished data). The study site at 3900 m had an average annual mortality rate of 29%, resulting almost entirely from a few bouts of elephant feeding (Mulkey *et al.* 1984). Post-reproductive mortality is discussed in detail in Smith & Young (1982).

Adult size class distribution for *Senecio keniodendron*

Senecio keniodendron tends to occur in stands of relatively uniform size, in which the individuals are probably of similar age. Data from a series of stratified random vegetation transects (Young & Peacock 1992) were analysed to produce height class histograms for all transects in which

Table 15.5 *Total leaf production (g dry weight) per rosette for* S. keniodendron *from May 1977 to August 1978, in relation to past reproduction (as evidenced by branching) and future reproduction (1979)*

Population	Total dry weight production				Plant height (cm)	
	Branched	Unbranched	Reprod. 1979	Veget. 1979	Reprod. 1979	Veget. 1979
3900 m Elephant damage prevented analysis						
4200 m						
Talus	1308 ± 591	1478 ± 1301	2105 ± 1316	1095 ± 461	314 ± 120	223 ± 83
	d.f. = 61, $F = 0.51$, $p > 0.50$		d.f. = 61, $F = 20.69$, $p < 0.001$		d.f. = 61, $F = 10.77$, $p < 0.005$	
North-facing talus	1179 ± 637	956 ± 710	1504 ± 885	931 ± 499	270 ± 84	229 ± 74
	d.f. = 84, $F = 1.73$, $p < 0.50$		d.f. = 87, $F = 14.19$, $p < 0.001$		d.f. = 87, $F = 3.25$, $p < 0.20$	

North-facing grassy	1298 ± 90	1951 ± 220	1955 ± 123	1048 ± 98	229 ± 15	246 ± 20
	d.f. = 62, F = 10.57, p < 0.005		d.f. = 64, F = 33.65, p < 0.001		d.f. = 64, F = 0.43, p > 0.50	
Valley floor	811 ± 468	1180 ± 668	1708 ± 942	815 ± 449	198 ± 73	202 ± 77
	d.f. = 51, F = 4.83, p < 0.05		d.f. = 52, F = 33.90, p < 0.001		d.f. = 52, F = 0.03, p > 0.50	
Ridge	1849 ± 970	1544 ± 964	2270 ± 882	1230 ± 676	227 ± 82	168 ± 59
	d.f. = 42, F = 0.92, p > 0.50		d.f. = 42, F = 19.46, p < 0.001		d.f. = 42, F = 6.58, p < 0.025	
4340 m						
North-facing talus	780 ± 530	1196 ± 1030	2351	1007 ± 870	225	192 ± 70
	d.f. = 39, F = 1.62, p < 0.50		n = 1	n = 48	n = 1	n = 48
4500 m						
Talus	2041 ± 1398	3075 ± 2490	4432 ± 2143	2438 ± 2200	183 ± 5	157 ± 42
	d.f. = 41, F = 2.08, p < 0.50		d.f. = 40, F = 2.98, p < 0.20		d.f. = 40, F = 1.07, p > 0.50	

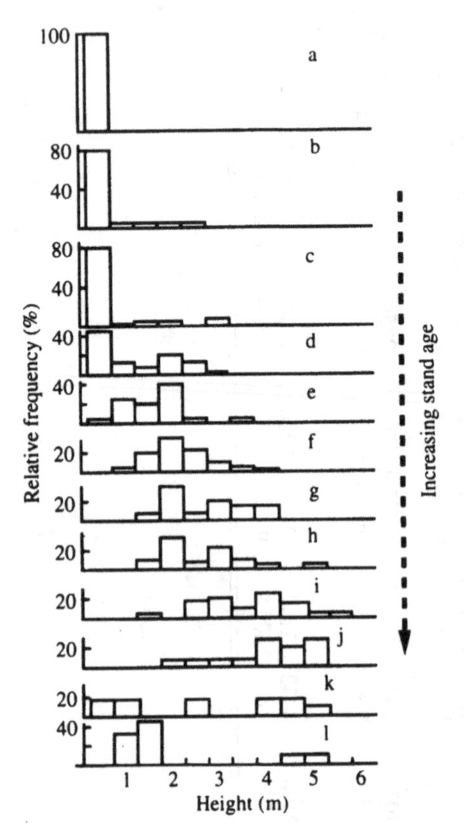

Figure 15.4. Adult size class distributions for 12 populations of *S. keniodendron* in the upper Teleki Valley.

there were caulescent individuals of *S. keniodendron* (Figure 15.4). The 30 m × 20 m transects were all located in the upper Teleki Valley at elevations of 3800–4200 m. The histograms in Figure 15.4 can be considered a sequence of increasingly older stands of even-aged plants, and are interpreted as follows.

Initially (Figure 15.4a–c), stands are composed mostly of individuals in the smallest height class (0.5 m tall). As these individuals grow, and do so at variable rates, mean plant height increases, as does variation in plant height. Over a broad range of mean stand heights (Figure 15.4f–j), there is apparently little or no recruitment of caulescent plants, although there may be acaulescent seedlings in abundance. As individuals reach maximum height (*c.* 5 m), the stands begin to senesce and recruitment into the smaller height classes commences anew (Figure 15.4k, l).

Bouts of recruitment into caulescent size classes are apparently rare, and may be limited by competition with adults. The experiments described above imply that although adults may increase the growth and survivorship of acaulescent seedlings, transplanted seedlings became caulescent only in the absence of adults, and then did so quickly. In addition, rare events of synchronous reproduction may contribute to even-aged stand structure.

Comparison of *S. keniodendron* with *S. brassica*

Senecio brassica appears to differ from *S. keniodendron* in reproductive biology. The capitula are upright rather than nodding, and have conspicuous bright yellow ray flowers; pollen is sticky and appears to be dispersed by a variety of butterflies, moths, wasps and flies. Flowering occurs every year, although the proportion of the population that flowers varies from year to year; between 1977 and 1985 peak flowering years in *S. brassica* coincided with flowering years in *S. keniodendron*. Pre-dispersal seed predation by hyrax is common in some populations (Young & Smith, Chapter 18). Seeds are wind dispersed. *Senecio brassica* seedlings are less abundant than those of *S. keniodendron*; it is uncommon to find more than 5–10 seedlings per m^2, whereas it is common to find over 100 *S. keniodendron* seedlings per m^2 in some microhabitats. *Senecio brassica* inflorescences are terminal, as in *S. keniodendron*; the terminal rosette dies after reproduction, and several lateral rosettes are produced. In *S. keniodendron* these eventually grow into lateral branches, forming an aerial crown of rosettes; in *S. brassica* the new rosettes develop their own root systems and the connections with the old plant base eventually break down. As a result, clones of rosettes develop, with immediately adjacent rosettes often maintaining physiological connections. Because of this clonal growth pattern, it is impossible to define physiological and genetic 'individuals' without extensive, destructive excavation or electrophoretic analyses. As a result, meaningful mortality rates were not calculated for comparison with *S. keniodendron*.

Senecio brassica and *S. keniodendron* hybridize commonly within the narrow zone of overlap between the two species (see Figure 15.2). Hybrids appear to show continuous morphological variation in sexual and vegetative characters. Nonetheless, the species distinctions remain unambiguous outside the narrow hybrid zone. Seeds from several hybrid individuals at 4200 m elevation showed 2% germination when kept in pots in the field at 4200 m for $6\frac{1}{2}$ months (100 seeds were sown on the soil surface, using soil from the grassy north-facing slope site, concurrently with treatments shown in Table 15.1).

Table 15.6 *Total leaf production (g dry weight) per rosette from May 1977 to August 1978 for* S. brassica, *in relation to reproduction in 1979*

Population	Reproduced in 1979	Vegetative in 1979	d.f.	F	p
3550 m					
Level site	447 ± 327	254 ± 170	49	6.28	<0.05
3900 m					
South-facing slope	550 ± 149	239 ± 120	48	28.79	<0.001
North-facing slope	633 ± 187	280 ± 161	48	17.29	<0.001
Valley floor	444 ± 121	251 ± 147	54	18.66	<0.001
Ridge	367 ± 286	325 ± 153	38	0.18	>0.50
4200 m					
Valley floor, well-drained	173 ± 56	87 ± 43	56	7.77	<0.01
Valley floor, flooded	392 ± 144	161 ± 90	56	53.13	<0.001
4340 m					
Valley floor	280 (n = 1)	82 ± 42 (n=37)			<0.001

Patterns of juvenile establishment and growth in *S. brassica* appear to be similar to those in *S. keniodendron*; most individuals show little or no increase in size from year to year; only occasional plants in favorable microsites grow rapidly. Fastest growth appears to occur in sites which are moist throughout the year, are protected from soil frost heaving and have few interspecific and intraspecific competitors.

Adult growth rates in *S. brassica* are not significantly correlated with rainfall, unlike *S. keniodendron.* One possible explanation for this is that *S. brassica* typically grows on wet valley bottoms which generally stay moist in the dry season due to runoff from melting glaciers at higher elevations. Nearest neighbor effects could not be evaluated for *S. brassica*, because the degree of past and present physiological interdependence among neighbors could not be determined. There was generally a correlation between vegetative growth rate and reproduction (Table 15.6), as in *S. keniodendron.* Total dry weight production between May 1977 and August 1978 was greater for rosettes which subsequently flowered than for rosettes which remained vegetative during the following year. A given rosette seldom grew more than 1 m tall before flowering and dying; production of new lateral rosettes results in lateral clone expansion rather than the vertical growth exhibited by *S. keniodendron.*

Comparison with *Espeletia* (Asteraceae) in the Andes

There are many parallels between *Senecio* on Mount Kenya and *Espeletia* in the northern Andes. Studies on representative *Espeletia* species in the Venezuelan Andes have been presented elsewhere (Smith 1974, 1979, 1980, 1981, 1984; Baruch & Smith 1979), and are only briefly summarized here. The original generic name *Espeletia* is used here in preference to the new generic names proposed by Cuatrecasas (1976) because of the high degree of hybridization that occurs among the proposed genera (Smith 1981; Berry & Calvo, Chapter 13).

Espeletia timotensis, a dominant species at elevations above 4100 m, shows synchronous flowering over its geographical range at intervals of approximately 4 years (Smith 1981); most species at lower elevations flower every year. *Espeletia* species at elevations above approximately 4000 m (e.g. *E. timotensis*, *E. moritziana*, *E. spicata*) typically have nodding capitula with greatly reduced ray flowers, similar to *S. keniodendron*, and are wind-pollinated, whereas species at lower elevations (e.g. *E. schultzii*, *E. floccosa*) have erect capitula with conspicuous ray flowers and are insect-pollinated (Berry & Calvo, Chapter 13), paralleling *S. brassica*.

Unlike *S. keniodendron* and *S. brassica*, many *Espeletia* species have lateral inflorescences, so that terminal rosettes do not die after reproduction, and branching typically does not occur. However, two species in treeline forests, *E. humbertii* and *E. neriifolia* do have terminal inflorescences, and have extensive post-reproductive branching; mortality appears to be greater among individuals that flowered than among vegetative individuals of the same size (A. P. Smith, unpublished data). *Espeletia atropurpurea*, a species common at elevations of 3400–3600 m in areas with very high rainfall, also has terminal inflorescence, and shows a pattern of post-reproductive rosette die-back and clonal growth similar to that in *S. brassica*. *Espeletia floccosa*, a species common on dry slopes between 3400 and 3700 m, also has terminal inflorescences, but the entire plant dies after first reproduction, without lateral shoot production (Smith 1981), a pattern similar to that in other tropical alpine rosette species (Smith & Young 1987; Young & Augspurger 1991), including *Lobelia telekii* (Young 1984, 1990 and Chapter 14), *Puya raimondii* (Bromeliaceae) in the Peruvian and Bolivian Andes, *Argyroxiphium sandwicense* (Asteraceae) in Hawaii (Rundel & Witter, Chapter 16) and *Echium wilprettii* (Boraginaceae) in the Canary Islands (Smith, Chapter 1).

Patterns of seedling establishment and growth in *E. schultzii* are similar in many ways to those in *S. keniodendron* (Smith 1980, 1981, 1984). Most seedlings die during the first dry season after germination. Individuals which survive this early stage typically spend many years as suppressed juveniles, showing little or no increase in height or diameter, and eventually die. Competition, frost heaving and drought stress are major factors contributing to mortality and reduced growth. Only a few juveniles in favorable microsites grow rapidly and survive to reproductive age (see also Perez 1991). As in *S. keniodendron*, adult *E. schultzii* plants appear to have a 'nurse tree' effect, improving juvenile establishment and growth in sites where interspecific competition is a significant factor (Smith 1984).

Growth in *E. schultzii* adults is reduced by competition with neighboring conspecific adults and with other species (Smith 1980, 1984) and by dry season stress (Smith 1981), as in *S. keniodendron*. Annual leaf production per rosette at 4200 m was 172 g dry weight per year for *E. schultzii* and 231 g d.w. per year for *E. timotensis* (Smith 1981; '*E. lutescens*' in Smith 1981 is a synonym for *E. timotensis*), compared to 1154 g d.w. per year for *S. keniodendron* (Smith & Young 1982) and 313 g d.w. per year for *S. brassica* (recalculated from data in Table 15.6). There is a distinct elevation gradient in plant size, both among and within *Espeletia* species (Smith 1980): within *E. schultzii* both mean and maximum adult height increase with increasing elevation. In addition, the species producing the tallest

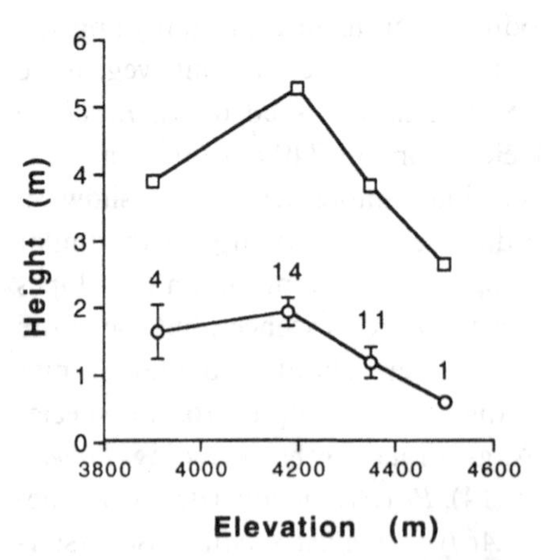

Figure 15.5. □, Plant height, to highest leaf tip, of the tallest plant within each of four *Senecio keniodendrpm* study populations at different elevations in Teleki Valley. ○, Mean height (±1 SE) of *S. keniodendron* plants found in vegetation transects in the Teleki Valley (cf. Young & Peacock 1992). Sample sizes are the number of transects in each elevational class.

plants tend to occur at the highest elevations (excluding those arborescent *Espeletia* species which grow in treeline forests). This trend appears to be related primarily to greater longevity at high elevations rather than to greater height growth rates. Height growth is generally 1–2 cm per year at all elevations. The within-species trend does not seem to apply to *Senecio* on Mount Kenya (Figure 15.5). However, the between-species trend does pertain: *S. brassica* is dominant at lower elevations and seldom grows more than 1 m tall, while *S. keniodendron* dominates at higher elevations, and grows several meters tall. Possible causes of such trends are discussed in Smith (1980, 1984) and Young & Smith (Chapter 18).

Espeletia schultzii populations are subject to extensive mortality due to fire, drought and unusually severe flooding (Smith 1981), but are not currently subject to the sort of intense herbivore attacks by mammals that can decimate *S. keniodendron* populations (Mulkey *et al.* 1984; Young & Smith, Chapter 18).

The distinct niche differentiation seen between *S. keniodendron* and *S. brassica* along elevational and topographic gradients has also been documented in *Espeletia* in Venezuela (Baruch & Smith 1979). *Espeletia atropurpurea* occurs only in high rainfall areas just above treeline, whereas *E. schultzii* extends into drier and higher elevation sites. In the zone of overlap between the two species *E. atropurpurea* abundance is positively correlated with slope steepness while *E. schultzii* is negatively correlated with steepness (compare Figure 15.2 here).

Similarities between *Espeletia* and *Senecio* with respect to population biology tend to support the suggestion, based largely on morphology, that the two groups are strongly convergent (Hedberg & Hedberg 1979; Smith & Young 1987). However, evaluation of this hypothesis will require a more synthetic approach. It is clear that morphological, physiological and life history characters covary in complex ways both within and among species. For example, life history is closely related to stem height in *E. schultzii* (Smith 1980), and stem height affects water relations (Goldstein *et al.* 1984). Moreover, these character syndromes have evolved in response to complex biotic as well as edaphic and climatic interactions. For example, herbivory (Young & Smith, Chapter 18) and interspecific competition (Smith 1984) probably influenced the evolution of life history, suggesting that variation in population processes and in other characters must be analysed in a community context. Finally, phylogenetic history must be considered, because *Espeletia* and *Senecio* are both in the Asteraceae, although in different tribes. Their similarities may therefore result in part from common origin, although similar character syndromes occur in unrelated genera (Smith, Chapter 1) supporting the idea of convergence. The challenge is to integrate these diverse fields in a comprehensive analysis of convergence among tropical alpine taxa.

Acknowledgements

Research was supported by grants from the Smithsonian Scholarly Studies Program, the Smithsonian Fluid Research Fund, the Smithsonian Research Opportunities Fund, The Smithsonian Tropical Research Institute, and the Explorer's Club. We are grateful to Martin Otieno, who collected many of the long-term data on adult growth and seedling dynamics. John Omirah Miluwi provided invaluable logistic support and advice. Additional assistance was provided by William Newmark, Josh Lincoln, Stephen S. Mulkey, Mary Peacock and Nate Stephenson. We thank Phil Snyder and Mr Oloo, former Wardens, Mt Kenya National Park, for their assistance. Dr S. K. Imbamba, former Chairman, Department of Botany, University of Nairobi, provided essential laboratory and herbarium facilities. We thank the Ministry of Tourism and Wildlife (now Kenya Wildlife Service) and the Office of the President, Republic of Kenya, for permission to carry out manipulative experiments on Mt Kenya.

References

Baruch, Z. & Smith, A. P. (1979). Morphological and physiological correlates of niche breadth in two species of *Espeletia* (Compositae) in the Venezuelan Andes. *Oecologia* **38**, 71–82.

Cuatrecasas, J. (1976). A new tribe in the Heliantheae (Compositae): Espeletiinae. *Phytologia* **35**, 43–61.

Hedberg, O. (1969). Growth rate of the East African giant Senecios. *Nature* **222**, 163–4.

Hedberg, I. & Hedberg, O. (1979). Tropical alpine life-forms of vascular plants. *Oikos* **33**, 297–307.

Goldstein, G., Meinzer, F. C. & Monasterio, M. M. (1984). The role of capacitance in the water relations of Andean giant rosette species. *Plant, Cell and Environment* **7**, 179–86.

Mulkey, S. S., Smith, A. P. & Young, T. P. (1984). Predation by elephants on *Senecio keniodendron* (Compositae) in the alpine zone of Mt. Kenya. *Biotropica* **16**, 246–8.

Perez, F. L. (1987). Soil moisture and the upper altitudinal limit of giant paramo rosettes. *Journal of Biogeography* **14**, 173–86.

Perez, F. L. (1989). Some effects of giant Andean rosette species on ground microclimate, and their ecological significance. *International Journal of Biometeorology* **33**, 131–5.

Perez, F. L. (1991). Soil moisture and the distribution of giant Andean rosettes on talus slopes of a desert paramo. *Climate Research* **1**, 217–31.
Pielou, E. C. (1962). The use of plant-to-neighbor distances for the detection of competition. *Journal of Ecology* **50**, 357–67.
Smith, A. P. (1974). Bud temperature in relation to nyctinastic leaf movement in an Andean giant rosette plant. *Biotropica* **6**, 263–6.
Smith, A. P. (1979). The function of dead leaves in an Andean giant rosette plant. *Biotropica* **11**, 43–7
Smith, A. P. (1980). The paradox of plant height in an Andean giant rosette species. *Journal of Ecology* **68**, 63–73.
Smith, A. P. (1981). Growth and population dynamics of *Espeletia* (Compositae) in the Venezuelan Andes. *Smithsonian Contributions to Botany* **48**, 1–45.
Smith, A. P. (1984). Post-dispersal parent–offspring conflict in plants: antecedent and hypothesis from the Andes. *American Naturalist* **123**, 354–70.
Smith, A. P. & Young, T. P. (1982). The cost of reproduction in *Senecio keniodendron*, a giant rosette species of Mt. Kenya. *Oecologia* **53**, 418–20.
Smith, A. P. & Young, T. P. (1987). Tropical alpine plant ecology. *Annual Review of Ecology and Systematics* **18**, 137–58.
Young, T. P. (1984). Comparative demography of semelparous *Lobelia telekii* and iteroparous *Lobelia keniensis* on Mount Kenya. *Journal of Ecology* **72**, 637–50.
Young, T. P. (1985). *Lobelia telekii* herbivory, mortality, and size at reproduction: variation with growth rate. *Ecology* **66**, 1879–83.
Young, T. P. (1990). The evolution of semelparity in Mount Kenya lobelias. *Evolutionary Ecology* **4**, 157–71.
Young, T. P. & Augspurger, C. K. (1991). Ecology and evolution of long-lived semelparous plants. *Trends in Ecology and Evolution* **6**, 285–9.
Young, T. P. & Evans, M. E. (1993). Alpine vertebrates of Mount Kenya, with particular notes on the rock hyrax. *Journal of the East Africa Natural History Society* **82** (202), 55–75.
Young, T. P. & Peacock, M. M. (1992). Giant senecios and the alpine vegetation of Mount Kenya. *Journal of Ecology* **80**, 141–8.

16

Population dynamics and flowering in a Hawaiian alpine rosette plant, *Argyroxiphium sandwicense*

PHILIP W. RUNDEL and M. S. WITTER

Introduction

A notable feature of tropical alpine floras in many parts of the world is the presence of rosette plants with monocarpic growth habits. These long-lived perennials flower only once in their lives, producing a giant inflorescence with large numbers of flowers before the parent plant dies. In the South American páramos of Venezuela, species of *Espeletia* (Asteraceae) provide classic examples of this life-form (Cuatrecasas 1986; Berry & Calvo, Chapter 13)). Giant species of Afroalpine *Lobelia* (Campanulaceae) reach to 5–6 m in height when flowering before dying, and form one of the most spectacular elements of the flora of the tropical African highlands (Mabberley 1974; Young, Chapter 14). On Tenerife in the Canary Islands, two alpine species of *Echium* (Boraginaceae) have also evolved rosette growth forms and a monocarpic habit (Carlquist 1974). In each of these groups, the monocarpic rosette plants have evolved from polycarpic ancestors with a shrubby growth form.

Argyroxiphium sandwicense (Asteraceae) provides another well-known example of the evolution of a monocarpic rosette plant in a tropical alpine environment. The genus *Argyroxiphium*, with five species, is one of three genera of Hawaiian tarweeds that have evolved from a monophyletic North American origin (Carr 1985; Witter & Carr 1988; Baldwin *et al.* 1991). This group displays a remarkable diversity of growth forms and adaptive morphologies. Monocarpic rosette plants have evolved not only in species of *Argyroxiphium* in bog and in alpine habitats, but also in *Wilkesia gymnoxiphium*, a stalked rosette plant in scrub and open forest habitats.

In this chapter, we review the demography and reproductive biology of *Argyroxiphium sandwicense* as a model for understanding the evolution of monocarpy in long-lived perennial plants, and the relationship between monocarpy and a rosette growth form. Our studies are based on a permanent plot of individuals established in Haleakala National Park, Maui, in 1982, and monitored for growth and flowering over a subsequent 5-year period.

Silversword

Argyroxiphium sandwicense, the famous Hawaiian silversword, is one of five species of this genus that have evolved in the Hawaiian Islands. Two subspecies are known. *Argyroxiphium sandwicense* ssp. *sandwicense* is restricted to the upper slopes of Mauna Kea on the island of Hawaii where it was once reported to have been widespread and common. The impact of feral animals has drastically reduced the abundance of this subspecies and it is restricted today to one small population in the Wailuku River drainage at about 2850 m elevation (Powell 1985).

The more abundant subspecies, *Argyroxiphium sandwicense* spp. *macrocephalum* occurs on the outer slopes and inner caldera of Haleakala Crater on East Maui where it grows on cinder cones and lava flows at 2200–3000 m elevation. The silversword is the focal point of public interest in Haleakala National Park and concern about its management was an important factor leading to the establishment of Haleakala as a section of Hawaii National Park in 1916 (and later as Haleakala National Park in 1961). Browsing by goats and cattle and human vandalism led to marked declines in populations of Haleakala silversword in the latter part of the 19th century and early part of this century. Following protection afforded by National Park status, silversword numbers have increased and now total about 50 000 individuals (Loope & Crivellone 1986).

While overall population size of Haleakala silversword increased only about 10% between censuses in 1971 (Kobayashi 1973) and 1982 (Loope & Crivellone 1986), populations on individual lava flows and cinder cones with Haleakala showed pronounced fluctuations with increases up to 70% in numbers and decreases up to 50%. Little is known about which factors control such apparent population shifts. These data suggest that silversword may well exist under dynamic conditions for growth. Kobayashi's studies (1973) indicated that establishment and survival of silversword are restricted to a relatively specific stage of physiographic succession on their volcanic soils. Substrates on which silverswords grow in Haleakala Crater all appear to be less than 2500 years old (Macdonald 1978). Biological competition with other plant species does not appear to be a factor since few other species grow with silversword. Very often, silverswords grow as monocultures in open stands on the slopes of cinder cones within the caldera.

Despite the lunar appearance of Haleakala Crater, precipitation is moderate in the silversword habitat. Median annual precipitation is about 1250 mm, with a range of about 1000–1500 mm per year for the silversword habitats. About 75% of this rainfall occurs from November to April, with summers being relatively dry (Blumenstock & Price 1967). When the tradewind inversion is high, dense clouds may rapidly move into Haleakala Crater and create conditions with abundant fog drip. Kobayshi (1973) suggested that such fog drip may be an important source of moisture for silverswords. As is characteristic of tropical alpine environments, mean daily temperature varies by 4 °C from January to July, but diurnal changes in temperature of 20 °C are common. Freezing temperatures may occur at any time of the year.

Flowering in Haleakala silversword

Flowering in the Haleakala silversword takes place with the formation of a large inflorescence 70–150 cm in height from the center of the acaulescent rosette. Flowering stalks begin to emerge in late May and June, with peak flowering in July and August. The number of silverswords flowering in any given year is highly variable. Counts carried out from 1969 to 1987 by the National Park Service and associated researchers have recorded years with no or virtually no flowering of silverswords, as in 1970, and 1972, and years with more than 2200 flowering individuals as in 1982 (Figure 16.1). Even heavier flowering occurred again in 1991. Despite this long-term record of flowering, however, the causal factors promoting flowering are still not understood. Demographic data show that flowering is not correlated with growth pulses in the population structure of larger rosettes. Simple correlates with precipitation totals in the current or preceding year or season are not significant. Qualitatively, there appears to be a pattern where regimes of constant moisture, followed by drought, are required for heavy flowering. No simple environmental correlates have been established, however.

The size class distribution of flowering rosettes was measured from 1982 to 1985 in a permanent plot and in a larger population nearby along Silversword Loop. Individuals which flower in a given year are normally the larger individuals in a population, typically with rosette diameters of 30 cm or more. Few vegetative rosettes exceed 60 cm in diameter. During 1982, one of the most prolific years on record for flowering, the modal

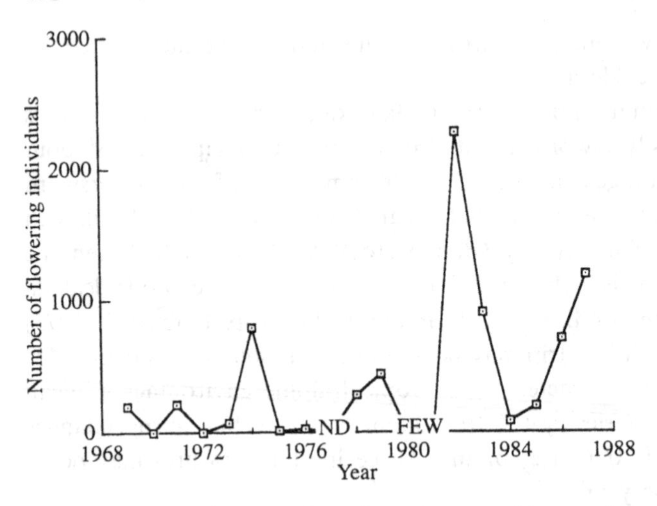

Figure 16.1. Number of flowering *Argyroxiphium sandwicense* in Haleakala National Park censused each year from 1969 to 1987. Data from Loope & Crivellone (1986) and L. L. Loope (personal communication).

size for flowering individuals at Silversword Loop and our nearby permanent plot in Haleakala National Park was 45–49 cm rosette diameter (Figure 16.2). Two individuals with rosettes smaller than 20 cm flowered, an unusually small size. Large individuals with rosettes over 50 cm diameter comprised 30% of the flowering population.

During 1983, a moderately good year for flowering, only 14 individuals flowered in this area, compared with 49 in the previous year. All of these flowering individuals had diameters between 30 and 49 cm, a much more constrained range than in 1982. This pattern suggests that conditions in 1982 were sufficient to cue physiologically a high percentage of large individuals to flower as well as to stimulate a select number of small individuals to produce inflorescences. It is not clear whether carbohydrate and/or nutrient reserves provide a cue to initiate flowering. It is interesting to note that silversword cultivated at the Haleakala National Park headquarters have flowered at relatively small size after only 3–4 years of growth. Favorable conditions for flowering in 1983 were moderated by a reduced population of large rosettes to cue for flowering after the previous year of extensive flowering. No individuals smaller than 35 cm rosette diameter flowered in 1984, a poor year, but 30% of the 10 flowering individuals sampled in the area of Silversword Loop were again 50 cm or more in diameter (Figure 16.2).

Inflorescence size showed a significant morphometric correlation with

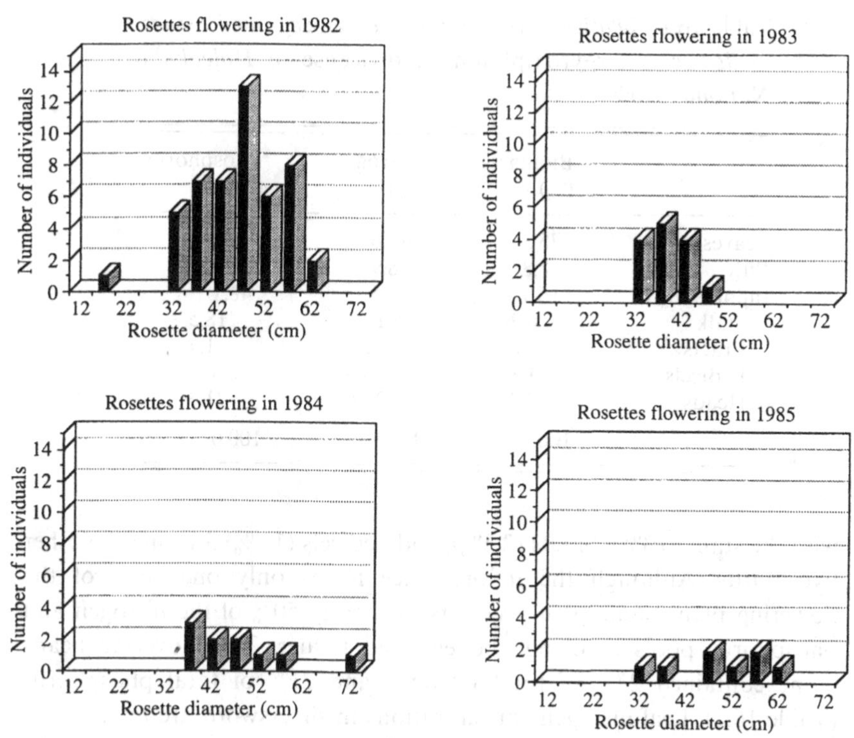

Figure 16.2. Number and rosette diameter of *Argyroxiphium sandwicense* flowering at Silversword Loop and Kobayashi No. 56 (1982), the site used in this study, from 1982 to 1989, in Haleakala National Park.

rosette diameter (Meyrat *et al.* 1983). Larger rosettes produced larger inflorescences with larger numbers of flowers and seeds. Thus if the number of seeds produced is an important component in successful re-establishment of silversword, then parent plants may well benefit from delayed reproductive maturity. However, since the parent genet dies completely following flowering, number of seeds may well be less important than timing of seed release into favorable microclimatic conditions. It is surprising, therefore, that flowering induction is not more clearly correlated with current season climatic conditions.

In a flowering silversword, living leaf tissues in the rosette comprised 75% of the above-ground biomass. The small internal pith added another 9% of this biomass leaving 16% for the inflorescence (Table 16.1). This is not an unusually large biomass allocation to flowering compared to other monocarpic perennials with large inflorescences. The flowering heads themselves accounted for one third of the total weight of the inflorescence,

Table 16.1 *Relative distribution (%) of aboveground tissues in flowering* Argyroxiphium sandwicense *in Haleakala National Park*

	Biomass (%)	Nitrogen (%)	Phosphorus (%)
Leaves	75.3	42.6	44.0
Pith	8.8	8.9	7.9
Inflorescence	15.9	48.5	48.1
Stalk	6.3	12.9	18.5
Bracts	2.6	3.1	1.4
Pedicels	1.8	5.9	6.9
Heads	5.2	26.6	21.3
	100.0	100.0	100.0

with the stalk (6.3%), bracts (2.5%), and pedicels (1.8%) forming the other two thirds. Although the inflorescence forms only one sixth of the flowering plant biomass, it accounts for nearly 50% of the nitrogen and phosphorus pools in living aboveground tissues. The flowering heads alone contained 27% of total nitrogen and 21% of total phosphorus (Table 16.1). Leaf nitrogen concentrations in silversword are remarkably low, at only 2.7 mg g^{-1} dry weight. This low concentration may be the result of dilution by the large relative biomass of carbon in the dense, non-living hairs which cover the leaf surface. The root/shoot ratio in the flowering individual was 0.10 for biomass, 0.06 for nitrogen, and 0.03 for phosphorus.

The largest single biomass and nutrient pool in a large silversword is formed by the dead marcescent leaves on the outer whorls of the rosette. This dead leaf mass comprises 73% of total aboveground tissues (Figure 16.3), 57% of total nitrogen and 42% of total phosphorus.

Demography

Detailed studies on population size of silversword were carried out in Haleakala by Kobayashi (1973), who divided individuals into three categories in his census: rosettes less than 20 cm diameter, rosettes over 20 cm in diameter, and flowering plants. This scale of sampling, however, was not sufficient to analyse the growth dynamics within a population.

In August 1982, we set up a permanently marked population of silverswords in Haleakala Crater near Silversword Loop. This site was

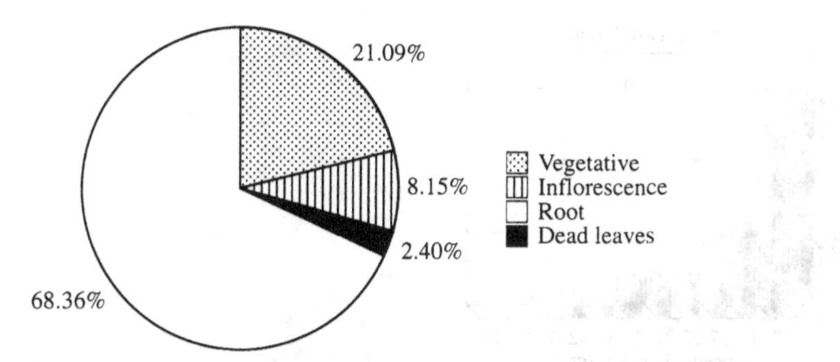

Figure 16.3. Distribution of living and dead biomass within tissues of a flowering rosette of *Argyroxiphium sandwicense* in Haleakala National Park.

termed population No. 56 in the census of Kobayashi (1973), and occurs on shallow cinders on the northeast flank of Puu o Maui at 2200 m elevation. A complete census was made of the 225 rosettes forming the major part of this population, and another 48 individuals were selectively added nearby to increase the sample size of larger rosette sizes to understand better the factors controlling flowering. The rosette diameter and flowering condition (if any) of each of these marked individuals was remeasured in the summers of 1983, 1984 and 1985, and the winter of 1987. Mortality of rosettes was recorded as well as recruitment of new seedlings into the population.

The initial population structure of our study site in 1982 is shown by size class in Figure 16.4. This careful census of 255 individuals included 78% with rosette diameters less than 20 cm. The overall distribution of our population was very similar to a census of 500 silversword rosettes carried out in 1983 at Silversword Loop nearby (Figure 16.4).

We have utilized our demographic data to develop a matrix model of population dynamics in Haleakala silversword. This model follows basic matrix structure for populations as first described by Leslie (1945), but modified to use size rather than age data (Hartshorn 1975). Applications of matrix models in population biology have been described in detail by Caswell (1989). We have utilized this model to look at changes in population structure over 3-year (1982–5) and 5.5-year (1982–7) intervals to estimate the mean age of plants in any growth state and to predict the distribution of plants by size class for a stable population structure. The first 3-year period was relatively dry and thus potentially stressful for silversword growth. For this reason we used an interval of 5.5 years to average out conditions of growth over a longer period.

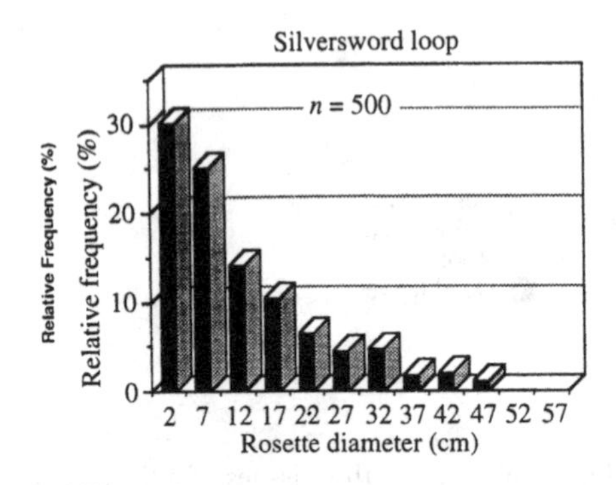

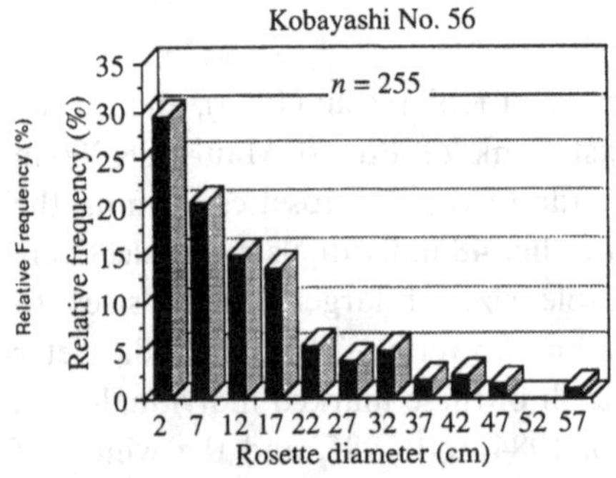

Figure 16.4. Population structure of *Argyroxiphium sandwicense* in 5 cm classes of rosette diameter for Silversword Loop (1983) and Kobayashi No. 56 (1982), the site used in this study. Data from Silversword Loop from L. Powell (personal communication).

In our model a silversword passes through 11 growth states, including emergence (germination) and death, as shown in Table 16.2. From our field data, a transition matrix was developed to predict the probability of change in growth state for a plant in a current condition (Table 16.3). Each element in the matrix is the probability (in per cent) of a plant moving to the next state. Unlike the size dynamics of many organisms, silversword rosettes have the potential to grow, stay the same, or actually shrank under stressful conditions. Even large rosettes up to 50 cm diameter shrank one size class over the 5.5 years of study. From the ninth state, flowering, it is only possible for a monocarpic rosette such as silversword to move into the tenth state, death. This final state is terminal and eventually accumulates all individuals entering the population.

The average age of plants within a given size class can be estimated by modelling the input of new seedlings into a hypothetical population

Table 16.2 *Growth states of* Argyroxiphium sandwicense

Emergence	0
1 cm rosette	1
2–4 cm rosette	2
5–9 cm rosette	3
10–19 cm rosette	4
20–29 cm rosette	5
30–39 cm rosette	6
40–49 cm rosette	7
50–59 cm rosette	8
Flowering	9
Death	10

Table 16.3 *Transition matrix for changes in growth states of* Argyroxiphium sandwicense *from 1982 to 1987 in Haleakala National Park, Hawaii. See Table 16.2 for description of growth states*

	Next state (j)									
Initial state (i)	1	2	3	4	5	6	7	8	9	10
0	0.061	0.091								0.848
1	0.061	0.091								0.848
2	0.225	0.375	0.100							0.300
3		0.365	0.538	0.058						0.038
4		0.013	0.184	0.645	0.105	0.026				0.026
5				0.125	0.525	0.300	0.025			0.025
6					0.087	0.632	0.158	0.070	0.035	0.018
7						0.077	0.538	0.077	0.307	
8								0.500	0.500	
9										1.00

followed by sequential growth periods. For our 5.5-year matrix, as shown in Table 16.3, the modal ages of plants, in states 4 and 5 above and below 20 cm diameter, were determined to be 22 and 33 years respectively. This suggests that it takes new seedlings 25–30 years to reach this size. Some plants, however, reached this size at 16.5 years in the model. Our 3-year matrix predicted a shorter time interval of 15–24 years to reach a 20 cm rosette, and some plants grew to such size in 12 years. Because statistical population models of this type may often overestimate mean ages of population size classes because they include small individuals that never grow to large size, we looked at the maximum growth rate in diameter

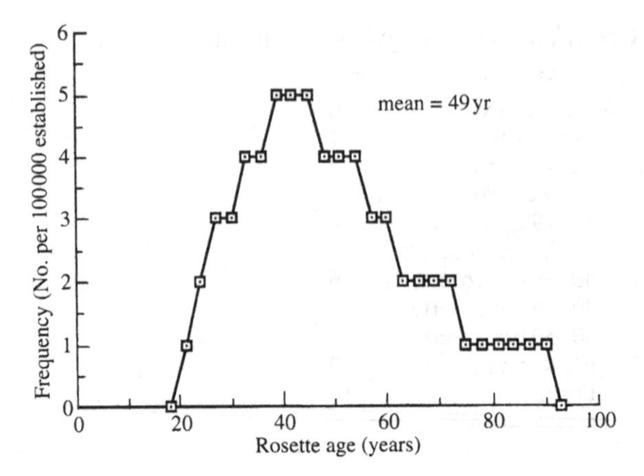

Figure 16.5. Model prediction of flowering in *Argyroxiphium sandwicense* in relation to rosette age for an initial population of 100 000 seedlings established. Data are based on a growth matrix for 1982–4.

for small rosettes. The maximum mean annual growth rate we observed in rosettes which were initially smaller than 5 cm diameter was 2 cm per year. At such a rate, a 2 cm diameter rosette would reach 20 cm in 10 years. Adding a few years for establishment, this is slightly younger than our model predicts. Even among large individuals of potential flowering size, it was rare to have mean annual growth rates of more than 2 cm. Single year growth rates in large rosettes were commonly 3–4 cm, with a record level of 10 cm, but such high rates were never maintained for multiple years.

Our 5.5-year matrix estimated a modal age for flowering of 49 years, very close to the prediction of 39–45 years for the 3-year model. The median age for a flowering plant is 63 and 49 years, respectively, in these two models as some plants continued to reproduce at relatively old age. For the 3-year model, silverswords begin flowering at an average age of 21 years and some individuals are predicted to survive up to 90 years before flowering (Figure 16.5). These estimates are considerably older than those previously suggested for silversword. Ruhle (1959) suggested that silverswords reach flowering age in only 7–20 years, less than half the time predicted by our model.

Conclusions

The adaptive significance of monocarpy in silversword is still poorly understood. Clearly the monocarpic growth habit allows the production of very large numbers of seeds in years in which flowering is abundant, similarly to the derived monocarpic condition in other tropical alpine rosette plants. Such a reproductive strategy could be expected to evolve under conditions where seedling survivorship is relatively high or adult mortality rates high. Under such conditions, long-term reproductive output is maximized by the monocarpic habit. Our data show that silverswords are long-lived, however, with some plants probably living up to nearly a century. A key factor in the silversword habitat may be the speed of physiographic weathering of their substrate. Studies by Kobayashi (1973) strongly suggested that silverswords colonize cinders and lava flows at a specific stage of substrate development, and furthermore, are absent from more deeply weathered substrates. If this proves to be true and the time frame for the weathering cycle can be measured in decades rather than centuries, an argument for the evolution of monocarpy may be made. Plants may face increasingly unfavorable conditions for survival as they age. Seed output might thus be maximized by monocarpic reproduction.

The basis for cueing for the initiation of flowering and the subsequent senescence of the vegetative rosette remains unknown. While physiological triggers related to carbohydrate and/or nutrient storage are the likely proximate cause of monocarpic flowering, environmental cues must be involved as well. Most flowering of thousands of rosettes may occur in favorable years and no flowering at all may occur in others. Attempts to correlate flowering with precipitation patterns in previous weeks, months or years has not been successful.

Flowering cues initiated a year or more in the future would seem a poor adaptation for silversword if seedling survivorship is variable between years. If conditions unfavorable for seed germination and seedling establishment occurred following flowering and seed production, maintenance of population size and structure could be significantly damaged. Careful studies of seedling survivorship are of critical importance in understanding the evolution of most flowering in silversword and other monocarpic rosette plants from tropical alpine habitats.

References

Baldwin, B., Kylos, D. W., Dvorak, J. & Carr, G. D. (1991). Chloroplast DNA evidence for a North American origin of the Hawaiian silversword alliance (Asteraceae). *Proceedings of the National Academy of Sciences* (*USA*) 88, 1840–3.

Blumenstock, D. I. & Price, S. (1967). Climates of the states: Hawaii.

Climatography of the United States, No. 60-51. U.S. Department of Commerce.
Carlquist, S. (1974). *Island Biology.* New York: Columbia University Press.
Carr, G. D. (1985). Monograph of the Hawaiian Madiinae (Asteraceae): *Argyroxiphium, Dubautia* and *Wilkesia. Allertonia* **4**, 1–123.
Caswell, H. (1989). *Matrix Population Models.* Sunderland, MA: Sinauer Associates.
Cuatrecasas, J. (1986). Speciation and radiation of the Espeletiinae in the Andes. In *High Altitude Tropical Biogeography*, ed. F. Vuilleumier and M. Monasterio, pp. 267–303. Oxford: Oxford University Press.
Hartshorn, G. S. (1975). A matrix model of tree population dynamics. In *Tropical Ecological Systems: Trends in Terrestrial and Aquatic Research*, ed. F. B. Golley and E. Medina, pp. 41–51. New York: Springer-Verlag.
Kobayashi, H. K. (1973). Ecology of the silversword *Argyroxiphium sandwicense* DC. (Compositae); Haleakala Crater, Hawaii. PhD dissertation, University of Hawaii, Honolulu.
Leslie, P. H. (1945). On the use of matrices in certain population mathematics. *Biometrika* **3**, 183–212.
Loope, L. L. & Crivellone, C. F. (1986). *Status of the Haleakala Silversword: Past and Present.* Cooperative National Park Resource Studies Unit, University of Hawaii, Honolulu, Tech. Rep. 58.
Mabberley, D. J. (1974). The pachycaul lobelias of Africa and St Helena. *Kew Bulletin* **29**, 535–85.
Macdonald, G. A. (1978). Geologic map of the crater section of Haleakala National Park, Maui. *U.S. Geol. Surv. Misc. Invest. Ser.*, Map I-1088.
Meyrat, A. K., Carr, G. D. & Smith, C. W. (1983). A morphometric analysis and taxonomic appraisal of the Hawaiian silversword *Argyroxiphium sandwicense* DC. (Asteraceae). *Pacific Science* **37**, 211–25.
Powell, L. (1985). The Mauna Kea silversword: a species on the brink of extinction. *Hawaiian Botanical Society Newsletter* **24**, 44–57.
Ruhle, G. C. (1959). *Haleakala Guide.* Honolulu: Hawaii Natural History Association.
Witter, M. S. & Carr, G. D. (1988). Adaptive radiation and genetic differentiation in the Hawaiian silversword alliance (Compositae: Madiinae). *Evolution* **42**, 1278–87.

17

Plant form and function in alpine New Guinea

R. J. HNATIUK

Introduction

The tropical alpine environment is surprising. Although it is predictable (Troll 1961; Humboldt, in Botting 1973), it is still remarkable to experience cool, misty mountains that rise out of the lowland tropics. These alpine places bear surprising resemblance in vegetation to the arctic, and to the alpine zones of temperate mountains. They also bear climatic and vegetational resemblance to remote islands of the Southern Ocean. The similarities of vegetation derive primarily from the changes in physical climate that occur as altitude increases on humid mountains. As temperatures decline, different plant species replace those adapted to growth in warmer conditions. The conspicuous changes to vegetation are a transition from tall, tree-dominated communities to low, shrub or herb-dominated communities. These transitions occur on a grand scale that has been expressed as a three-dimensional model of climate and vegetation of the earth (Troll 1961; Humboldt, in Botting 1973).

There is controversy about the terms that should be applied to the regions. The argument turns on the degree of recognition given to thermal seasonality, which is weak or absent on the high tropical mountains and profound in the temperate alpine and near polar regions. There are many regional names: alpine, páramos, pahonales, mountain grassland, oro-subantarctic, gras maunten, ais, etc. Each has its utility for its own zone – none is without misleading connotations when applied outside its region of origin. In this chapter, the phrase 'tropical alpine' is being used to conform to the usage of this book.

This chapter aims to summarize the sparse information on plant form and function that is available for the tropical alpine areas of New Guinea. Most of the data come from Mount Wilhelm (lat. 5° 47′ S, long 140° 01′, alt. 4510 m) which is the highest point in Papua New Guinea (Figure 17.1).

Figure 17.1. Mt Wilhelm with light ephemeral snow cover in early morning.

To understand the form and function of tropical alpine plants it is necessary to know the climate of these high humid mountain regions. The major influences are perpetually cool temperatures, frequent diurnal frosts throughout the year, little or no seasonality of temperatures, some seasonality of precipitation, generally high UV, very variable levels of photosynthetically active radiation in relation to diurnal cloud formation, and generally little wind. To varying degrees these aspects of climate and microclimate have been documented in New Guinea (Hnatiuk *et al.* 1976; Smith 1977; Korner *et al.* 1983) and are discussed on a world perspective in Chapters 1 and 4 of this volume.

Tropical alpine zone of New Guinea

Tropical alpine environments of the island of New Guinea occur on isolated peaks that rise from the 2000 km long highland spine which runs along the centre of the island. The western peaks rise highest and culminate in the 4884 m Mount Jaya. These western peaks lie in Irian Jaya, a province of Indonesia, and are very little known biologically (Hope *et al.* 1976). The eastern peaks, occurring in Papua New Guinea, are somewhat better known (Brass 1964; Wade & McVean 1969; Smith

1975a, b, 1977; Hnatiuk *et al.* 1976; Hope 1976; Hnatiuk 1978; Corlett 1984). Of these Mt Wilhelm is best known, largely because of a semi-permanent research facility that was established in 1966 and for many years run by the Australian National University.

Mount Wilhelm is geologically young, being Miocene to Pliocene in age. It was formed by uplift associated with the northward movement of the Australian plate, in terms of plate tectonics and the breakup of Gondwana. The rocks are derived from a large granodiorite batholith (Noakes 1939) and are known as Bismarck granodiorite (Rickwood 1955). The summit areas have more basic rocks such as pyroxenite and hornblendite (Dow & Dekker 1964) and a variety of other accessory minerals are locally abundant. The topography of the tropical alpine and parts of the subalpine zones has been formed by Pleistocene glaciation (Hope *et al.* 1976). The soils derived from these rocks, and the peats overlying them are generally low in nutrients (Wade & McVean 1969).

The palaeontological history from Pleistocene to the Present of areas of the New Guinea tropical alpine zone are discussed by Hope (1976) and Corlett (1984). These studies indicate the dynamic nature of the vegetation as glaciers came, grew, and decayed on the highest peaks. They also document the effects which man has had in modifying the vegetation. In particular, the use of fire has extended the 'alpine' shrub, grass and herblands at the expense of forested land over the past 700 years (Corlett 1984).

On Mount Wilhelm, the true tropical alpine zone is believed to begin at about 3800–4200 m, but treeless grassland mires, that are floristically related to and in part derived from the tropical alpine flora, extend in valley bottoms down to 3510 m (Wade & McVean 1969; Hnatiuk 1978). The upper limit of tree growth in the humid mountains of south and eastern Asia correlated closely with the mean temperatures of the warmest and coldest months (Ohsawa 1990).

Plant form and function

Most detailed studies of plant form and function in tropical alpine regions have dealt with its most stunning plant forms – the giant rosettes and giant herbs (Hedberg 1964; Smith 1974; Mabberley 1975; Young 1984) which characterize some of these environments and those of some sub-Antarctic islands. These forms do not occur in the tropical alpine of New Guinea, except to the extent that stout tree ferns grow abundantly in grassland at the lower end of otherwise subalpine valleys.

Table 17.1 *Number of plant associations recognized on Mount Wilhelm (Wade & McVean 1969), and Mount Jaya (Hope 1975)*

Association	Mount Wilhelm	Mount Jaya
Subalpine grasslands	7	3
Mires	5	3
Landslips	4	?3
Alpine grasslands and fern meadows	3	2
Alpine heath and tundra	4	4

The non-arboreal plants of the tropic alpine–subalpine zones of Mount Wilhelm exhibit a range of plant forms common on other high mountains. These are low herbs, dwarf and tall shrubs, tufted and tussocked grasses, mosses, and lichens, to name the most common. Unusual growth forms are found in the stands of treeferns (*Cyathea* spp.), about 2 m in height, in the subalpine grassland, and the finger-like ferns (*Papuapteris linearis*) of the alpine grasslands. Hard cushion bogs, which show no indentation when stepped on, occur on the western peaks of New Guinea. Individual cushion plants occur in the central mountains of that island. Similar growth forms also occur in the Andes, New Zealand, Fuegia, Tasmania, Africa, and the sub-Antarctic islands.

Plant forms, in combination with floristic composition, have produced 23 plant associations as determined by Wade & McVean (1969) (Table 17.1) for the subalpine and tropic alpine areas of Mount Wilhelm. On Mount Jaya, a less extensive survey revealed 17 plant associations (Hope 1976). Hope comments that the plant communities are significantly different between the two areas, apart from the feature of extensive grassland–shrubland vegetation. The non-forested areas are of very different ages post glaciation – Mount Jaya is much younger. Surveys that are both more extensive and more detailed are required of most of the other high mountain areas of New Guinea.

A characteristic feature of tropical alpine–subalpine vegetation is the zone of tall shrubs found between the subalpine forests and the low shrubland–grasslands (Wade & McVean 1969; Smith 1975b; Hope 1976). A range of species inhabit this zone. In form and location they are reminiscent of *krummholz* but the similarity of form and function needs further study. They appear to be especially prone to destruction by burning (Smith 1975b; Hope 1976).

The floras of Mount Wilhelm and Mount Jaya are not particularly rich.

About 160 species of vascular plants, 100 species of mosses, perhaps fewer than 100 species of lichens and an unknown number of liverworts were reported in the only comprehensive floristic survey of this area (Wade & McVean 1969). On Mount Jaya, Hope (1976) reported 169 vascular plant species, 55 species of mosses, 27 liverworts and 31 lichens. Although several species have been discovered since, the general situation still attains. There is a mingling of genera with origins in the north and south temperate zones which give a distinctive character to the landscape (e.g. southern genera: *Astelia, Oreomyrrhis, Drapetes*; northern genera: *Ranunculus, Potentilla, Gentiana*).

Very few species of the tropical alpine of New Guinea have been studied in ecological detail. Korner *et al.* (1983) have reported on altitudinal variation in leaf diffusive conductance and leaf anatomy, whilst Smith (1975a) has studied introduced plant species. The most detailed study of plant form is that of Hnatiuk (1979) which only reported on the growth and primary productivity of *Deschampsia klossii*, the dominant tussock grass of the tropical alpine and subalpine communities. The remainder of this chapter summarizes primarily the latter work.

The *D. klossii* tussock grasslands are native to the tropical alpine zone (Figure 17.2). In this zone they are composed of small (0.5 m high, 0.5 m diameter: Figure 17.3), closely spaced tussocks intermixed with dwarf shrubs, and mosses. They are extensive below 4270 m on slopes of less than 30° but from 4100 to 4400 m they are confined to areas of deep, moisture retaining soils (Wade & McVean 1969). In the subalpine zone, from 3510 to 3750 m, *D. klossii* dominates extensive tussock grasslands on well-drained valley floors with slopes of less than 10°. In these areas, the grassland comprises large tussocks (1 m high, 0.9 m diameter; Figure 17.3) spaced such that crowns generally overlap, but with low shrubs being moderately common.

The tussock form is characterized by a central pedestal formed as a dense, often cylindrical mass of stems, roots and leaf bases. On Mount Wilhelm, the development of pedestals is best at mid to low alpine–subalpine altitudes and weakest at high alpine altitudes. Growing upwards from the pedestal is a mass of leaves which can be long or short, abundant or sparse, rigid or flexuose. Growing downwards and outwards from the base are roots, but very rarely stems. Only some of these roots anchor the pedestal to the ground; the rest remain within the pedestal. The sides of the pedestal may be swathed in marcescent leaves. They may also provide support for small vascular or non-vascular plants that grow on their sides or amongst the bases of the green leaves.

Figure 17.2. Lake Aunde, Mt Wilhelm. Alpine and subalpine grasslands clothe the slopes.

It is not known whether the roots within the pedestal and those in the underlying peat and mineral soil function identically in these two environments. The underlying soil is prone to waterlogging following heavy showers, but the raised pedestals are generally free of waterlogging. The pedestal may become relatively dry during extended rainless periods. At such times, deep roots may provide essential moisture.

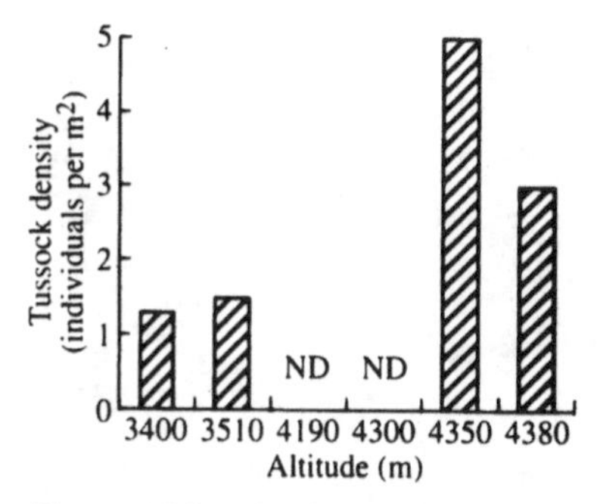

Figure 17.3. The density of tussocks of *Deschampsia klossii* in relation to altitude on Mt Wilhelm. Sample sites and altitudes: Waterfall, subalpine tussock grassland, 3400 m; Field Station, subalpine grassland, 3510 m; Bivouac Gap, alpine tussock grassland, 4190 m; Wilhelm Track, alpine tussock grassland, 4300 m; Upper Valley 1 alpine tussock grassland, 4350 m; Upper Valley 2 alpine tussock grassland, 4380 m.

The distribution of roots of *D. klossii* between pedestal and 'soil' varied with locality. At low altitude sites (3400–3510 m) there was about twice the mass of roots in the 'soil' as in the pedestal (35 g m^{-2} vs 17 g m^{-2}). At high altitude (4350 m) there was about eight times more in the 'soil' than in the pedestal (41 g m^{-2} vs 5 g m^{-2}). This contrast was despite the four-fold increase in tussock density over the same altitude. The difference is largely the result of much poorer development of pedestals at high altitudes (23% of tussocks had distinct pedestals at 4350 m compared with 47% at 3400 m). The total live root mass did not vary greatly from high to low altitude sites: at 4350 m there were 46 g m^{-2} while at 3400 m there were 50 g m^{-2}.

The leaves of tussock grasses are variable in size and shape. Although broad, flat-leaved species of grass occur in tussock grassland on Mount Wilhelm (e.g. *Hierochloe redolens*), it appears that narrow or tightly rolled leaves are most common. Stomata are sunken and occur in narrow papillose channels inside tightly rolled leaves of *D. klossii*. New regrowth from tussocks, clipped to the top of the pedestal during the dry season, was always of soft flat leaves. Gradually the rolled leaf form took over as time from clipping increased. The relative availability of water and nutrients may be partly responsible for the changed leaf morphology.

The leaves of *D. klossii* occurred in two forms with respect to altitude. Below 3510 m, the leaves were long (50–70 cm), flexuose, erect to spreading. With increasing altitude, modal leaf length declined, reaching a minimum of 20–25 cm at 4350 m. Leaf densities appeared to fall into two

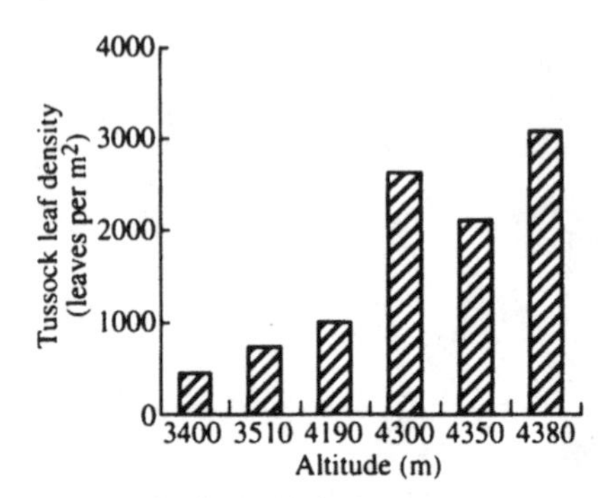

Figure 17.4. Density of tussock leaves in relation in altitude on Mt Wilhelm. Sample sites as in Figure 17.3.

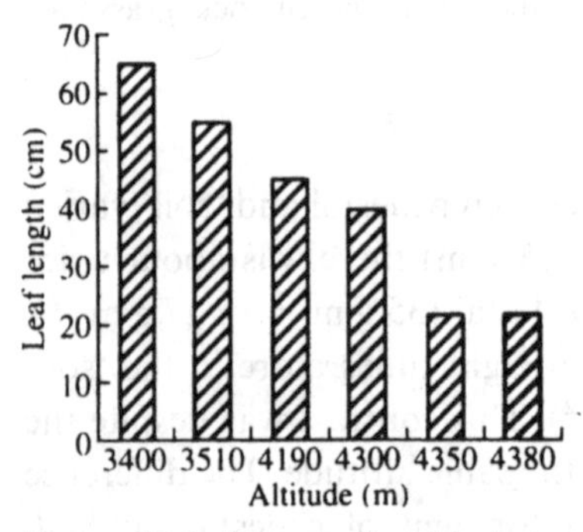

Figure 17.5. The relationship between leaf length and altitude of tussock grasses on Mt Wilhelm. Lengths are the mid points of modal classes at each site: 3400 m: 60–70 cm; 3510 m: 50–60 cm; 4190 m: 40–50 cm; 4300 m: 30–40 cm; 4350 and 4380 m: 20–25 cm. Sample sites as in Figure 17.3.

classes (Figure 17.4). There were 2000–3000 leaves per m^2 above 4300 m and 400–1000 leaves per m^2 below 4200 m. Leaf area index was variable and not correlated with altitude but ranged from 1.6 to 2.4 (differences not statistically significant).

The change in length of leaves was not as abrupt as the change in density of tussocks with respect to altitude (Figure 17.5). Leaf length is largely controlled by cell length, not by the number of cells in the leaf (Stukey 1942; Evans *et al.* 1964). As temperatures decline, cell elongation declines. With mean temperatures tending to decline steadily with increasing altitude, leaf length can be expected to decrease fairly steadily in the one species. Overall the trend was for tussocks to get larger and less numerous with decreasing altitude.

Data on leaf phenology and longevity are fragmentary. Pronounced seasonal death, characteristic of many alpine plants at high latitudes, has neither been seen nor been reported for any time of year on Mount Wilhelm or elsewhere in New Guinea. New leaves are initiated throughout at least the dry season. Leaf death is centripetal along the leaf, and the measured rates are variable and often interrupted for periods of a week to more than a month. Whilst less than 5% of young leaves present early in the dry season die by the onset of the next wet season, all leaves initiated during one dry season are dead by the middle of the next dry season (i.e. a maximum longevity of 7–16 months). The rate of initiation and death during the wet season has not been recorded. It is estimated that leaf biomass takes 17–18 months to accumulate, if growth is assumed to be constant. There is no rapid increment in leaf numbers from year to year, as would be implied by an 18-month longevity. It must be, then, that growth during the dry season (when rates were measured) is less than during the wet season.

Productivity, or the rate of accumulation of dry matter, of tussock grasses in the tropical alpine has not been extensively documented around the world. The environment is harsh, despite an apparently year-round growing season. The high incidence of night-time radiation frosts, high insolation in morning, and heavy cloud in the afternoon, all mitigate against high productivity.

Net productivity (of roots, leaves and leaf bases, but not stems) is calculated to be 0.69 g m^{-2} per day at 3510 m, and 1.21 g m^{-2} per day on 4350 m (Figure 17.6). It is highest at high elevations, for these tussock grasses.

The figures for net primary production of *D. klossii* tussock grassland are similar to those for tundra, summarized on a worldwide basis (Leith

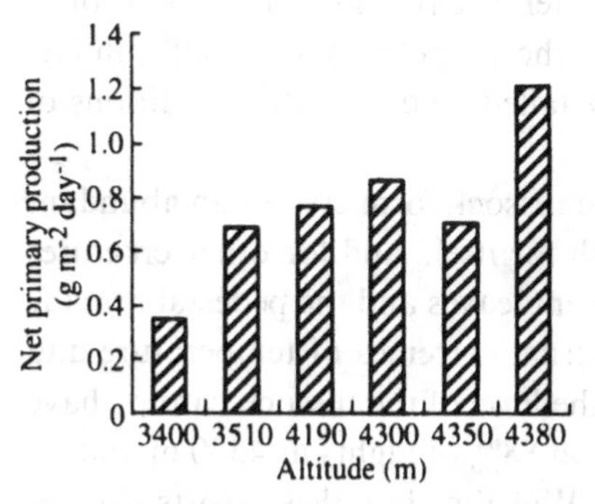

Figure 17.6. Net primary productivity of tussock grasslands in relation to altitude on Mt Wilhelm. Sample sites as in Figure 17.3.

1972). Using Leith's numerical models relating temperature and precipitation to net primary productivity, the calculated (expected) production is 10 times greater than the measured value at 3510 m and twice that for the 4350 m site. The estimates based on temperature give better estimates than those based on precipitation, indicating that temperatures may be the major factor limiting growth of *D. klossii*. It appears that the limitation is greater at lower, and therefore warmer, altitudes, suggesting that *D. klossii* is a cool-adapted species that is invading low altitude sites. The palaeontological evidence (Corlett 1984) suggests that the lower altitude grasslands became available and extensive only during the past 700 years, which corroborates the proposal that *D. klossii* is spreading from high to low altitudes. The altitudinal breaking point for the species seems to be about 4200 m judging from the changes in tussock and leaf densities. This is also the upper altitude suggested for the natural treeline now.

The rate of biomass accumulation in reproductive structures was not measured but is not expected to be high on an areal basis. The production of inflorescences occurs during the dry season but at no time is a mass of flowers visible and only a small proportion of tussocks appears to produce inflorescences during one season.

The greatest proportion of biomass appears to be in the leaves and leaf bases and not in the roots. There is from 8 to 12 times more aboveground than belowground live matter in the tussocks of *D. klossii*. Similar ratios have been reported by Smith & Klinger (1985) for tropical alpine sites in Venezuela. These ratios are not consistent with those from temperate alpine and arctic environments (Billings & Mooney 1968; Bliss 1971; Wielgolaski 1972). Bliss has argued that the distribution of biomass above and below ground in the arctic-alpine reflects the severity of the aerial environment and the perennial nature of the species. In tropical alpine habitats, plants are perennial, but the aerial environment is harsh on a diurnal, not an annual basis. Perhaps the perpetually cool soil inhibits root growth relative to leaf growth. Detailed studies under conditions of controlled environments are needed.

Within the tropical alpine region, the tussock form creates an abundant microenvironment that is suitable both for itself, and for other creatures (Coe 1967, 1969). The girdle of marcescent leaves and the pedestal provide a habitat that is buffered from the diurnal extremes of temperature and humidity fluctuations. Night frosts at the level of the tussock canopy have been recorded, during the dry season, on 88% of nights at 4300 m and on 72% of nights at 3480 m on Mount Wilhelm, but these frosts do not penetrate more than a few millimeters into the soil or pedestal. The growing points the grass leaves are located within the upper pedestal and are thus subject to less frost than are exposed parts of the plants. In contrast, the long, narrow, cylindrical leaves of *D. klossii* and other grasses and some shrubs mean that leaf temperatures closely follow air temperatures (Korner *et al.* 1983).

Conclusion

The New Guinea tropical alpine zone has recently become known to western science. Some of its flora and vegetation are known, but they only emphasize the great diversity between mountain peaks. Extensive surveys are still required for many of the upland areas. Recent glaciation and the advance of human occupation complicate the study of regional variation. Variations in rock type, gradation in precipitation and cloudiness, and differences in soil-forming properties all need investigation. The nature of the adaptations of the tropical alpine plants to their environments needs examination in detail. How these plants respond to changing environments brought about by human actions is also in need of study.

References

Billings, W. D. & Mooney, H. A. (1968). The ecology of the arctic and alpine plants. *Biological Reviews* **43**, 481–529.

Bliss, L. C. (1971). Devon Island, Canada, High Arctic ecosystem. *Biological Conservation* 3, 229–31.

Botting, D. (1973). *Humboldt and the Cosmos*. London: Sphere Books.

Brass, L. J. (1964). Results of the Archibold Expeditions No. 86. Summary of the Sixth Archibold Expedition to New Guinea (1959). *Bulletin of the American Museum of Natural History* **127**, 145–216.

Coe, M. J. (1967). The ecology of the alpine zone of Mount Kenya. *Monographiae Biologicae 17*. The Hague: Junk.

Coe, M. J. (1969). Microclimate and animal life in the equatorial mountains. *Zoologica Africana* **4** (2), 101–28.

Corlett, R. T. (1984). Human impact on the subalpine vegetation of Mt Wilhelm, Papua New Guinea. *Journal of Ecology* **72**, 841–54.

Dow, D. B. & Dekker, F. E. (1964). Geology of the Bismarck Mts, New Guinea. *Rep. Bureau Min. Resources, Geol. Geophys. Aust. 76.*

Evans, L. T., Wardlow, I. F. & Williams, C. N. (1964). Environmental control of growth. In *Grasses and Grasslands*, ed. C. Barnard, pp. 102–25. London: Macmillan.

Hedberg, O. (1964). Features of Afroalpine plant ecology. *Acta Phytogeographica Suecica* **49**, 1–144.

Hnatiuk, R. J. (1978). The growth of tussock grasses on an equatorial high mountain and on two sub-Antarctic islands. In *Geoecological Relations between the Southern Temperate Zone and the Tropical Mountains*, ed. C. Troll and W. Lauer, pp. 159–90. Erdwissenschaftliche Forschung 11. Wiesbaden: Franz Steiner.

Hnatiuk, R. J., McVean, D. N. & Smith, J. M. B. (1976). *The climate of Mt. Wilhelm. Mt Wilhelm Studies 2.* Research School of Pacific Studies Publication BG/4. Canberra: Australian National University.

Hope, G. S. (1976). The vegetational history of Mt Wilhelm, Papua New Guinea. *Journal of Ecology* **64**, 627–61.

Hope, G. S., Peterson, J. A., Allison, I. & Tadok, U. (ed.) (1976). *The Equatorial Glaciers of New Guinea.* Rotterdam: A. A. Balkema.

Korner, C., Allison, A. & Hilscher, H. (1983). Altitudinal variation of leaf diffusive conductance and leaf anatomy in heliophytes of montane New Guinea and their interrelation with microclimate. *Flora* 174, 91–135.

Leith, H. (1972). Modelling of the primary productivity of the world. *Nature and Resources* **8**, 5–10.

Mabberley, D. J. (1975). The giant lobelias: pachycauly, biogeography, ornithophily and continental drift. *New Phytologist* **74**, 365–74.

Noakes, L. C. (1939). Geological report of the Chimbu-Hagen area, Territory of New Guinea. *Geol. Surv. Terr. New Guinea.*

Ohsawa, M. (1990). An interpretation of latitudinal patterns of forest limits in South and East Asian mountains. *Journal of Ecology* **78**. 326–39.

Rickwood, F. K. (1955). Geology of the Western Highlands of New Guinea. *Journal of the Geological Society of Australia* **2**.

Smith, A. P. (1974). Bud temperature in relation to nyctinastic movement in Andean giant rosette plants. *Biotropica* **6**, 263–6.

Smith, J. M. B. (1975a). Notes on the distributions of herbaceous angiosperm species in the mountains of New Guinea. *Journal of Biogeography* **2**, 87–101.

Smith, J. M. B. (1975b). Mountain grasslands of New Guinea. *Journal of Biogeography* **2**, 87–101.

Smith, J. M. B. (1977). Vegetation and microclimate of east- and west-facing slopes in the grasslands of Mt Wilhelm, Papua New Guinea. *Journal of Ecology* **65**, 39–53.

Smith, J. M. B. & Klinger, L. (1985). Above-ground:below-ground phytomass ratios in Venezuelan paramo vegetation. *Arctic and Alpine Research* **17**, 189–98.

Stukey, I. H. (1942). Some effects of photoperiod on leaf growth. *American Journal of Botany* **29**, 92–7.

Troll, C. (1961). Klima und Pflanzenkleid der Erde in dreidimensionaler Sicht. *Die Naturwissenschaften* **9**, 332–48.

Wade, L. K. & McVean, D. N. (1969). *Mt Wilhelm Studies I. The alpine and subalpine vegetation.* Research School of Pacific Studies Publication BG/1. Canberra: Australian National University.

Wielgolaski, F. E. (1972). Vegetation types and primary production in tundra. In *Meeting on the Biological Productivity of Tundra*, Leningrad, October 1971, ed. F. E. Wielgolaski and T. H. Rosswall, pp. 9–34. Tundra Biome Steering Committee, Stockholm.

Young, T. P. (1984). The comparative demography of semelparous *Lobelia telekii* and *Lobelia keniensis* in Mount Kenya. *Journal of Ecology* **72**, 637–50.

18
Alpine herbivory on Mount Kenya

TRUMAN P. YOUNG and ALAN P. SMITH

Introduction

The savannas of East Africa, perhaps more than anywhere in the world, are known for dramatic plant–animal interactions. Although large herbivores are less common above timberline than below it in East Africa, herbivory is also a powerful force in Afroalpine plant ecology. Many vertebrates and invertebrate herbivore species occur above treeline on Mount Kenya (Moreau 1944; Coe 1967; Jabbal & Harmsen 1968; Coe & Foster 1972; Mulkey *et al.* 1984; Young 1991; Young & Evans 1993). Herbivores influence the distribution of plant species, the size structure of populations, and the success of individual plants on Mount Kenya.

Long-term studies of giant rosette *Lobelia* and *Senecio* species have brought to light a number of interesting patterns of herbivory that have major impacts on the biology of these species.

Lobelia

The two high alpine giant rosette lobelias on Mount Kenya are *Lobelia telekii* and *L. keniensis*. Mount Kenya lobelias are subject to herbivory from a variety of animal species (Table 18.1). None of the herbivores is host-specific, except perhaps the coleopteran larvae associated with *L. telekii* roots. Most of these herbivores have only minor effects. However, *Lobelia* populations near hyrax colonies have suffered severe predation.

The leaves and stems of both *Lobelia* species are protected by a bitter latex containing anti-herbivore compounds (Mabberley 1975). Although this may limit the activity of their herbivores, it far from renders the plants invulnerable. Hyrax elsewhere are known to eat the toxic leaves of other plants, including those of *Ficus* species, which also contain latex (Dorst & Dandelot 1970).

Table 18.1 *Herbivores of alpine giant rosette lobelias on Mount Kenya. The coleopteran larvae are probably* Seneciobius basirufus (*Jabbal & Harmsen 1968*)

Host and herbivore	Effect
Lobelia telekii	
Hyrax (*Procavia johnstoni mackinderi*)	Eat and often kill rosettes and inflorescences, particularly during drought
Various invertebrates	Limited feeding on old inflorescences
Coleopteran larvae	Found in association with roots of larger rosettes
Rodents (probably *Otomys otomys*)	Occasionally feed on small rosettes
Lobelia keniensis	
Hyrax	Feed on rosettes and inflorescences near colonies
Duikers (*Sylvicapra grimmia*)	Occasional feeding on rosettes and inflorescences
Various invertebrates	Extensive feeding on old inflorescences
Undetermined lepidopteran larvae	Occasional outbreaks on single large rosettes
Snails (probably *Helaxaryon* spp.)	Occasional feeding on leaves
Rodents (probably *Otomys otomys*)	Occasional feeding on small rosettes

Hyrax predation may periodically restrict the upper limit of *L. keniensis* on Mount Kenya (see Young, Chapter 14). In recent drought years, hyrax have eliminated the few remaining adult *L. keniensis* populations in the upper Teleki Valley (T. P. Young and A. P. Smith, personal observations). However, hyrax herbivory has been strongest, and best documented, on *L. telekii*, which occurs in habitats most used by hyrax.

Hyrax predation on *Lobelia telekii* rosettes

Lobelia telekii populations often occur near hyrax colonies on dry rocky slopes. The inner fleshy pith of giant lobelias provides an apparently rich food source, but it is protected both by latex-bearing leaves and by latex-bearing cells in the outer layer of the stem itself.

At the beginning of the *Lobelia* study (1977–9), the weather was wet and the *Lobelia* populations were healthy; mortality was low and growth rates were high (Young 1984). Although there was some hyrax herbivory on *L. telekii* during this wet period, it was usually limited to nipping of the leaf-bud tips. This would later appear as a ring of truncated leaves in

Figure 18.1. A ring of truncated leaves on a *Lobelia telekii* rosette, the result of an earlier nipping of the central leaf bud, probably by hyrax.

the rosette (Figure 18.1), but did not appear to adversely affect *L. telekii* growth or survivorship.

Hyrax predation on *L. telekii* began in earnest as the weather became dry in 1980, and a multi-year drought ensued. In these cases, the hyrax would individually remove, but not eat, the outer leaves of the rosette (Figure 18.2*a*). When they reached the innermost bud and the stem, the hyrax would feed, concentrating on the inner, unprotected portion of the stem (Figure 18.2*b*). This would invariably kill the plant, because *L. telekii* plants are unbranched and have only a single shoot meristem.

Hyrax attacked reproductive and large vegetative rosettes almost exclusively. This predation became so severe that by the middle of 1983, all of the larger *L. telekii* rosettes in the drier sites and many of those in the wetter sites had been killed, drastically altering population structure (see Figure 14.7, p. 263). As mentioned by Young (Chapter 14), the size structure of smaller rosettes remained virtually unchanged in both *L. telekii* populations; predation was clearly directed toward larger rosettes.

Probable sources of *L. telekii* mortality are shown in Table 18.2. One quarter of the larger rosettes (largest leaf $\geqslant 12.0$ cm) that died between

Figure 18.2. (*a*) Hyrax predation on a vegetative *L. telekii* rosette. (*b*) Close-up of the partially eaten inner stem.

February 1978 and July 1983 showed signs of hyrax herbivory at the time of death. These records are minimum estimates. Plants were surveyed every 6–8 weeks, and for many plant deaths it was not possible to identify associated factors. Hyrax attack was listed in 85% (71–100%) of the larger *L. telekii* deaths where an associated factor was identified. This confirms

Table 18.2 *Factors associated with* Lobelia telekii *rosette death on Mount Kenya between February 1978 and July 1983*

H, hyrax herbivory; I, invertebrate herbivory; M, herbivory by unspecified mammal; B, bud damage or sickness not associated with herbivory; W, leaf wilting; L, burial under small mudslides; U, no associated factor recorded

Rosette size (leaf length, cm)	Factors associated with rosette death									
	H	H & I	I	I & B	M	B	W	L	U	Total
⩽2.0	0	0	0	0	1	0	1	0	63	65
2.5–3.5	0	0	0	0	1	1	1	6	45	54
4.0–5.5	1	0	1	0	0	0	1	5	44	52
6.0–8.5	2	0	1	2	1	2	0	1	30	39
9.0–11.5	6	1	2	1	0	3	0	0	77	90
12.0–14.5	15	0	0	0	1	3	2	0	51	72
15.0–17.5	19	1	0	0	2	0	1	0	48	71
18.0–20.5	10	0	0	0	0	0	0	0	36	46
⩾21.0	5	0	0	0	0	0	0	0	10	15
Total	58	2	4	3	6	9	6	12	404	504

the impressions of both T. P. Young and our assistant, Martin Otieno, that hyrax were the major proximate cause of death in larger *L. telekii* plants.

Although hyrax often killed *L. telekii* rosettes, the underlying cause of death was apparently drought stress. Hyrax predation was severe only after the drought began. Similar interactions between climate and herbivory on trees have been shown with insects (White 1976, 1979) and with elephants (Laws 1970; Western & Van Praet 1973). Increased herbivory on a particular plant species during drought could be due to any of the following factors (from Young 1985). (i) Herbivore needs change during drought, increasing their preference for plant species that supply those needs. Such may be the case for elephants that obtain water from baobab trees during drought. (ii) As more palatable plant foods become exhausted in a drought, herbivores begin feeding on less favored species. (iii) Plants put under stress during a drought become inherently more palatable. White (1976) suggested that stressed plants contain more nitrogen and so are more attractive. A reduction in the defenses of stressed plants is a more likely cause.

Increased hyrax herbivory on *L. telekii* was probably due to a combination of increased *Lobelia* palatability and the exhaustion of alternate food plants. It is unlikely that hyrax dietary needs change dramatically during drought. Water was available throughout even the driest years on Mount Kenya. It is more likely that the switch to *L. telekii* was at least partly due to the disappearance of more palatable food plant species. Growth rates of all alpine plants studied on Mount Kenya decreased significantly during dry weather (Young 1984 and Chapter 14; Smith & Young, Chapter 15), and this would reduce the production of hyrax food, forcing them to rely on less palatable food plants. Unfortunately, we do not have data on relative abundance of other hyrax food plants to test this.

There is indirect evidence that stressed *L. telekii* plants become inherently more palatable during drought. Hyrax prefer to eat stressed, slow-growing *L. telekii* plants. During drier weather or in drier habitats, plants grew more slowly, had higher mortality independent of hyrax attacks, and tended to flower at a smaller size than did plants during wetter weather or in wetter habitats (Young 1984). We suggest that these patterns are demographic responses to drought stress. Increased hyrax herbivory was associated with drier sites and drier weather.

More revealing were patterns of hyrax herbivory within a population within a weather period. During dry season attack on a population of *L. telekii* plants, hyrax preferred slower growing plants to faster growing plants (Table 18.3). Paralleling large-scale patterns, slower growing plants in this population also exhibited higher non-hyrax mortality and smaller reproductive size than faster growing plants (Young 1985). These differences among plants within a site strongly imply microsite differences, with hyrax preferentially attacking stressed plants in less favorable microsites.

Table 18.3 *Mean growth (change in leaf length, cm) from February 1978 to May 1981 of vegetative* Lobelia telekii *rosettes that were either killed by hyrax or not killed, from May to November 1981*

Size class (leaf length, cm)	Rosettes killed by hyrax $X \pm$ SE (n)	Rosettes not killed by hyrax $X \pm$ SE (n)	t	p
12.0–14.5	0.45 ± 0.60 (11)	1.36 ± 0.26 (44)	1.48	0.07
15.0–17.5	0.15 ± 0.53 (10)	1.85 ± 0.26 (37)	2.98	<0.01
18.0–20.5	0.72 ± 0.32 (20)	2.23 ± 0.30 (22)	3.45	<0.01
⩾21.0	−0.83 ± 1.45 (3)	1.25 ± 0.32 (4)	1.59	0.09

From Young 1985.

Neither the depletion of alternate food sources nor changes in basic hyrax nutritional requirements alone can explain these local herbivory differences. Therefore, the increase in hyrax herbivory on *L. telekii* during drought was at least partly due to increased palatability of these plants when under stress.

In summary, the chances that a vegetative *L. telekii* rosette will be devoured by hyrax is a function of its size, its proximity to hyrax colonies, the dryness of its habitat (both spatially and temporally), and the quality of its microsite.

Hyrax herbivory on reproductive *Lobelia telekii*

Hyrax also attack *L. telekii* inflorescences, eating flowers and pods at various stages of maturity. There was differential hyrax floral/pod herbivory on *L. telekii* inflorescences of different heights. Figure 18.3 shows that smaller inflorescences lost proportionately more flowers and pods to hyrax than did larger inflorescences (Spearman rank correlation = -0.47, $n = 166$, $p < 0.01$). This pattern was attributed to two factors.

First, hyrax preferentially attacked slower growing rosettes (see above), and slower growing rosettes tended to flower at a smaller size (Young 1985). Differences in inflorescence size were strongly correlated with differences in rosette size at reproduction, a trait probably more influenced by environment than by genetics (Young 1984). Therefore hyrax preference for stressed (slow-growing) plants also exhibited itself as a preference for smaller inflorescences. During the height of the drought many smaller

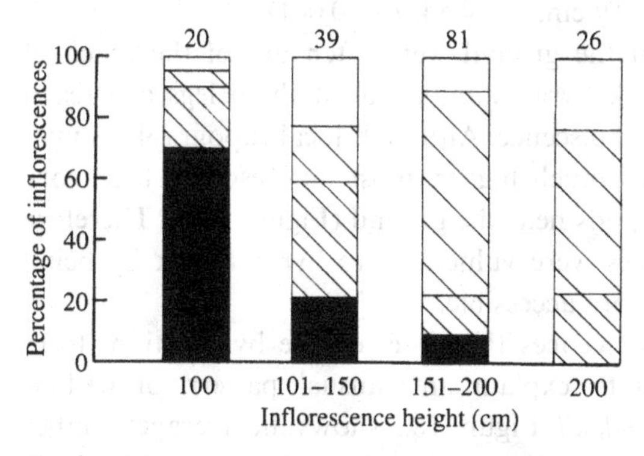

Figure 18.3. Relative hyrax herbivory on flowers and pods of *L. telekii* inflorescences of different heights. Open = less than 5% of the flowers and/or pods eaten; wide hatch = 5–20% eaten; narrow hatch = 20–50% eaten; solid = more than 50% eaten. The number of inflorescences surveyed in each height class are given at top.

Figure 18.4. Hyrax attack on a *L. telekii* inflorescence limited to pods near the ground, within reach.

flowering plants lost their entire inflorescence to hyrax herbivory. The average size (leaf length) of flowering rosettes killed by hyrax was significantly less than the size of flowering rosettes not killed by hyrax (19 ± 0.4 cm (SE) vs 22.4 ± 0.9 cm, $t = 4.44$, $p < 0.001$).

Second, flowers far from the ground were often out of the reach of hyrax. A Mount Kenya rock hyrax standing on its hind legs can reach only about 50 cm up an inflorescence. Although local topography sometimes allowed hyrax to feed much higher, most inflorescence herbivory was limited to flowers and pods near the ground (Figure 18.4). Therefore small *L. telekii* inflorescences were vulnerable to hyrax attack by being both more palatable and more accessible.

Flowers higher on inflorescences therefore 'escape by position' from herbivory. This may help to explain an unusual pattern of within-inflorescence seed set in *L. telekii*. Figure 18.5 shows the average number of seeds per pod by relative height on unattacked inflorescences of both *L. telekii* and *L. keniensis*. Inflorescences of both species mature from the bottom up. In *L. keniensis* the lower, earlier maturing pods produced the most seeds, and the average number of seeds in later (distal) pods was

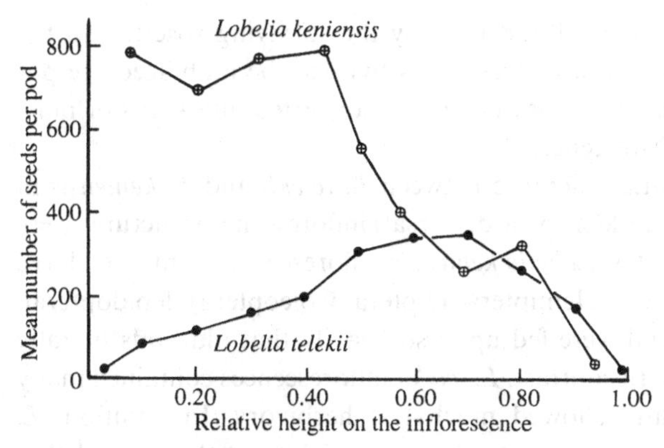

Figure 18.5. Average seed set of pods at different relative heights from 10 inflorescences each of *L. telekii* and *L. keniensis*.

less. This pattern appears to be common in plants (Stephenson 1981). By contrast, in *L. telekii* seed set per pod was initially low, and later pods higher on inflorescence had the greatest seed set.

We suggest that the unique pattern of seed set within *L. telekii* inflorescences has evolved to favor those pods most likely to escape hyrax attack. Resources put into seeds high on inflorescences would be less likely to be lost to herbivory. Because *L. keniensis* plants usually occur far from hyrax colonies, they are not subject to nearly as much hyrax herbivory, and so have not evolved this pattern. Care must be taken with this interpretation, however, given the very different life histories of these two species, which are reflected in other aspects of allocation (Young 1990). A possible test of this hypothesis would be to examine patterns of seed set in the lobelias on the nearby Aberdare Mountains, where there are no alpine hyrax.

Invertebrate herbivory on *Lobelia* inflorescences

The mature seeds of both species apparently have some protection from invertebrate herbivores. Although old pods were often eaten by invertebrates, especially in *L. keniensis*, there was no evidence of the seeds themselves being eaten before dispersal.

Lobelia life history differences may affect patterns of inflorescence herbivory by invertebrates. In the iteroparous *L. keniensis* resources are partitioned at reproduction between the inflorescence and a clone of vegetative rosettes. After reproduction, some of the inflorescence resources can presumably be recovered and used by the surviving rosettes. In the semelparous *L. telekii*, all of the resources that can be mobilized are put into the inflorescence, where they either are converted into seeds or must remain in the old inflorescence.

There was a dramatic difference between *L. telekii* and *L. keniensis* in the degree to which the older (post-dispersal) inflorescence structures were fed upon by invertebrates. *Lobelia keniensis* inflorescences contained large numbers of adult insects (Hemiptera, Diptera, Coleoptera), lepidopteran larvae, and spiders, and were fed upon so heavily that old pods literally fell apart in the hands. In contrast, *L. telekii* inflorescences contained many fewer invertebrates and showed much less herbivory. In addition, *L. keniensis* inflorescences decayed much more quickly on the ground that did *L. telekii* inflorescences.

We suggest that these two *Lobelia* species differed in their re-allocation of defensive compounds from mature inflorescences. The latex-borne compounds that protected the flowers and pods during maturation in iteroparous *L. keniensis* could be recovered after pod maturity, having served their purpose. The recovered compounds could then be re-allocated to surviving parts of the clone, leaving the old inflorescence structures more susceptible to herbivory. In contrast, semelparous *L. telekii*, which dies after reproduction, has nowhere to re-allocate resources from aging inflorescences. Defensive compounds initially invested in flowers and pods would therefore remain there even after they were no longer needed. This scenario of differential re-allocation would explain why herbivory on old inflorescences was much less in *L. telekii* than in *L. keniensis*. Differences in re-allocation of resources, including defensive compounds, in relation to plant life history is likely to be important for our understanding of herbivory on different plant species and different life stages.

Summary of *Lobelia* herbivory

The giant rosette plants *Lobelia telekii* and *Lobelia keniensis* have a variety of interactions with their invertebrates and vertebrate herbivores. These interactions have occurred on evolutionary as well as ecological time scales. Most of these interactions appear to have limited effects on Mount Kenya lobelias. However, drought-mediated hyrax herbivory may limit the distribution of *L. keniensis*, and has so altered the population structure of *L. telekii* at surveyed sites on Mount Kenya in recent years that it will be decades before a diverse size or age structure is regained.

Senecio

Lepidopteran larvae (Cochylidae, possibly *Trachybyrsis* sp.) occasionally burrow into the inflorescence stalks of *Senecio keniodendron* and *S. brassica*. This can cause entire inflorescences or inflorescence branches to break off. Adult weevils (*Aparasystales elongatus* Hust., Curculionidae) occasionally feed on *S. keniodendron* leaves, and unidentified weevils are seen occasionally in mature seed heads of *S. keniodendron*. In general, insect herbivory on achenes and leaves appears to be a minor factor for *Senecio* compared to the effects of mammals.

Hyrax regularly fed on the flower stalks of *Senecio brassica* plants occurring near their colonies. Typically the entire inflorescence was eaten, so that the affected plants produced no flowers or fruit that year. To illustrate this effect, a series of 10 m wide transects was established at 10, 25, 40 and 200 m from the edge of a hyrax colony at 4200 m in the Teleki Valley. The first 100 reproductive *S. brassica* plants in each transect were surveyed for hyrax damage. Results (Figure 18.6) indicate that inflorescence loss drops sharply with distance from the colony. Most *S. brassica* populations occur below the lower limit of hyrax (about 3800 m); those that occur above this elevation tend to occur on valley floor sites far from hyrax colonies, and so escape damage.

Hyrax also feed on the leaves and inflorescences of *S. keniodendron*. This species' elevational range is similar to that of hyrax; it is most common on slopes rather than valley floor sites, and is abundant on rocky sites where hyrax colonies are most common. It is therefore readily

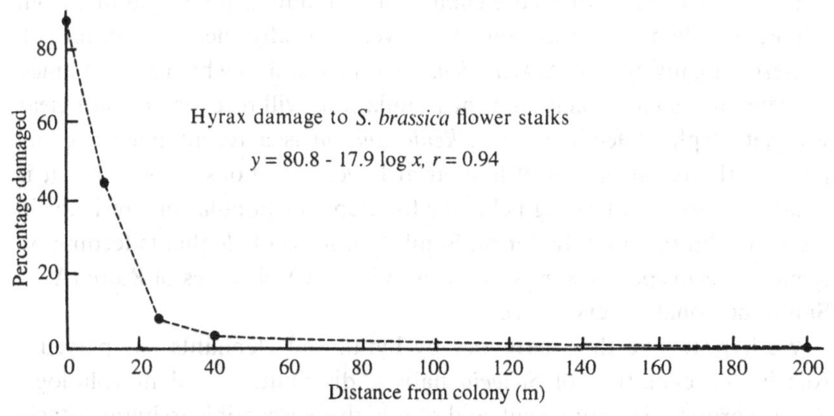

Figure 18.6. Proportion of inflorescences eaten by hyrax in *Senecio brassica* populations at different distances from a hyrax colony.

accessible to hyrax. However, access to rosettes and inflorescences is limited by plant height; very tall plants are less likely to be attacked than short plants. To quantify this effect a 4 m wide transect was established along a west facing slope at 4200 m in the Teleki Valley. Thirty reproductive adults with hyrax damage to inflorescences were encountered. Height to the highest leaf was recorded for each of these plants; height of the nearest reproductive neighbor lacking hyrax damage was also recorded. Height of the damaged plants was 174 ± 7 cm, compared with 274 ± 23 cm for undamaged neighbors.

Hyrax feed on *S. keniodendron* leaves primarily during the dry season, and especially during unusually severe periods of drought. During normal years they generally eat only a small section of the midrib of each leaf they attack. This causes the leaves to droop, but does not kill them immediately. During the severe drought of 1984 hyrax ate entire rosettes on short plants.

Hyrax also feed on the dead leaves retained on *S. keniodendron* stems. This sometimes eliminates the sheath of dead leaves near the base of the stem. Experimental elimination of these dead leaves caused reduced dry season production of new leaves and decreased minimum night-time stem temperatures (A. P. Smith, unpublished data), suggesting that feeding on these leaves by hyrax may be detrimental to the plants.

Elephants periodically feed on large adults of *S. keniodendron*, killing the plants (Mulkey *et al.* 1984). They appear to ignore plants shorter than approximately 1.5 m. This herbivory is common below 3800 m, where it caused extensive elimination of adults in several valleys (Liki North, MacKinder's, Teleki), with less extensive losses in other valleys. Elimination of adults causes dramatic change in community physiognomy, from 'alpine woodland' to grassland. However, typically there are abundant, fast-growing juveniles of *S. keniodendron* on these sites which are not killed by elephants, so it is likely that the populations will recover. It is not clear whether elephant feeding on *S. keniodendron* is a recent phenomenon, perhaps the result of loss of habitat at lower elevations, or whether it is a natural aspect of feeding behavior for elephant populations resident in the surrounding montane forest. Similar patterns of elephants feeding on alpine *Senecio* species were seen in the Virunga Volcanoes of Zaïre (A. P. Smith, personal observation).

We hypothesize that herbivory by hyrax and elephants has played a role in the evolution of *Senecio* habitat distribution and morphology. *Senecio brassica* is acaulescent, and so is highly susceptible to hyrax attack. This may have imposed selection for growth in microsites and at elevations far from hyrax colonies. The acaulescent habit may reduce susceptibility to elephant feeding, since elephants fed almost exclusively on *Senecio* plants over 1.5 m tall. Elephant feeding may have selected against *S. keniodendron* at lower elevations, because tall reproductive plants are preferentially eaten and killed. Hyrax herbivory at higher elevations may have imposed selection for tall plants, since greater height confers escape from attack. Hypotheses of this sort are not testable, and are presented simply as a way of re-emphasizing the potentially great role that herbivores can play in tropical alpine plant communities.

It has been suggested that hyrax and elephant herbivory on giant *Lobelia* and *Senecio* species threaten the future of these plants, and that the intense herbivory described above is mainly due to human interference, such as the discouragement of hyrax predators in the heavily visited Teleki Valley, and habitat loss forcing elephants into the alpine zone (Rehder *et al.* 1988). We do not agree. Hyrax attacks on *Lobelia* species are also common in valleys rarely visited by humans. Elephant destruction of *Senecio keniodendron* was limited to older individuals, lower elevations, and a few valleys. There is evidence for past large-scale die-offs of these species, and considerable current potential for the recovery of present populations (see Chapters 14 and 15). Although herbivory has strongly affected Mount Kenya's giant rosette plants in the past, and will probably continue to do so in the future, there is not sufficient evidence for alarm over short-term trends in these long-lived plants.

Other hyrax herbivory

The most visible mammalian herbivore above treeline on Mount Kenya is the rock hyrax, *Procavia johnstonii mackinderi*, common in moraines and rock outcrops from 3800 to 4500 m. Rock hyrax do not occur above treeline on other East African mountains. However, tree hyrax are found in the alpine zone of the Ruwenzori and Virunga Mountains. The vegetation near hyrax colonies of Mount Kenya is very different from the surrounding vegetation. Hyrax feed intensively near the rocky outcrops in which they live. Analysis of hyrax dung indicates that they feed considerably on grasses (Mahaney & Boyer 1986). They crop vegetation almost to the ground, leaving a carpet-like mat of grasses and mosses. This cropped area is much greener than the surrounding uncropped vegetation; this pattern allows reliable identification of hyrax colonies at a distance. Closer inspection always reveals abundant hyrax dung and an appropriate rock shelter. The dung may affect the nearby vegetation; the adjacent dry rocky slopes are generally low in nitrogen (Young 1984). Several plant species occur mostly at hyrax colonies and are rare elsewhere: *Carduus keniensis*, *Valeriana abyssinica*, *Arabis alpina*, *Anthoxanthum nivale*, and *Sedum ruwenzoriense*. We do not know whether these plants' distributions are due to different soil chemistry or to the fact that hyrax do not feed on these species, giving them a local competitive edge.

Carduus keniensis (Asteraceae) appears to be most common in and around hyrax colonies. It forms clones composed of spiny-leaved rosettes, each to up 1 m tall. The inflorescences grow to 1.5 m tall, and are also densely spiny. Neither leaves nor inflorescences are eaten by hyrax under natural conditions. The following experiment was carried out to test the importance of *Carduus keniensis* spines in deterring hyrax herbivory. Spines were removed from sections of leaves and inflorescence, and the sections were placed on the ground near a hyrax colony at 4200 m in the Teleki Valley. Control sections with spines retained were randomly intermixed with the experimental sections. All sections with spines removed were eaten within one hour; none of the control sections were eaten (A. P. Smith, unpublished data). This suggests that spines represent an effective defense against hyrax herbivory.

Carduus platyphylla, which commonly co-occurs with *C. keniensis* in and around hyrax colonies, produces small flattened rosettes with leaves appressed against the ground, and with an acaulescent inflorescence. Neither leaves nor inflorescences are densely spiny. It appears that both leaves and inflorescences are generally inaccessible to hyrax because hyrax have difficulty feeding on plants appressed to the ground. When rosette leaves were propped up at an angle of approximately 45°, they were eaten by hyrax within 24 hours, while adjacent appressed leaves were untouched (A. P. Smith, unpublished data), suggesting an alternative form of defense against hyrax herbivory.

Alchemilla argyrophylla (Rosaceae) is a low shrub common in the alpine zone of Mount Kenya up to approximately 4300 m, except near hyrax colonies. There is a significant negative correlation between frequency of hyrax dung and frequency of *Alchemilla* plants in the Teleki Valley at 4200 m (J. Lincoln and A. P. Smith, unpublished data). In order to test the hypothesis that this distribution is enforced by hyrax herbivory rather than by habitat requirements of the plant, an exclosure experiment was initiated (J. Lincoln and A. P. Smith, unpublished data). Twenty-eight adult *A. argyrophylla* plants were transplanted to wire enclosures at 4200 m in the Teleki Valley immediately adjacent to a hyrax colony, and 34 plants were transplanted to areas just outside the enclosures. Hyrax

Table 18.4 *Effects of hyrax herbivory on* Alchemilla argyrophylla *inside and outside exclosures*

Final heights were taken after one year. Confidence limits are ±1 SE.

	Initial height (cm)	Final height (cm)	Survivorship
Exclosure	19 ± 1	22 ± 1	
	$n = 28$	$n = 24$	0.86
Open	15 ± 1	5 ± 1	
	$n = 34$	$n = 21$	0.62

began to feed on the plants in the open as soon as they were transplanted. After one year the protected plants were taller and had higher survivorship than the unprotected controls (Table 18.4).

Rodents

Mount Kenya mole rats (*Tachyoryctes rex*) also alter the vegetation around their burrows. These rats are common on the drier alpine slopes, and form mounds of soil above their burrows (Jarvis & Sale 1971). The vegetation on these mounds has fewer grasses and more woody vegetation. Mole rats eat plant roots. We do not know whether the vegetation associated with their burrows is due to belowground herbivory or to edaphic differences in mound soil.

The rat *Otomys otomys* is abundant throughout alpine East Africa. Coe (1967) suggests that it is by far the most abundant herbivore on Mount Kenya, and our observations do not contradict this. Mahaney & Boyer (1986) suggest that these rodents eat more browse and roots than do hyrax. These rats are found in virtually all vegetated alpine habitats on Mount Kenya. In high elevation valley bottoms their extensively browsed runs are ubiquitous. The community effects of these herbivores have not been studied, but are likely to be great.

Ungulates

Large mammal herbivory is most evident on the dry northern slopes. Here, an otherwise continuous band of forest is broken up to form a 'treeless gap', allowing (until recently) the movement of animals between the plains and alpine areas. Herbivore species that may have once regularly migrated between the lowlands and the mountain, such as Burchell's zebras, are now resident in the alpine scrub and grasslands, cut off from the plains by human settlement. Elands are long-term residents above treeline on Mount Kenya, Kilimanjaro, and the Aberdare Mountains, and both elephants and buffaloes regularly visit the alpine areas of these mountains. Bush duikers are common in all of East Africa's alpine areas (Young 1991; Young & Evans 1993).

The treeless gap on the northern slopes of Mount Kenya is probably maintained by an interaction between fire and herbivory. The drier northern and northwestern slopes burn regularly. The flush of fresh growth after a burn attracts numbers of large herbivores, including zebras, buffaloes, elands, and steinbucks (T. P. Young, personal observation). The unburned tussocks of *Festuca pilgeri* and *Carex* spp. are exceedingly coarse. The lack of regular fires and of subsequent grazing on the wetter slopes of Mount Kenya results in vast expanses of living tussocks consisting mostly of dead leaves. This means that the drier parts of alpine Mount Kenya are often greener than the wetter parts. Although it is clear that herbivory and fire play dominant roles in the plant ecology of Mount Kenya's drier slopes, their effects have not yet been studied in detail.

Acknowledgements

We thank Martin Otieno for many years of invaluable assistance in the field. Insect identifications were done by R. Madge (Coleoptera) and J. D. Bradley (Lepidoptera) at the Commonwealth Institute of Entomology, British Museum.

References

Coe, M. J. (1967). Ecology of the alpine zone of Mount Kenya. *Monographiae Biologicae 17*. The Hague: Junk.

Coe, M. J. & Foster, J. B. (1972). Notes on the mammals of the northern slopes of Mount Kenya. *Journal of the East Africa Natural History Society* **131**, 1–18.

Dorst, J. & Dandelot, P. (1970). *A Field Guide to the Larger Mammals of Africa*. Boston: Houghton Mifflin.

Jabbal, I. & Harmsen, R. (1968). Curculionidae (weevils) of the alpine zone of Mount Kenya. *Journal of the East Africa Natural History Society* **117**, 141–54.

Jarvis, J. U. M. & Sale, J. B. (1971). Burrowing and burrow patterns of East African mole-rats. *Journal of Zoology (London)* **163**, 451–79.

Laws, R. M. (1970). Elephants as agents of habitat and landscape change in East Africa. *Oikos* **21**, 1–15.

Mabberley, D. J. (1975). The giant lobelias: toxicity, inflorescence and tree building in the Campanulaceae. *New Phytologist* **75**, 289–95.
Mahaney, W. C. & Boyer, M. G. (1986). Small herbivores and their influence on landform origins in the Mount Kenya afroalpine belt. *Mountain Research and Development* **6**, 256–60.
Moreau, R. E. (1944). Mount Kenya: a contribution to the biology and the bibliography. *Journal of the East Africa Natural Society* **18**, 61–92.
Mulkey, S. M., Smith, A. P. & Young, T. P. (1984). Predation by elephants on *Senecio keniodendron* (Compositae) in the alpine zone of Mount Kenya. *Biotropica* **16**, 246–8.
Rehder, H., Beck, E. & Kokwaro, J. O. (1988). The afroalpine plant communities of Mount Kenya (Kenya). *Phytocoenologia* **16**, 433–63.
Stephenson, A. G. (1981). Flower and fruit abortion: proximate causes and ultimate functions. *Annual Review of Ecology and Systematics* **12**, 253–79.
Western, D. & Van Praet, C. (1973). Cyclical change in the habitat and climate of an East African ecosystem. *Nature* **241**, 104–6.
White, T. C. R. (1976). Weather, food and plagues of locusts. *Oecologia* **22**, 119–34.
White, T. C. R. (1979). An index to measure weather-induced stress of trees associated with outbreaks of psyllids in Australia. *Ecology* **72**, 905–9.
Young, T. P. (1984). Comparative demography of semelparous *Lobelia telekii* and iteroparous *Lobelia keniensis* on Mount Kenya. *Journal of Ecology* **72**, 637–50.
Young, T. P. (1985). *Lobelia telekii* herbivory, mortality, and size at reproduction: variation with growth rate. *Ecology* **66**, 1879–83.
Young, T. P. (1991). The ecology, flora and fauna of Mount Kenya and Kilimanjaro. In *Guidebook to Mount Kenya and Kilimanjaro*, ed. A. Allen, pp. 37–49. Nairobi: Mountain Club of Kenya.
Young, T. P. (1990). Evolution of semelparity in Mount Kenya lobelias. *Evolutionary Ecology* **4**, 157–71.
Young, T. P. & Evans, M. E. (1993). Alpine vertebrates of Mount Kenya, with particular notes on the rock hyrax. *Journal of the East Africa Natural History Society* **82** (202), 55–79.

19

Biotic interactions in Hawaiian high elevation ecosystems

LLOYD L. LOOPE and ARTHUR C. MEDEIROS

Introduction

High elevation shrublands occur above the tradewind inversion (*c.* 2000 m), on the geologically young (<1 000 000 years old) volcanoes of Maui (Haleakala volcano) and Hawaiian (Mauna Kea, Mauna Loa and Hualalai volcanoes). Except for Mauna Loa, which recently erupted at the 3000 m level in 1984, these volcanoes are quiescent. Vegetation of the high elevation Hawaiian volcanoes consists of shrubs (*Styphelia, Coprosma, Vaccinium, Dubautia, Dodonaea, Geranium*), small trees (*Sophora, Myoporum, Santalum*), perennial graminoids (*Deschampsia, Agrostis, Trisetum, Luzula, Carex*), and other perennial herbaceous species (*Pteridium aquilinum, Asplenium, Silene, Sanicula*, etc.). With increasingly severe climatic conditions at higher elevations and with proximity to a mountain's summit, vegetation becomes more sparse and smaller in stature. Vegetation at the uppermost limit consists of prostrate shrubs, grasses, and ferns (Hartt & Neal 1940; Fosberg 1959; Mueller-Dombois & Krajina 1968; Whiteaker 1983). The fresh lava substrates of Mauna Loa (elevation 4170 m) result in much less development of soil and vegetation there than on older Mauna Kea (4207 m) and Haleakala (3056 m), both of which have extensive outcrops of cinder and ash deposits (Fosberg 1959). Hualalai Volcano (2522 m) barely reaches into the high elevation zone.

Mean temperature in the Hawaiian Islands decreases upward on the volcanoes at a rate of about 0.53 °C per 100 m. The upper limit of cloud forest and the lower limit of shrubland on windward East Maui is at *c.* 1800–2000 m, an elevation which corresponds to the mean level of the base of a subsidence inversion (the 'tradewind inversion') present about 70% of the time (Blumenstock & Price 1967; Giambelluca & Nullet 1991; Leuschner & Schulte 1991). Moist air and often a cloud layer typically occur below the inversion. Above the inversion, dry air dominates with low relative humidity and high solar radiation (Giambelluca & Nullet 1991). Mean annual precipitation generally decreases upslope from the tradewind inversion; it totals 800 mm for the Haleakala summit (Kobayashi 1973) and 146.5 mm for the Mauna Kea summit (Noguchi *et al.* 1987). The difference in mean temperature between the warmest and coolest months is about 4 °C (Whiteaker 1983). Snow falls only occasionally on upper Haleakala, but winter snow is relatively frequent on Mauna Kea and Mauna Loa. Freeze–thaw processes result in a variety of stone stripes and other active patterned ground phenomena near Haleakala's summit (Noguchi *et al.* 1986). Whereas upper Haleakala occupies a marginal periglacial environment, permanent ice exists less than one meter below the surface in the cinder of the summit cones of Mauna Kea (Woodcock 1976).

The Hawaiian biota has originated from the infrequent establishment of plants and animals from other lands. The difficulties of long-distance dispersal and establishment on these islands, isolated 4000 km from the nearest seed sources, have resulted in few colonizations. Many plant and animal groups failed to reach the islands prior to man's arrival. There are no native reptiles, amphibians, gymnosperms and, with the exception of a single bat species, no native land mammals. The native insect fauna consists of only 15 of the world's 26 orders and 15% of the world's insect families (Simon *et al.* 1984). New arrivals often speciated prolifically, exploiting ecological niches and food resources unavailable to continental relatives and producing a unique biota. An estimated 272 original immigrants are ancestral to some 1000 known species of native flowering plants (Fosberg 1948; Wagner *et al.* 1990). The nearly 6500 described species of native insects originated apparently from only 150–250 ancestral colonizers (Zimmerman 1948; Simon *et al.* 1984).

The characteristic Hawaiian high elevation shrubland flora comprises 29 genera of native flowering plants (21 dicots and eight monocots) and four genera of native ferns. The relatively simple vegetation of this zone consists of three typical fern genera, one tree-fern genus, eight graminoid and four herb genera, 13 shrub genera, and three genera of small trees.

Most high elevation Hawaiian species have related taxa found at lower elevations, some in middle to upper elevation windward forests (*Coprosma, Geranium, Rumex, Styphelia, Vaccinium*), others in drier leeward forests (*Artemisia, Gnaphalium, Myoporum, Santalum, Osteomeles, Schiedea, Sophora, Tetramolopium*). A distinctly alpine element occurs only in Hawaiian high elevation zones, perhaps dispersed from boreal regions by birds such as the winter resident, the lesser golden plover (*Pluvialis dominica*). The best examples of this alpine element are the grasses *Trisetum* and *Agrostis*.

Some high elevation taxa occur on both Maui and Hawaii as distinct species (*Dubautia*), some as distinct subspecies (*Tetramolopium humile*, *Argyroxiphium sandwicense*, *Geranium cuneatum*), and others as undifferentiated species (*Styphelia tameiameiae*, *Sophora chrysophylla*, *Stenogyne microphylla*). Though lacking some floristic components, pioneer vegetation on new lava surfaces at low elevations on Hawaii island, such as at Kilauea volcano area and south Kona, closely resembles many structural and floristic aspects of high elevation shrubland, most notably in abundance of *Vaccinium* and *Styphelia*.

Endemic fauna

The native insect fauna

Perturbations by introduced ant and parasitoid species and destruction of habitat have decimated much of the native Hawaiian insect fauna, especially in the lowlands. Nearly all (about 98%) of the 6500 described native Hawaiian insects are endemic to the islands. The Hawaiian insect fauna has high levels of local endemism; a high percentage of the species are phytophagous and restricted to certain native host plants (Zimmerman 1948).

Beardsley (1980) recorded 389 insect species in Haleakala National Park's Crater District. Of these, 60% were endemic to the Hawaiian Islands, 21% restricted to upper Haleakala volcano, and 40% introduced. Despite some serious threats, the native insect faunas of Hawaii's high elevation areas are still relatively intact. The relative harshness of the environment has apparently provided a buffer against the perturbations that have decimated Hawaii's lowland native insect fauna.

Definitive work on breeding systems and pollination in Hawaiian natural systems is lacking. Many native plant species of Hawaii's high elevation shrublands are apparently entomophilous and are pollinated by native bees (*Nesoprosopis*), flies (*Pipunculus*, *Neotephritis*, *Trupanea*), and moths (numerous genera). The strong tendency toward bird pollination of the native Hawaiian flora seems to be augmented by insect pollination in high elevation shrublands.

Levin and Anderson (1970) pointed out that when plant species rely on the same pollinators, selection will favor the separation of flowering times. Flowering times of major pollen- and nectar-producing native species of upper Haleakala are spread over the February–October period.

The abundance of *Styphelia* flowers during the spring months of March–May on Haleakala provide an ample food source heavily used by the native solitary bee (*Nesoprosopis*), supporting buildup of large populations by early summer. In July–October, *Styphelia* flowers are sparse, but *Argyroxiphium*, *Geranium* and *Dubautia* are sequentially in flower. *Vaccinium reticulatum* apparently does not require pollen transfer between plants for successful fruit set (S. P. van der Kloet, personal communication). The pollen of dioecious (and probably wind-pollinated) *Coprosma montana* as well as that of *Vaccinium reticulatum* may serve as a pollinator food source.

A remarkable example of coevolved plant–insect interactions is found in the Haleakala silversword, *Argyroxiphium sandwicense* subsp. *macrocephalum* (Asteraceae). The insect fauna associated with this endemic plant species includes a variety of unique endemic forms. Silverswords are monocarpic, flowering once after years of vegetative growth, and are self-incompatible, requiring pollinator transfer of pollen for seeds to be set (Carr *et al.* 1986). The flowering silversword sends an aromatic central flowering stalk up to 1.5 m tall with hundreds of composite heads. These attract many species of arthropods including pollinators, herbivores (especially on seeds), predators and parasitoids. Native flies, wasps, bees and moths, attracted in large numbers to flowering silverswords, all probably act as pollinators. The normally dark-colored predacious native wolf spider (*Lycosa hawaiiensis*) is cryptically silver-colored when hunting among whorled silversword rosettes. Mirid bugs (*Nesosydne argyroxiphii*), found only on the silversword, are plant-suckers on leaves, but do no serious damage. Long-horned beetle larvae (Cerambycidae) of the endemic Hawaiian genus and species *Plagithmysus terryi* bore in the lower stems and roots of flowering plants. Members of this beetle family usually exploit wood of native hardwood trees, yet three closely related species of the genus (all known only on upper Haleakala Volcano) feed on aborescent composites (Gressit 1978). The larvae of *Rhynchephestia rhabdotis*, the sole species of an endemic pyralid moth genus, feed in the inflorescence of Haleakala silversword consuming the interior of the central stalk and pedicels as well as the developing seeds (Swezey 1932). The larvae of tephritid flies (*Trupanea cratericola*) and lygaeid bugs (*Nysius* spp.) are also seed predators. In some years, most viable seeds are consumed by these organisms. However, in years of abundant flowering, silversword seed predation may be less than 50% (Kobayashi 1974; L. L. Loope unpublished data). Alternating years of abundant synchronized flowering with years of little flowering (Loope & Crivellone 1986) of this self-incompatible species may be an adaptation to minimize seed predation and maximize outcrossng and seed set. The surviving insect fauna of the Mauna Kea silversword (*Argyroxiphium sandwicense* subsp. *sandwicense*) of the island of Hawaii is comparatively meager, due at least in part to the severe reduction of plant populations because of ungulate damage (Elizabeth Powell, personal communication).

Leaf herbivory by native insects is less conspicuous in high elevation shrubland than in adjacent cloud forests, where reduction of leaf area by native insects is often obvious. In the warm summer months, however, foliage herbivory by lepidopterans in the high elevation zone can be substantial, especially for *Dubautia*.

Seed predation is common in the shrubland zone. For Haleakala, Beardsley (1980) recorded seven endemic species of the tephritid fly genus *Trupanea*, whose seed-feeding larvae destroy much of the seed crop of native plant species of the Asteraceae. The larva of tortricid moths (*Cydia* spp.) feed in the seeds of *Sophora chrysophylla*, a dominant species over much of the shrubland; damage in some areas destroys 50–70% of the seed crop (Zimmerman 1978). Several species of microlepidopterans of the genus *Carposina* infest the fruits of the shrubland genera *Styphelia* and *Vaccinium*.

For high elevaton Haleakala, Beardsley (1980) recorded 16 species of endemic seed-eating lygaeid bugs (*Nysius* spp.) that feed in the flowers and fruits of a wide variety of native as well as introduced species. Two species (*N. coenosulus* and *N. nemorivagus*) develop primarily on lowland weeds, feeding on plants in the Chenopodiaceae and Amaranthaceae that flush after spring rains. In the warm summer months, these insects migrate upslope to form large aggregations on the summit of Haleakala. Populations of these insects gather in such multitudes that operation of a solar observatory on Haleakala's summit was abandoned because of interference caused by the swarming flights of *Nysius* (Beardsley 1966).

The seasonal upslope movements of insects to the summits of high mountains in Hawaii may be critical in the biology of native predacious and scavenger specialist insect species of aeolian ecosystems. The term 'aeolian' in Hawaii is primarily applied to non-weathered lava substrates, mostly but not exclusively at high elevations, where arthropods feed on organic matter blown in from elsewhere (e.g. Howarth & Mull 1992). These areas are characterized by little vegetation (generally 0–5% cover), edaphically and climatically xeric conditions, wide daily temperature fluctuations, and a sparsity of food. Thermal and moisture regulation are critical factors in the adaptation of arthropods to this unique habitat. An endemic flightless lygaeid bug, *Nysius wekiuicola*, known only from the summit region of Mauna Kea, appears to feed on moribund and dead arthropods (Ashlock & Gagne 1983). The worldwide genus *Nysius* otherwise consists of seed and leaf feeders. The bug is often found in large numbers among boulders that offer shelter and concentrate wind-borne organic debris. Three species of spiders share this habitat (Howarth & Montgomery 1980). Larvae of a noctuid moth species (*Agrotis* sp.), abundant in areas without vegetation on upper Haleakala, are also apparently scavengers or predators belonging to a primarily herbivorous order (A. C. Medeiros and F. R. Cole, unpublished).

Ten species of native ground beetles have been recorded within the aeolian zone of upper Haleakala; nine (including two monotypic genera) are endemic to the volcano. Of these, five are entirely restricted to the upper 150 m elevation occurring just below the summit. *Pseudobroscus lentus* is the sole member of an endemic genus which has been collected only at the very summit of the mountain at 3030 m (G. A. Samuelson, in litt.). The five aeolian carabid beetle species are flightless scavenger–predator specialists that are extremely rare; little is known of their current status or biology.

The phenomenon of flightlessness occurs in diverse orders, including lacewings (Neuroptera: Hemerobiidae), moths (Lepidoptera: Xyloryctinae) carabid beetles (Coleoptera: Carabidae), true bugs (Hemiptera: Nabidae), and flies (Diptera: Dolichopodidae). The endemic flightless lacewing genera *Pseudopsectra* and *Nesothauma*, both present on upper Haleakala, were regarded by Zimmerman (1957) as 'among the marvels of insular creation'. *Pseudopsectra cookeorum*, not encountered since 1945, has been collected only in the summit area of Haleakala (Zimmerman 1957).

The native avifauna

Recent exploration and study of Hawaii's fossil avifauna (James & Olson 1991; Olson & James 1991) has revealed much greater bird diversity than previously suspected, including flightless forms. Subfossil evidence verifies newly described species of eagles, owls, ibis, geese, rails, honey-creepers, honeyeaters, and crows. Some of these, such as the large goose-like birds, the crows, and certain honeycreepers, were probably important in pollinating and dispersing the seed of native plants, including those of high elevations. The reduction of native birds and resultant loss of pollination and dispersal may have profound ecosystem-level consequences.

The native Hawaiian goose or nene (*Nesochen sandvicensis*) is primarily a grass grazer, but also eats the fruits and disperses the seeds of native species of the upper shrubland, *Vaccinium reticulatum*, *V. berberifolium* and *Styphelia tameiameiae* (Baldwin 1947; Banko & Manuwal 1982). Hillebrand (1988), referring to the nene, stated that the fruits of *Vaccinium reticulatum* are 'the principal food of the wild Hawaiian goose'. The large extinct native goose-like birds, termed moa-nalo, up to 1.2 m tall, weighing 20 kg, and with 'toothed' dentitions on the bill, presumably for grinding fruits and seeds, were likely important vectors in dispersing seeds, especially large fleshy seeds (Olson & James 1991).

The omao or Hawaiian thrushes (Turdidae: *Myadestes* spp.) were the principal known frugivores of the Hawaiian avifauna. These birds consume a wide variety of native plant fruits including the high elevation genera *Coprosma*, *Rubus*, *Styphelia*, *Vaccinium* and *Myoporum* (van Riper & Scott 1979). Fossil remains of thrushes are known from most main Hawaiian islands (Olson & James 1984) and were abundant in lava tube deposits on leeward Haleakala (H. L. James, personal communication). Hawaiian thrushes are still extant on Kauai and Hawaii, but commonly encountered only on the latter island, there ranging as high as 2300 m (Scott *et al.* 1986).

Among the most important species radiations in the Hawaiian Islands is the Hawaiian honeycreeper alliance (Fringillidae: Drepanidinae), likely derived from a long-distance dispersal of finch-like birds from the Americas. Evolution over millions of years produced many species with long, decurved bills and infolded siphon-like tongues adapted for nectar-feeding (Raikow 1977), primarily on the flowers of *Metrosideros polymorpha*, *Sophora*, lobelias, mints, and numerous other plant genera. These birds also feed on invertebrates, especially while rearing young.

On the southern slopes of Mauna Kea at 1800–2900 m, the Federally listed Endangered palila (*Psittarostra bailleui*) feeds on the immature legumes of *Sophora*, occasionally dispersing seeds of this species (van Riper 1980) and *Myoporum sandwicense*.

At Haleakala, the amakihi (*Loxops virens*) and the apapane (*Himatione sanguinea*) commonly visit sandalwood (*Santalum haleakalae*) for nectar. When the *Sophora* trees flower intensely from February to April, apapane and 'i'iwi (*Vestiaria coccinea*) move from their resident *Metrosideros* rain forest habitat upslope to subalpine shrublands. Systematic trapline feeding by these strong-flying pollinators may be important in the reproductive biology of certain high elevation plant species.

Some plant species, endemic to high elevations of Maui and Hawaii, including *Sanicula sandwicensis* and *Uncinia uncinata*, have seeds with recurved hooks presumably adapted for dispersal by birds, since there are no native ground-dwelling mammals.

Impacts of alien species

Effects of alien invertebrates

Although over 2000 alien insect species have become established in Hawaii (Howarth 1985) and about 20 additional new species establish each year (Beardsley 1979), relatively few herbivorous insects have become established in Hawaiian high elevation areas. This may be largely because most inadvertent introductions take place in the lowlands. However, the Hawaiian flora in general has proved scarcely more susceptible to alien herbivorous insects than has the flora of continental areas, presumably because it has evolved under rigorous pressure from herbivorous insects. Alien insects may have a greater indirect effect on native vegetation by attacking native insects, such as lepidopterans, some of which are involved in crucial pollination of native plant species.

Two immigrant hymenopteran species – the western yellowjacket (*Vespula pensylvanica*) and the Argentine ant (*Iridomyrmex humilis*) – pose substantial threats to the high elevation biota of Haleakala, and probably to Mauna Kea and Mauna Loa as well. There are no social insects that are native to the Hawaiian Islands. As such, the native biota is not adapted to the intensity of foraging by predacious social insects such as ants and wasps. Both threaten the conservation of native ecosystems by their reduction of native insect pollinators.

Vespula pensylvanica, native to western North America, became established in the Hawaiian Islands in 1978. Since, then, populations have often built up to the point where they are considered a significant hazard to humans by the Hawaii State Health Department. They have been found on East Maui at 100–2900 m elevation, but thrive best at middle to high elevations. During years with mild winters, yellowjacket nests which are typically annual, have overwintered producing large nests and large number of both foragers and reproductives. It is estimated that 20 times the number of workers are produced in a nest that overwinters versus one that does not (Gambino *et al.* 1990). One overwintered nest at 2470 m elevation on Haleakala measured 1.2 m × 1.5 m × 2.1 m. The annual nests of most continental areas are, in contrast, the size of a volleyball. Overwintering of this normally annual species is known also from mild climate areas such as parts of central and coastal California within its native range.

Workers of *Vespula pensylvanica* are opportunistic predacious foragers with a wide dietary range. At high elevations, *Vespula* nest preferentially under the native shrub *Styphelia* presumably because of the substantial accumulated leaf litter. *Styphelia* also supports large populations of the native mealy bug *Pseudococcus nudus* which produces a honeydew, an important food source for adult yellowjackets (Gambino *et al.* 1990). Recent work at Haleakala National Park has shown that these yellowjackets also consume a wide variety of alien and native insects, including endemic pollinators and highly localized and/or flightless species (Gambino *et al.* 1987).

The Hawaiian insect fauna lacks endemic ant species and has not developed adaptations to cope with the fierce predation of this group (Zimmerman 1978). Introduced ants have had a devastating impact on the endemic insect fauna of low elevations in the Hawaiian Islands. The Argentine ant (*Iridomyrmex humilis*), a serious pest in California, South Africa and Australia, was first recorded in the 1960s in Haleakala National Park. This alien species poses a serious threat to native invertebrates of high elevation Hawaiian areas, especially rare and/or flightless species. Areas occupied by the Argentine ant showed lower numbers of individuals of many native insect species, including Lepidoptera larvae, carabid beetles (three genera), ground-nesting bees and wasps, collembola, lygaeid bugs, and various spiders (Medeiros *et al.* 1986; Cole *et al.* 1992). This alien ant species, if allowed to spread unchecked, may radically alter the composition of native invertebrates in high elevation systems and might ultimately cause failure of seed set among pollinator-dependent native plants.

The alien moth (*Uresiphita polygonalis*) is seasonally the most conspicuous insect herbivore in Haleakala's shrubland. Defoliation by this moth was a major seasonal source of mortality of *Sophora* at three sites on the island of Hawaii (Conant 1975). On Haleakala, the partial defoliation of *Sophora* by this species is conspicuous primarily during the warmer summer months. During the cold, wet winters, damage by *U. polygonalis* is either absent or inconspicuous.

Another threat to native biota is posed by parasitic wasps, some of which have been introduced to control agricultural pests. These insects have often attacked native species (especially lepidoterans), with disastrous results (Zimmerman 1978; Howarth 1985). The polyembryonic encyrtid wasp *Capidosoma bakeri* first appeared in Hawaii in 1985 (Beardsley & Kumashiro 1989). It is a parasite of a wide generic range of moth larvae of the family Noctuidae and has the potential to attack many endemic taxa. In 1988, it was found to be more abundant than all other Hymenoptera combined in yellow pan traps set out in the Kula area of Maui (Beardsley 1990). Beardsley stated that since *C. bakeri* is widely distributed in North America, 'the cool temperatures which are characteristic of the upper elevations of Haleakala on Maui, and on Hawaii where many endemic agrotine moths occur, probably would not inhibit its spread. This parasitoid could have significant impact on populations of these endemic Noctuidae'.

On Haleakala, the introduced honeybee (*Apis mellifera*) visits the flowers of alien and native plant species for pollen and nectar. Native species visited included *Santalum haleakalae*, *Sophora chrysophylla*, *Dubautia menziesii*, *Vaccinium berberifolium*, *Argyroxiphium sandwicense* and *Styphelia tameiameiae*. Honeybees probably compete significantly for food resources with native bees and wasps and this competition may result in limiting population levels of native pollen-dependent species such as the native yellow-faced bees (*Hylaeus* subgenus *Nesoprosopis*). Though good data are lacking, pollination by *Apis mellifera* may aid seed set in both alien and native plants.

Effects of alien ungulates

Mammalian herbivores were lacking in the Hawaiian Islands until their introduction at Cook's arrival in 1778. Many Hawaiian plant species, including those of high elevation areas, have proved susceptible to grazing and browsing of introduced mammalian ungulates. Carlquist (1974) and others have noted the absence in Hawaiian plants of many of the anti-herbivore strategies utilized by plants of continental areas. Hawaiian plants are notably non-poisonous, free from armament, and relatively lacking in oils, resins, stinging hairs, and coarse texture. After 200 years of increased fire frequency, browsing by large mammalian herbivores, and abundant alien plant introductions, the native Hawaiian biota has survived primarily in the harshest habitats, including high elevation rain forest and shrubland, as well as xeric leeward shrubland. Most lowland native flora and fauna have disappeared, often with minimal documentation.

Feral goats (*Capra hircus*) have been highly damaging to island biotas throughout the world (Coblentz 1978) and especially so in the Hawaiian Islands. Established on all Hawaiian islands by the 1780s, feral goats multiplied rapidly, concentrating in open high elevation shrublands (Tomich 1986). On Haleakala, Wilkes (1845) observed goats at the summit in 1841, and feral goat populations were likely to have been near carrying capacity on upper and leeward East Maui by that time.

Efforts to remove goats from Haleakala National Park through hunting by park personnel and citizen hunters began in the 1930s. Yocum (1967) stated that Haleakala's goat population may have exceeded 4000 as recently as the early 1950s and estimated the 1963 population at 600. In the early 1980s, numbers of goats at Haleakala were still estimated at 2000–4000. In 1986, Haleakala National Park completed a goat-proof fence encircling the high elevation section of the park. By 1988, goat populations had been reduced to several hundred animals; as of mid-1992, less than a dozen goats survive within fenced units.

Domestic cattle (*Bos taurus*) were grazed on Haleakala's upper slopes and crater until the 1930s when they were excluded from Haleakala National Park. During that period, goats and cattle eliminated several unique plant species endemic to the upper slopes of Haleakala: a silversword relative (*Argyroxiphium virescens*), a shrubby composite (*Tetramolopium arbusculum*), and a unique arborescent lobelioid (*Clermontia haleakalensis*).

Haleakala's high elevation shrub species are consumed by goats in a consistent order of preference. The dominant native high elevation shrubs from least- to most-browsed are *Styphelia* < *Dodonaea* < *Vaccinium* < *Coprosma* < *Santalum* < *Dubautia* < *Sophora*. Many rare high elevation species have become extinct or are now largely confined to cliff faces and ledges (e.g. *Schiedea haleakalensis*, *Plantago princeps*, *Artemisia mauiensis*) or to sites protected from browsing by encompassing shrubs or trees (e.g. *Stenogyne crenata*, *Sanicula sandwicensis*). In areas of high goat concentrations, the relatively unpalatable *Styphelia* is often the only surviving native shrub. In other areas, even *Styphelia* has been largely eliminated by goat browsing, with only dead stumps remaining. Although nearly all native species quickly decline in areas subjected to very intense goat browsing, certain alien herbaceous species thrive despite heavy cropping and trampling (e.g. *Sporobolus africanus*).

Cattle and sheep (*Ovis aries*) were introduced into the Hawaiian Islands in the 1790s by Captain Vancouver and others (Tomich 1986). The establishment of cattle and sheep, especially on Hawaii island, was greatly aided by the native '*kapu*' or restriction against killing these animals. By the 1820s, huge herds of these introduced ungulates soon roamed Mauna Kea and other areas. Mouflon sheep (*Obis musimon*) were introduced to Mauna Kea for hunting in the 1960s and have multiplied while other ungulates were removed as a result of a Federal Court order (Tomich 1986).

Middle to high elevation (1600–2800 m) *Sophora* forests of Mauna Kea have been destroyed or highly modified by the feeding feral sheep, cattle, goats and mouflon. These animals eat the shoots, leaves, bark and flowers of *Sophora* (Scowcroft & Sakai 1983). Analysis of aerial photographs spanning a 21-year period in the heavily sheep-browsed Mauna Kea Forest Reserve showed significant loss of tree cover in *Sophora* forest near treeline (Scowcroft 1983). As on Haleakala, most other native species of the high elevation shrublands of Mauna Kea have also declined dramatically (Hartt & Neal 1940). The Mauna Kea silversword (*Argyroxiphium sandwicense sandwicense*) was severely reduced in numbers in the 19th century and is currently on the verge of extinction (Powell 1987). High elevation portions of Mauna Kea are now dominated by introduced grasses due to grazing damage, whereas native grasses still predominate on the younger lava flows of Mauna Loa (Mueller-Dombois & Krajina 1968). After 16 years of protection from browsing, vegetation in exclosures showed substantial recovery and increase in cover of native species including *Sophora*, *Agrostis sandwicensis*, *Trisetum glomeratum* and *Deschampsia nubigena* (Scowcroft & Giffin 1983).

Though feral pigs (*Sus scrofa*) are more abundant in the low to middle elevation rain forests, they also have substantial impacts in high elevation shrublands and grasslands by their digging and feeding. Pigs occur commonly as high as 2600 m and have been seen near 3000 m on Haleakala. Seasonal movements of pigs occur with movement to higher elevations during prolonged dry and warm periods. Gosmore (*Hypochoeris radicata*) leaves and roots, and bracken fern (*Pteridium aquilinum*) rhizomes are preferred foods of pigs in Haleakala's high elevation grasslands (Jacobi 1981). Chronic pig digging damages native vegetation and results in a progressive increase in alien species, especially pioneer grasses and forbs.

The impact of large mammalian herbivores along with other perturbations is directly responsible for the destruction of most lowland Hawaiian native ecosystems. Continued damage has the potential for eliminating native vegetation from many of the upland areas of the Hawaiian Islands where it still remains. In many areas, only protection from large herbivores will preserve much remaining native vegetation. In other areas, especially on leeward slopes, the damage has already been too pervasive and only scattered native species remain. Steps have recently been taken to improve this situation, especially in the national parks. With continued efforts, substantial areas of native vegetation may be preserved in a near natural state for the foreseeable future. Feral animals have been drastically reduced on portions of State land on upper Mauna Kea and Haleakala, allowing some vegetation recovery.

Effects of alien rodents and small mammals

Introduced rodents are ubiquitous in the Hawaiian Islands. Colonizing Polynesians introduced the Polynesian rat (*Rattus exulans*) in about the 4th century AD. For nearly 1500 years, the Polynesian rat was the only rodent until the arrival of the house mouse (*Mus domesticus*) soon after western contact (Tomich 1986). The black or roof rat (*Rattus rattus*) was accidently introduced in the 1870s (Atkinson 1977).

Before the arrival of European man and the two Old World rodents, Polynesian rats (*Rattus exulans*) may have had a substantial impact on native biota; currently, they are only rarely if ever found in high elevation shrubland. House mice and black rats are important consumers in natural environments as least as high as 2900 m on Haleakala, where Cole *et al.* (1986 and unpublished data) conducted snap-trapping at 3-month intervals along elevational transects beginning in November 1984 to determine their population dynamics, food habits, and possible impact on native arthropod populations. Black rat populations remained low and stable throughout 2 years of study. Populations of mice were more variable, attaining peak densities in November 1984 and 1985 when availability of fruits and seeds was high. The rodent diets consisted primarily of grass seeds and various fruits when abundant, shifting to arthropods when these were scarce. Arthropods were most important as food items at the highest elevations. Araneida, Lepidoptera (primarily larvae), Homoptera and Coleoptera were the main arthropod taxa taken. *Rattus rattus*, and especially *Mus domesticus*, appear to exert a strong impact on local arthropod species, many of which are endemic.

By their consumption of plant seeds, rodents may reduce reproduction of some native plant species. Seed predation by seasonally high populations of rodents in the high elevation shrubland may threaten uncommon plant species (e.g. *Sanicula sandwicensis*, *Tetramolopium humile*) especially those that produce relatively few large-sized seeds (e.g. *Santalum haleakalae*). The black rat, more prolific and arboreal than the Polynesian rat, may have been crucial in the decline and loss of certain native bird species that acted as pollinators and seed dispersers of native plants (Atkinson 1977).

The introduced mongoose (*Herpestes auropunctatus*), feral cat (*Felis catus*), and possibly the black rat, are serious threats to the long-term survival of the endangered Hawaiian dark-rumped petrel (*Pterodroma phaeopygia sandwicensis*). This seabird nests only in caves burrowed deep into the sides of rocky lava cliffs on upper Haleakala and on the volcanoes of the island of Hawaii, having been eliminated from lower elevations (Simons 1985). The Hawaiian dark-rumped petrel, abundant and ubiquitous throughout the Hawaiian Islands prior to arrival of Polynesians in the 4th century AD (Olson & James 1984), was formerly undoubtedly an important agent in transport of nutrients from marine to terrestrial ecosystems.

Effects of alien birds

Introduced game birds, the chukar (*Alectoris chukar*) and ring-necked pheasant (*Phasianus colchicus*), disperse seeds of native and alien plant species in high elevation shrubland. Chukars occur on upper Haleakala at densities of up to 100 birds per 100 ha, most abundant at 2500 m elevation, but found as high as 2830 m. Pheasants occur at densities of 25–75 birds per 100 ha at elevations between 2000 and 2500 m. Seeds of native shrubs (*Styphelia tameiameiae*, *Vaccinium reticulatum*, *Coprosma ernodioides*, *C. montana*, *Geranium cuneatum*) and several alien grass and herb species have been germinated from pheasant droppings. The native shrubs mentioned above are difficult to germinate in cultivation and seedlings of these species are quite rare in natural environments. Game birds frequently disturb seedlings by scratching in the soil. Their most negative effects, however, may be both in carrying diseases and in sustaining greater numbers of alien predators than would otherwise be the case by providing a larger prey base. Both factors threaten rare, ground-nesting native birds such as the Hawaiian goose and dark-rumped petrel (F. R. Cole, L. L. Loope and A. C. Medeiros, unpublished data).

Conservation of Hawaiian high elevation ecosystems

The high mountains of Hawaii are not immune to direct impacts of man (e.g. extensive observatories on Mauna Kea and Haleakala) and are threatened by man's introductions of alien species. Though high elevation ecosystems are less vulnerable than lowland ecosystems, they have nonetheless suffered much damage despite recent conservation efforts.

Disappointingly little is known regarding biotic interrelationships in tropical alpine ecosystems throughout the world. The Hawaian areas are perhaps better known than most because of their relative accessibility to biologists. Hawaiian mountains provide outstanding opportunities for both the short- and long-term studies so urgently needed. Much progress in conservation has been made in recent years. The worst impacts of feral herbivores has been removed from many natural areas, with subsequent partial recovery of native species. Cautious optimism may be warranted. If current trends in research and management continue, it is possible that large areas of high elevation native ecosystems can be maintained nearly intact for future generations.

References

Ashlock, P. D. & Gagne, W. C. (1983). A remarkable new micropterous *Nysius* species from the aeolian zone of Mauna Kea, Hawaii island (Hemiptera: Heteroptera: Lygaeidae). *International Journal of Entomology* **25**, 47–55.

Atkinson, I. A. E. (1977). A reassessment of the factors, particularly *Rattus rattus* L., that influenced the decline of endemic forest birds in the Hawaiian Islands. *Pacific Science* **31**, 109–33.

Baldwin, P. H. (1947). Foods of the Hawaiian goose. *Condor* **49**, 108–20.

Banko, P. C. & Manuwal, D. A. (1982). *Life History, Ecology, and Management of Nene* (Branta sandvicensis) *in Hawaii Volcanoes and Haleakala National Parks*. Final Rept. to NPS.

Beardsley, J. W. (1966). Investigations of *Nysius* spp. and other insects at Haleakala, Maui during 1964 and 1965. *Proceedings of the Hawaiian Entomological Society* **19**, 187–200.

Beardsley, J. W. (1979). New immigrant insects in Hawaii: 1962 through 1976. *Proceedings of the Hawaiian Entomological Society* **23**, 35–44.

Beardsley, J. W. (1980). *Haleakala National Park Crater District Resources Basic Inventory: Insects*. Coop. Natl Park Resources Studies Unit, Univ. of Hawaii/Manoa, Dept. of Botany, Tech. Rept. 31.

Beardsley, J. W. (1990). Notes and exhibitions. *Capidosoma* sp., probably *bakeri* (Howard) (Hymenoptera: Encyrtidae). *Proceedings of the Hawaiian Entomological Society* **30**, 10–11.

Beardsley, J. W. & Kumashiro, B. (1989). *Capidosoma* sp., probably *bakeri* (Howard). *Proceedings of the Hawaiian Entomological Society* **29**, 4–5.

Blumenstock, D. I. & Price, S. (1967). Climates of the States: Hawaii. *Climatography of the States*, No. 60–51. U.S. Department of Commerce.

Carlquist, S. (1974). *Island Biology*. New York: Columbia University Press.

Carr, G. D., Powell, E. A. & Kyhos, D. W. (1986). Self-incompatibility in the Hawaiian Madiinae (Compositae): an exception to Bakers's rule. *Evolution* **40**(2), 430–4.

Coblentz, B. E. (1978). The effects of feral goats (*Capra hircus*) on island ecosystems. *Biological Conservation* **13**, 279–86.

Cole, F. R., Loope, L. L. & Medeiros, A. C. (1986). Population biology and food habits of introduced rodents in high-elevation shrubland of Haleakala National Park, Maui, Hawaii. (Abstract) *Program of the IVth International Congress of Ecology*, p. 220. International Association for Ecology, Syracuse, New York, 10–16 August 1986.

Cole, F. R., Medeiros, A. C., Loope, L. L. & Zuelke, W. W. (1992). Effects of the Argentine ant on the arthropod fauna of Hawaiian high-elevation shrubland. *Ecology* **73**, 1313–22.

Conant, M. (1975). *Seasonal Abundance of the Mamane Moth, its Nuclear Polyhedrosis Virus, and its Parasites.* Univ. Hawaii, Dept. Botany, US/IBP Island Ecosystems IRP, Tech. Rept. No. 64.

Fosberg, F. R. (1948). Derivation of the flora of the Hawaiian Islands. In *Insects of Hawaii, Vol. 1. Introduction*, ed. E. C. Zimmerman, pp. 107–10. Honolulu: University of Hawaii Press.

Fosberg, F. R. (1959). The upper limits of vegetation on Mauna Loa. *Ecology* **40**, 144–6.

Gambino, P., Medeiros, A. C. & Loope, L. L. (1987). Introduced *Paravespula pensylvanica* (Saussure) yellowjackets prey on Maui's endemic arthropod fauna. *Journal of Tropical Ecology* **3**, 169–70.

Gambino, P., Medeiros, A. C. & Loope, L. L. (1990). Invasion and colonization of upper elevations on East Maui (Hawaii) by *Vespula pensylvanica* (Hymenoptera: Vespidae). *Annals of the Entomological Society of America* **83**, 1088–95.

Giambelluca, T. W. & Nullet, D. (1991). Influence of the trade-wind inversion on the climate of a leeward mountain slope in Hawaii. *Climate Research* **1**, 207–16.

Gressitt, J. L. (1978). Evolution of the endemic Hawaiian Cerambycid beetles. *Pacific Insects* **18**, 137–67.

Hartt, C. E. & Neal, M. C. (1940). The plant ecology of Mauna Kea, Hawaii. *Ecology* **21**, 237–66.

Hillebrand, W. (1888). *Flora of the Hawaiian Islands*, 3rd edition. Reprinted in 1973 by Lubrecht & Cramer, Monticello, NY.

Howarth, F. G. (1985). Impacts of alien land arthropods and molluscs on native plants and animals in Hawaii. In *Hawaii's Terrestrial Ecosystems: Preservation and Management*, ed. C. P. Stone and J. M. Scott, pp. 149–79. Coop. Natl. Park Resources Studies Unit, University of Hawaii, Honolulu.

Howarth, F. G. & Montgomery, S. L. (1980). Notes on the ecology of the high altitude aeolian zone on Mauna Kea. *Elepaio, Journal of the Hawaiian Audubon Society* **41**(3), 21–2.

Howarth, F. G. & Mull, W. P. (1992). *Hawaiian Insects and Their Kin.* Honolulu: University of Hawaii Press.

Jacobi, J. D. (1981). *Vegetation Changes in Subalpine Grassland in Hawaii following Disturbance by Feral Pigs.* Coop. Natl. Park Resources Studies Unit, Univ. Hawaii/Manoa, Dept. of Botany, Tech. Rept. 41, pp. 29–52.

James, H. F. & Olson, S. L. (1991). Descriptions of thirty-two new species of birds from the Hawaiian Islands: Part II. Passeriformes. *Ornithological Monographs* No. 46, The American Ornithologists Union, Washington, DC.

Kirch, P. V. (1982). The impact of the prehistoric Polynesians on the Hawaiian ecosystem. *Pacific Science* **36**, 1–14.

Kobayashi, H. K. (1973). Ecology of the silversword *Argyroxiphium sandwicense* DC (Compositae), Haleakala Crater, Hawaii. PhD dissertation, University of Hawaii, Honolulu.

Kobayashi, H. K. (1974). Preliminary investigations in insects affecting the reproductive stage of the silversword (*Argyroxiphium sandwicense* DC) Compositae, Haleakala Crater, Maui, Hawaii. *Proceedings of the Hawaiian Entomological Society* **21**(3), 397–402.

Lammers, T. & Freeman, E. (1986). Ornithophily among the Hawaiian Lobelioideae (Campanulaceae): evidence from floral nectar sugar compositions. *American Journal of Botany* **73**, 1613–19.

Leuschner, C. & Schulte, M. (1991). Microclimatological investigations in the tropical alpine scrub of Maui, Hawaii: evidence for a drought-induced alpine timberline. *Pacific Science* **45**, 152–68.

Levin, D. A. & Anderson, W. W. (1970). Competition for pollinators between simultaneously flowering species. *American Naturalist* **104**, 455–67.

Loope, L. L. & Crivellone, C. F. (1986). *Status of the Haleakala Silversword: Past and Present*. Coop. Natl. Park Resource Studies Unit, Dept. of Botany, University of Hawaii, Honolulu, Tech. Rept. 58.

Medeiros, A. C., Loope, L. L. & Cole, F. R. (1986). Distribution of ants and their effects on endemic biota of Haleakala and Hawaii Volcanoes National Parks: a preliminary assessment, pp. 39–51. In *Proc. 6th Conf. in Natural Sciences, Hawaii Volcanoes National Park*, ed. C. W. Smith and C. P. Stone. pp. 39–51. Coop. Natl. Park Resources Studies Unit, Dept. of Botany, University of Hawaii, Honolulu.

Mueller-Dombois, D. & Krajina, V. J. (1968). Comparisons of east-flank vegetations on Mauna Loa and Mauna Kea, Hawaii. In *Proceedings of the Symposium on Recent Advances in Tropical Ecology*, ed. R. Misra and B. Copal, pp. 508–20. Varanasi, India: International Society for Tropical Ecology.

Noguchi, Y., Tabuchi, H. & Hasegawa, H. (1987). Physical factors controlling the formation of patterned ground on Haleakala, Maui. *Geografiska Annaler* **69**A(2), 329–42.

Olson, S. L. & James, H. F. (1984). The role of Polynesians in the extinction of the avifauna of the Hawaiian Islands. In *Quaternary Extinctions: a Prehistoric Revolution*, ed. P. S. Martin and R. G. Klein, pp. 768–80. Tucson: The University of Arizona Press.

Olson, S. L & James, H. F. (1991). Descriptions of thirty-two new species of birds from the Hawaiian Islands: Part I. Non-Passeriformes. *Ornithological Monographs No. 45*. Washington, DC: The American Ornithologists Union.

Powell, E. A. (1987). Population structure and conservation of the Mauna Kea silversword, an endangered plant of Hawaii. (Abstract). *Bulletin of the Ecological Society of America* **68**(3), 388–9.

Raikow, R. J. (1977). The origin and evolution of the Hawaiian honeycreepers (Drepanididae). *Living Bird* **15**, 95–117.

Scott, J. M., Mountainspring, S., Ramsey, F. L. & Kepler, C. B. (1986). *Forest Bird Communities of the Hawaiian Islands: Their Dynamics, Ecology, and Conservation.* Studies in Avian Biology No. 9. Cooper Ornithological Society.

Scowcroft, P. G. (1983). Tree cover changes in mamane forests grazed by sheep and cattle. *Pacific Science* **37**, 109–19.

Scowcroft, P. G. & Giffin, J. G. (1983). Feral herbivores suppress mamane and other browse species on Mauna Kea, Hawaii. *Journal of Range Management* **36**, 638–45.

Scowcroft, P. G. & Sakai, H. F. (1983). Impact feral herbivores on mamane forests of Mauna Kea, Hawaii: bark stripping and diameter class structure. *Journal of Range Management* **36**, 495–8.

Simon, C. M., Gagne, W. C., Howarth, F. G. & Radovsky, F. J. (1984). Hawaii: a natural entomological laboratory. *Bulletin of the Entomological Society of America* **30**, 8–17.

Simons, T. R. (1985). Biology and behavior of the endangered Hawaiian Dark-rumped Petrel. *Condor* **87**, 229–45.
Swezey, O. H. (1932). Notes on Hawaiian Lepidoptera, with descriptions of new species. *Proceedings of the Hawaiian Entomological Society* **8**, 197–202.
Tomich, P. Q. (1986). *Mammals in Hawaii: a Synopsis and Notational Bibliography*, 2nd edition. B. P. Bishop Museum Spec. Pub. 57, Honolulu.
van Riper, C. (1980). Observations on the breeding of the Palila *Psittarostra bailleui* of Hawaii. *Ibis* **122**, 462–75.
van Riper, C. & Scott, J. M. (1979). Observations on distribution, diet, and breeding of the Hawaiian thrush. *Condor* **81**, 65–71.
Wagner, W. L., Herbst, D. R. & Sohmer, S. H. (1990). *Manual of the Flowering Plants of Hawaii* (2 vols.). Honolulu, Hawaii: University of Hawaii Press and Bishop Museum Press.
Whiteaker, L. D. (1983). The vegetation and environment in the Crater District of Haleakala National Park. *Pacific Science* **37**, 1–24.
Wilkes, C. (1845). *Narrative of the United States Exploring Expedition during the Years 1838, 1839, 1840, 1841, 1842. Vol. IV.*
Woodcock, A. H. (1976). Permafrost and climatology of a Hawaii volcano crater. *Arctic and Alpine Research* **6**, 49–62.
Yocum, C. F. (1967). Ecology of feral goats in Haleakala National Park, Maui, Hawaii. *American Midland Naturalist* **77**, 418–51.
Zimmerman, E. C. (1948). *Insects of Hawaii*. Vol. 1. *Introduction*. Honolulu: University of Hawaii Press.
Zimmerman, E. C. (1957). *Insects of Hawaii*, Vol. 6. *Ephemeroptera–Neuroptera– Trichoptera*. Honolulu: University of Hawaii Press.
Zimmerman, E. C. (1978). *Insects of Hawaii*, Vol. 9. *Microlepidoptera*, Part II. Honolulu, University of Hawaii Press.

20

Tropical alpine ecology: progress and priorities

PHILIP W. RUNDEL, F. C. MEINZER and A. P. SMITH

As physiological plant ecologists and population biologists continue to work in tropical alpine environments, our knowledge of the form and functional relationships will surely grow rapidly. These ecosystems present unusual challenges for plant establishment and survival, but remain poorly studied. In this closing chapter, we briefly review what we feel are the important accomplishments to date in tropical alpine ecology and the challenges that remain. We focus on plant growth forms, plant demography, physiological convergence, ecosystem function and global climate change.

Plant growth forms

The most striking aspect of tropical alpine habitats is the diversity of plant growth forms, and the apparent convergence between geographically disjunct tropical mountains with respect to these forms (Hedberg & Hedberg 1979; Rauh 1978). Much progress has been made in understanding the ecological and physiological significance of the giant rosette form, perhaps the most conspicuous and typical form of high tropical mountains. However, the majority of tropical alpine growth forms have not been subject to quantitative and experimental analysis. Sclerophyllous-leaved shrubs, cushion plants, tussock grasses and acaulescent rosette forms have received minimal attention. Species in each of these forms may respond differently to changing environmental conditions, and may contribute in very different ways to edaphic, microclimatic and biotic environments.

Classical ecological paradigms suggest the evolution will lead toward convergence in adaptive traits of morphology, phenology, and physiology which provide the 'best' ecological solution to similar environmental stresses in disjunct habitats. Thus broad-leaved deciduous trees predominate in the cold winter, non-seasonal rainfall regimes of the Northern

Hemisphere, conifers or evergreen trees in nutrient-limited regions with similar climates, and evergreen sclerophyllous shrubs in Mediterranean-type ecosystems. Tropical alpine ecosystems, however, show a remarkable mix of successful growth forms. Does this suggest that the environmental stresses of diurnal freezing cycles, low growing season temperatures, variable levels of seasonal water stress, and low nutrient availability are sufficiently complex that no single suite of adaptive traits is 'best'? An alternative hypothesis is that tropical alpine ecosystems provide mosaics of microhabitats, each associated with individual dominant growth forms. If such mosaics exist, we have not been able to develop strong correlations between the adaptive traits of individual growth forms and soil/microclimate gradients across landscape gradients.

The adaptive traits which characterize specific growth forms other than giant rosette plants should provide an important theme for study. These should include not only physiological and ecological traits but anatomy as well (Carlquist, Chapter 6). While Hnatiuk (Chapter 17) has provided a good description of the growth dynamics of perennial bunch grasses in alpine New Guinea, comparable studies have not been carried out in East Africa or in the tropical Andes. Differences in dominant growth forms between regions can also provide important areas for thoughtful study, as can convergences. In the high Andes, for example, both the páramo and puna habitats show a high diversity of woody cushion plants (Heilborn 1925; Rauh 1939). While this growth form is occasionally present in the other tropical alpine regions and in lowland temperature regions as well (Godley 1978; Rauh 1978; Raunkiaer 1934) the convergent evolution of this growth form in so many families in the high Andes is remarkable. Almost nothing is known of the ecological significance of the abiotic and perhaps biotic stresses that have led to the convergent evolution of cushion growth forms in this single region.

Even within the familiar giant rosette growth form, there are large gaps in our knowledge of form/functional relationships. The convergent evolution of giant rosettes from a polycarpic shrubby ancestor has occurred many times in tropical alpine ecosystems. Frequently monocarpic reproductive systems are associated with the rosette growth form. Species of *Espeletia* in the northern Andes, *Argyroxiphium sandwicense* in the Hawaiian Islands, *Echium wildpretii* in the Canary Islands, and *Lobelia* and *Senecio* in the Afroalpine region all show morphological convergence to varying degrees. Despite relatively intensive ecological studies that have been made on *Espeletia schultzii* and *E. timotensis* in Venezuela (Smith 1974, 1979; Baruch 1979; Goldstein & Meinzer 1983; Goldstein *et al.* 1984,

1985; Meinzer & Goldstein 1985; Meinzer *et al.* 1985; Monasterio 1986), and *Senecio keniodendron* and *S. brassica* in East Africa (Hedberg 1964; Coe 1967; Beck *et al.* 1980, 1982, 1984; Smith & Young 1982), we continue to know surprisingly little about the comparative ecophysiology of giant rosette growth forms.

Plant demography

Detailed demographic data are now available for a number of giant rosette taxa. These include *Espeletia* (Smith 1981, 1984), *Lobelia* (Young, Chapter 14), *Senecio* (Smith & Young, Chapter 15) and *Argyroxiphium* (Rundel & Witter, Chapter 16). Such data provide a linkage between environmental studies and physiological ecology in helping to understand the relative importance of factors which influence growth and mortality at critical stages of life history. New demographic studies are needed to investigate establishment, growth and reproduction in other growth forms in the same habitats as these rosette plants. Microenvironment (Meinzer *et al.*, Chapter 3) may have quite a different impact on seedlings than on mature plants, which are often buffered against short-term environmental change.

Statistical analysis of plant demographic data not only leads to an improved understanding of life history dynamics and reproductive strategies, but also may be used to generate testable hypotheses concerning ecological factors controlling plant growth and survivorship. Such hypotheses can provide an important research tool through the use of controlled and replicated field experiments in which resource availability and competition are manipulated. Such experiments (see Smith & Young, Chapter 15) can reveal the relative importance of individual biotic and abiotic factors which control local patterns of germination and establishment in the seedling stage, and growth and reproductive output in mature plants (see Hnatiuk, Chapter 17; Young & Smith, Chapter 18). Related to such demographic studies is the need for better information on the breeding systems and pollination biology of tropical alpine floras (see Miller, Chapter 10; Beck, Chapter 11)

Physiological convergence

As pointed out by Smith in Chapter 1, the existence of qualitatively similar environments in widely separated tropical alpine areas has apparently resulted in a high degree of morphological convergence among different plant growth forms, especially within the giant rosettes (Monasterio 1986; Monasterio & Sarmiento 1991). Nevertheless, important physiological differences among plants of the same growth form in geographically isolated areas suggest either that these environments exhibit some critical quantitative differences, or that colonizing taxa are genetically canalized, imposing evolutionary constraints on physiological adaptation. A case in point is the different frost resistance mechanisms exhibited by African and Andean giant rosette plants. Leaves of the former are able to tolerate the extracellular freezing that occurs in the field (Beck, Chapter 5), while leaves of the latter apparently rely on supercooling and are irreversibly damaged upon extended freezing (Rada *et al.* 1985). Additional studies, involving a larger number of species and growth forms occurring in a wide range of tropical alpine environments, are needed to assess the relative roles of environmental and evolutionary histories in the occurrence of specific physiological adaptations. In frost resistance studies, the application of powerful tools such as NMR-imaging should prove valuable for ascertaining the state of water in plant tissues at low temperatures (Goldstein & Nobel 1991).

Because tropical alpine ecosystems present pronounced gradients in the seasonality of rainfall (see Rundel, Chapter 2), they should provide the opportunity for natural experiments to assess the influence of seasonal water stress on patterns of plant growth and development. Water relations studies to date have been focused largely on tropical alpine areas of Venezuela and Hawaii where rainfall is sharply seasonal (Rundel, Chapter 2). It would be interesting to carry out comparable studies of seasonal water relations in a variety of growth forms in wet páramo habitats or in the wet East African mountains or New Guinea. Relationships between soil frost cycles and plant water uptake should also be explored in more detail. Capacitance in giant rosette plants, for example, may be equally important in wet and dry tropical alpine habitats if soil freezing strongly affects the early morning availability of soil moisture.

The photosynthetic responses of tropical alpine plants have received little study (Schulze *et al.* 1985). Such studies could help identify the relative importance of stomatal and non-stomatal controls on net assimilation in these habitats, and define the significance for carbon balance of individual growth form strategies. If a rosette growth form has evolved in part to maximize radiant heating of photosynthetic tissues, gas exchange studies should reveal relationships between leaf energy balance and photosynthetic assimilation.

Although numerous hypotheses have been proposed to relate key environmental factors determining treeline and the upper elevation limits for vascular plants in tropical mountains, relatively little information is available on environmental factors limiting the lower altitude distribution of tropical alpine species (Young 1993). This question will become increasingly important in the context of predicting the effects of climate change on vegetation zonation in tropical mountains. Tropical alpine plants have evolved in environments in which diurnal temperature variations are much larger than seasonal ones (see Rundel, Chapter 2). It is thus conceivable that they possess very limited temperature acclimation capabilities compared with temperate zone plants. At low elevations, carbon balance of tropical alpine plants may be adversely affected by a limited capability to raise the optimum temperature for photosynthesis and/or limit high dark respiration rates at the higher night temperatures prevalent at lower elevations (see Goldstein *et al.*, Chapter 7). This possibility should be tested in field transplant experiments and in growth chamber studies.

Most studies of the environmental physiology of tropical alpine plants have focused on individual species or conspicuous growth forms occurring at a given site. Future research would profit from comparing physiological responses over the entire range of life forms growing in a particular site. For example, in a survey of frost resistance mechanisms in high tropical Andean plants, Squeo *et al.* (1991) found that all ground-level plants (i.e. cushions and small rosettes) exhibited freezing tolerance as the main mechanism of frost resistance while arborescent forms exhibited freezing avoidance mechanisms. Intermediate growth form (i.e. shrubs and perennial herbs) exhibited both freezing avoidance and tolerance. Similar comparative approaches applied to other aspects of plant function such as water and carbon balance would also help in understanding the trade-offs involved in possessing one particular morphological or physiological trait versus another.

Ecosystem function

Physiological ecology and population biology can provide important bases to begin to understand community-level responses of plants to their abiotic and biotic environment. Many physiological functions can be best understood in an ecosystem context of biogeochemical pools and fluxes of water and limiting nutrients for plant growth. Research to date has provided little insight concerning such processes. Data collected on net primary production and nutrient cycling in the alpine East Africa (Beck, Chapter 11; Rehder, Chapter 12) represent the first of such studies. More detailed comparative data from all of the major tropical alpine regions would be of great value.

One of the greatest gaps in our understanding of the ecological relations of tropical alpine plants lies in the form and function of root systems. Root architecture is critical to both water and nutrient uptake as well as providing support. The geometry, depth and horizontal spread of root systems associated with individual growth form, or even species, are largely undocumented. Root phenology, growth dynamics and physiology are similarly unknown. Given characteristic patterns of low soil nutrient availability and severe soil frost cycles in these ecosystems, rooting ecology should receive a high priority for future study.

Related to the need for studies of form and function in root systems, is limited knowledge of the biology of the soil environment in tropical alpine ecosystems. There has been significant work to date, for example, on the ecological or physiological relationships of mycorrhizal associations in a variety of ecosystems. Do the unusually harsh environmental conditions in these tropical alpine ecosystems provide special problems for mycorrhizal growth and inoculation? Soil biota should also be studied to understand how they interact with root growth and development and associated patterns of uptake for water and nutrients. Limited studies of soil biotic processes and decomposition in the Venezuelan páramo suggest that highly evolved systems of nutrient cycling are present for *Espeletia* (Monasterio 1986). Unusual patterns of sediment-based carbon nutrition in the high Andes are described by Keeley *et al.* in Chapter 9.

Global climate change

There has been virtually no research on the potential impact of global and regional climatic change and oscillations on the vegetation of tropical alpine zones. With regard to global and regional warming it can be hypothesized that the lower altitude limits of tropical alpine vegetation would rise more rapidly than the upper altitude limits, leading to a net reduction in area covered, even when the inevitable negative correlation between increasing altitude and available land surface area is taken into account. Colonization of high altitudes would be limited by suitability of substrate and rates of soil formation in previously unvegetated sites. A rise in the lower altitude limits of tropical alpine vegetation would also lead to isolation of previously contiguous breeding populations. This would influence genetic diversity within populations and ultimately rates of speciation.

Changes in the frequency and severity of climatic oscillations such as droughts may have a more immediate impact on tropical alpine vegetation than changing temperature regimes. Andean giant rosette species, for example, are well adapted to short-term frost-induced physiological drought, but apparently are less resistant to prolonged periods of low soil moisture during the dry season (Meinzer *et al.*, Chapter 4). This seems to be particularly true for juvenile plants among which widespread wilting has been observed during the unusually severe dry seasons of El Niño years (F. Meinzer and G. Goldstein, unpublished observations). Prolonged droughts such as those associated with the El Niño oscillations may thus exert a long-term impact on species composition and abundance in páramo vegetation if seedling establishment and mortality of giant rosette species are differentially affected.

Coupled with the need for understanding influences of global climate change are obvious problems of characterizing the impacts that human activities are having on tropical alpine ecosystems. Vegetation destruction, grazing, fire and introductions of exotic species are all having dramatic impacts today on tropical alpine systems throughout the world (Loope & Medeiros, Chapter 19; Balslav & Luteyn 1992).

References

Balslav, H. & Luteyn, J. L. (1992). *Páramo: An Andean Ecosystem Under Human Influence*. London: Academic Press.

Baruch, Z. (1979). Elevational differentiation in *Espeletia schultzii* (Compositae), a giant rosette plant of Venezuelan paramos. *Ecology* **60**, 85–98.

Beck, E., Scheibe, R., Sensar, M. & Miller, W. (1980). Estimation of leaf and stem growth of unbranched *Senecio keniodendron* trees. *Flora* **170**, 68–76.

Beck, E., Schütter, I., Schreibe, R. & Schulze, E.-D. (1984). Growth rates and population rejuvenation of East African giant groundsels (*Dendrosenecio keniodendron*). *Flora* **175**, 243–8.

Beck, E., Sensar, M., Scheibe, R., Steiger, H.-M. & Pongratz, P. (1982). Frost avoidance and freezing tolerance in afroalpine 'giant rosette' plants. *Plant, Cell and Environment* **5**, 215–22.

Coe, M. J. (1967). The ecology of the alpine zone of Mount Kenya. *Monographiae Biologicae 17*. The Hague: Junk.

Godley, E. J. (1978). Cushion bogs. In *Geoecological Relations between the Southern Temperate Zone and the Tropical Mountains*, ed. C. Troll and W. Lauer, pp. 141–58. Wiesbaden: Franz Steiner.

Goldstein, G. & Meinzer, F. C. (1983). Influence of insulating dead leaves and low temperatures on water balance in an Andean giant rosette plant. *Plant, Cell and Environment* **6**, 649–56.

Goldstein, G., Meinzer, F. C. & Monasterio, M. (1984). The role of capacitance in the water balance of Andean giant rosette species. *Plant, Cell and Environment* 7, 179–86.

Goldstein, G., Meinzer, F. C. & Monasterio, M. (1985). Physiological and mechanical factors in relation to size-dependent mortality in an Andean giant rosette species. *Oecologia Plantarum* **6**, 263–75.

Goldstein, G. & Nobel, P. (1991). Changes in osmotic pressure and mucilage during low-temperature acclimation of *Opuntia ficus-indica*. *Plant Physiology* **97**, 954–61.

Hedberg, O. (1964). Features of Afroalpine plant ecology. *Acta Phytogeographica Suecica* **49**, 1–144.

Hedberg, O. & Hedberg, I. (1979). Tropical-alpine life-forms of vascular plants. *Oikos* **33**, 197–307.

Heilborn, O. (1925). Contributions to the ecology of the Ecuadorian páramos with special reference to cushion plants and osmotic pressure. *Svensk Botanisk Tidskrift* **19**, 153–70.

Meinzer, F. & Goldstein, G. (1985). Leaf pubescence and some of its consequences in an Andean giant rosette plant. *Ecology* **66**, 512–20.

Meinzer, F., Goldstein, G. & Rundel P. W. (1985). Morphological changes along an altitudinal gradient and their consequences for an Andean giant rosette plant. *Oecologia* **65**, 278–83.

Monasterio, M. (1986). Adaptive strategies of *Espeletia* in the Andean desert páramo. In *High Altitude Tropical Biogeography*, ed. F. Vuilleumier and M. Monasterio, pp. 49–80. New York: Oxford University Press.

Monasterio, M. & Sarmiento, L. (1986). Adaptive radiation of *Espeletia* in the cold Andean tropics. *Trends in Ecology and Evolution* **6**, 387–91.

Rada, F., Goldstein, G., Azócar, A. & Meinzer, F. (1985). Freezing avoidance in Andean giant rosette plants. *Plant, Cell and Environment* **8**, 501–7.

Rauh, W. (1939). Über polsterförmiger Wuchs. *Nova Acta Leopoldina* **7**, 267–508.

Rauh, W. (1978). Die Wuchs- und Lebensformen tropischer Hochgebirgs-regionen und der Subantarktis – ein Vergleich. In *Geoecological Relations between the Southern Temperate Zone and the Tropical Mountains*, ed. C. Troll and W. Lauer, pp. 62–92. Wiesbaden: Franz Steiner.

Raunkiaer, C. (1934). *The Life Forms of Plants and Statistical Plant Geography*. Oxford: Clarendon Press.

Rundel, P. W. & Gibson, A. C. (1994). *Ecological Communities and Processes in a Mojave Desert Ecosystem; Rock Valley, Nevada*. Cambridge: Cambridge University Press (in press).

Schulze, E.-D., Beck, E., Scheibe, R. & Ziegler, P. (1985). Carbon dioxide assimilation and stomatal response of afroalpine giant rosette plants. *Oecologia* **65**, 207–13.

Smith, A. P. (1974). Bud temperature in relation to nyctinastic leaf movement in an Andean giant rosette plant. *Biotropica* **6**, 263–6.

Smith, A. P. (1979). Function of dead leaves in *Espeletia schultzii* (Compositae), an Andean caulescent rosette species. *Biotropica* **11**, 43–7.

Smith, A. P. (1981). Growth and population dynamics of *Espeletia* (Compositae) in the Venezuelan Andes. *Smithsonian Contributions to Botany* **48**, 1–45.

Smith, A. P. (1984). Postdispersal parent–offspring conflict in plants: antecedent and hypothesis from the Andes. *American Naturalist* **123**, 354–70.

Smith, A. P. & Young, T. P. (1982). The cost of reproduction in *Senecio keniodendron*, a giant rosette species of Mt. Kenya. *Oecologia* **55**, 243–7.

Squeo, F. A., Rada, E., Azócar, A. & Goldstein, G. (1991). Freezing tolerance and avoidance in high tropical Andean plants: Is it equally represented in species with different plant height? *Oecologia* **86**, 378–82.

Young, K. R. (1993). Tropical timberline: changes in forest structure and regeneration between two Peruvian timberline margins. *Arctic and Alpine Research* **25**, 167–74.

Index

Page numbers in bold refer to tables; page numbers in italic refer to figures.